Methods in Molecular Biology™

Series Editor
John M. Walker
School of Life Sciences
University of Hertfordshire
Hatfield, Hertfordshire, AL10 9AB, UK

For further volumes:
http://www.springer.com/series/7651

Nanotechnology for Nucleic Acid Delivery

Methods and Protocols

Edited by

Manfred Ogris

Department of Pharmacy, Ludwig-Maximilians-University Munich, Munich, Germany

David Oupicky

Department of Pharmaceutical Sciences, Wayne State University, Detroit, MI, USA

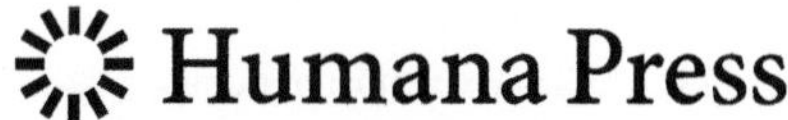

Editors
Manfred Ogris
Department of Pharmacy
Ludwig-Maximilians-University Munich
Munich, Germany

David Oupicky
Department of Pharmaceutical Sciences
Wayne State University
Detroit, MI, USA

ISSN 1064-3745 ISSN 1940-6029 (electronic)
ISBN 978-1-62703-139-4 ISBN 978-1-62703-140-0 (eBook)
DOI 10.1007/978-1-62703-140-0
Springer New York Heidelberg Dordrecht London

Library of Congress Control Number: 2012948771

Printed on acid-free paper

Humana Press is a brand of Springer
Springer is part of Springer Science+Business Media (www.springer.com)

Preface

Nanotechnology and nucleic acid-based therapies are two emerging fields in science whose combination has the potential to improve quality of life for patients suffering from various diseases that can so far be treated only in an unsatisfactory way either due to excessive side effects, such as in the case of malignant diseases, or due to the missing availability of therapeutically active compounds. Nucleic acids offer the potential for highly selective treatment of such diseases, for example the specific expression of therapeutically active proteins with gene vectors, or the highly specific modulation of gene expression with RNA interference. Within gene therapy approaches, the DNA delivered should act in principle as a completely inactive prodrug, which is activated only at the target site when being finally translated into the therapeutically active protein. The same applies for approaches based on small interfering RNA (siRNA), where rather short pieces of RNA have to be delivered to the appropriate site in the organism.

A key issue for successful nucleic acid therapies is the availability of a suitable delivery system. Viral vectors, e.g., recombinant adenovirus or lentiviral vectors, are highly efficient on the cellular level after local delivery but have to be optimized for systemic gene delivery approaches. In order to reduce residual biological risks of insertional mutagenesis, inflammation, or other adverse reactions after virus injection, numerous synthetic nucleic acid delivery devices have been developed within the last two decades. Here, the field of nanotechnology offers a multitude of possibilities to develop nanosized delivery vectors tailor-made for various local and systemic approaches.

As for many novel fields in modern medicine, safety aspects are of great concern in the development of nanocarriers for biomedical application. In Chapter 1, Maria Dusinska and colleagues report on safety aspects of engineered nanomaterials, including concerns related to biodistribution and bioavailability and cellular aspects, as well as regulatory issues. Chapters 2 and 3 by Moein Moghimi et al. describe the application of two assays to measure cellular toxicity of polycationic nucleic acid carriers, namely lactate dehydrogenase release as a direct measure of cell membrane disruption, and the quantification of caspase activation, which leads to apoptosis. Nanoparticle properties are known to strongly influence their in vivo behavior, like biodistribution or their interaction with the body defense system. Immunological aspects of the in vivo use of nucleic acid delivery are addressed in Chapter 4 by Tatsuhiro Ishida with a method to measure the induction of IgM by polyethyleneglycol, a polymer often used for surface coating of nanoparticles. In Chapter 5, Jean-Luc Coll and colleagues present a set of methods to determine the biodistribution of fluorescently labeled nucleic acid carriers in the living organism utilizing near infrared imaging.

Delivery systems based on lipids or cationic polymers are widely used for nucleic acids forming the so-called lipoplexes or polyplexes, respectively. In Chapter 6, Nathalie Mignet et al. describe the synthesis of charge neutral, thiourea-based lipids and the formation of liposomal nanoparticles with plasmid DNA. Antoine Kichler and colleagues in Chapter 7 give insights on the use of cationic peptides for nucleic acid delivery, and in Chapter 8, Rödl et al. report on the synthesis of conjugates of the polycation polyethylenimine and cell targeting peptides for tumor cell-directed gene delivery with polyplexes. Polycation topology is known to affect their toxicity and transfection activity. Several synthetic methods of controlling topology of bioreducible polycations is by Ye-Zi You et al. in Chapter 9. Nucleic acid-based nanoparticles are often made fresh by mixing nucleic acid and carrier molecule. Availability of stable formulations is crucial for preclinical and clinical applications. In Chapter 10, Julia Kasper and colleagues present a method for freeze-drying of polyplexes. Angela Pannier and Tatjana Segura (Chapter 11) report a system based on polyplexes, which are incorporated into a proteinous layer forming a hydrogel for surface-mediated gene delivery. Miriam Breunig and colleagues in Chapter 12 describe the potential of layer-by-layer films of therapeutic nucleic acids deposited on gold nanoparticles.

Disassembly of nanoparticles and subsequent release of nucleic acids is an important step for intracellular bioavailability of the delivered nucleic acids. In Chapter 13, Mao and colleagues describe an in situ atomic force microscopy method to study the disassembly of polyplexes under various conditions in liquid.

Biological activity of nucleic acids in local or systemic administration can be improved by combination with a range of physical methods. Niek Sanders et al. use focused ultrasound for site-directed disruption of gas-filled microbubbles, which enhance local gene delivery (Chapter 14). Dialechti Vlaskou et al. describe the synthesis and application of magnetic nanoparticles for DNA and RNA delivery, where targeting is achieved by applying a magnetic field (Chapter 15). Subsequent Chapter 16 by Yoshihisa Namiki describes the synthesis of "lipomag," a magnetic, lipidic DNA and RNA delivery system.

Chapters 17–19 present nanosized delivery systems and tissue-specific delivery strategies. In Chapter 17, Mark Zabel describes the synthesis and in vivo application of lipid- and peptide-based nanoparticles for siRNA delivery into the brain, which readily cross the blood–brain barrier after systemic injection. After systemic delivery, positively charged lipoplexes or polyplexes often end up in the first vascular bed encountered, namely the lung leading to transfection if lung tissue. Merkel and colleagues (Chapter 18) present a detailed protocol to analyze cell types in lung transfected after transgene delivery with the help of flow cytometry. Delivering nucleic acids to the liver can be achieved either by systemic administration or by direct infusion into the portal vein. Mao and colleagues present an alternative approach by transferring nanoparticulate systems via retrograde intrabiliary infusion (Chapter 19).

Taken together, this book covers the area of nanoparticulate delivery of nucleic acids in terms of biosafety, particle synthesis, and also its application in cell culture and after different routes of administration in vivo.

Munich, Germany *Manfred Ogris*
Detroit, MI, USA *David Oupicky*

Contents

Contributors

AMR S. ABU LILA • *Department of Pharmaceutics and Industrial Pharmacy, Faculty of Pharmacy, Zagazig University, Zagazig, Egypt*
HELENE ANDERSEN • *Centre for Pharmaceutical Nanotechnology, University of Copenhagen, Copenhagen Ø, Denmark*
BURKHARD BECHINGER • *Institut de Chimie, Université de Strasbourg/CNRS UMR 7177, Strasbourg, France*
MICHEL BESSODES • *Unité de Pharmacologie Chimique et Génétique, CNRS, UMR 8151, Paris, France; Inserm, U 1022, Paris, France; Faculté des Sciences Pharmaceutiques et Biologiques, Université Paris Descartes, Paris, France; ENSCP, Paris, France*
JENIFER BLACKLOCK • *Materials Reliability Division, National Institute of Standards and Technology, Boulder, CO, USA*
ANNE-LAURE BOLCATO-BELLEMIN • *Polyplus-transfection, Illkirch, France*
MIRIAM BREUNIG • *Department of Pharmaceutical Technology, University of Regensburg, Regensburg, Germany*
JEAN-LUC COLL • *INSERM U823, Institut Albert Bonniot, Grenoble, France; Université Joseph Fourier, Grenoble, France*
STEVEN K. COOL • *Laboratory for Gene Therapy, Faculty of Veterinary Medicine, Ghent University, Merelbeke, Belgium; Ghent Research group on Nanomedicines, Laboratory of General Biochemistry and Physical Pharmacy, Faculty of Pharmaceutical Sciences, Ghent University, Ghent, Belgium*
STEFAAN C. DE SMEDT • *Ghent Research group on Nanomedicines, Laboratory of General Biochemistry and Physical Pharmacy, Faculty of Pharmaceutical Sciences, Ghent University, Ghent, Belgium*
GILLES DIVITA • *CNRS UMR5237, Montpellier, France*
MARIA DUSINSKA • *NILU, CEE, Health Effects Group, Kjeller, Norway*
ASMAA ELBAKRY • *Department of Pharmaceutics, Faculty of Pharmacy, Al-Azhar University, Cairo, Egypt*
LISE MARIE FJELLSBØ • *NILU, CEE, Health Effects Group, Kjeller, Norway*
WOLFGANG FRIESS • *Department of Pharmacy, Pharmaceutical Technology and Biopharmaceutics, Ludwig-Maximilians-Universität, Munich, Germany*
HOLGER GARN • *Institute of Laboratory Medicine and Pathobiochemistry, Molecular Diagnostics, Philipps-Univeristät Marburg, Marburg, Germany; Sterna biologicals GmbH & Co. KG, Marburg, Germany*
BART GEERS • *Ghent Research group on Nanomedicines, Laboratory of General Biochemistry and Physical Pharmacy, Faculty of Pharmaceutical Sciences, Ghent University, Ghent, Belgium*
ACHIM GÖPFERICH • *Department of Pharmaceutical Technology, University of Regensburg, Regensburg, Germany*

JULIEN GRAVIER • *CEA Grenoble/LETI-DTBS, Grenoble, France*
ARNALDUR HALL • *Centre for Pharmaceutical Nanotechnology and Nanotoxicology, University of Copenhagen, Copenhagen Ø, Denmark*
MASAKO ICHIHARA • *Department of Pharmacokinetics and Biopharmaceutics, Subdivision of Biopharmaceutical Science, Institute of Health Biosciences, The University of Tokushima, Tokushima, Japan*
TATSUHIRO ISHIDA • *Department of Pharmacokinetics and Biopharmaceutics, Subdivision of Biopharmaceutical Science, Institute of Health Biosciences, The University of Tokushima, Tokushima, Japan*
XUAN JIANG • *Department of Materials Science and Engineering, Whiting School of Engineering, Johns Hopkins University, Baltimore, MD, USA; Translational Tissue Engineering Center and Whitaker Biomedical Engineering Institute, Johns Hopkins School of Medicine, Baltimore, MD, USA*
VÉRONIQUE JOSSERAND • *INSERM U823, Institut Albert Bonniot, Grenoble, France; Université Joseph Fourier, Grenoble, France*
JULIA CHRISTINA KASPER • *Department of Pharmacy, Pharmaceutical Technology and Biopharmaceutics, Ludwig-Maximilians-Universität, Munich, Germany*
ANTOINE KICHLER • *Laboratoire de Pharmacologie Chimique et Génétique et d'Imagerie, U1022 INSERM, Paris, France; Faculté de Sciences Pharmaceutiques et Biologiques, UMR 8151 CNRS, Paris, France*
THOMAS KISSEL • *Department of Pharmaceutics and Biopharmacy, Philipps-Universität Marburg, Marburg, Germany*
HIROSHI KIWADA • *Department of Pharmacokinetics and Biopharmaceutics, Subdivision of Biopharmaceutical Science, Institute of Health Biosciences, The University of Tokushima, Tokushima, Japan*
SARAH KÜCHLER • *Department of Pharmacy, Pharmaceutical Technology and Biopharmaceutics, Ludwig-Maximilians-Universität, Munich, Germany*
ANNA K. LARSEN • *Centre for Pharmaceutical Nanotechnology and Nanotoxicology, University of Copenhagen, Copenhagen Ø, Denmark*
INE LENTACKER • *Ghent Research group on Nanomedicines, Laboratory of General Biochemistry and Physical Pharmacy, Faculty of Pharmaceutical Sciences, Ghent University, Ghent, Belgium*
ZHIPING LI • *Department of Medicine, Johns Hopkins University School of Medicine, Baltimore, MD, USA*
HENRIK LUNDSGART • *Centre for Pharmaceutical Nanotechnology and Nanotoxicology, University of Copenhagen, Copenhagen Ø, Denmark*
ZUZANA MAGDOLENOVA • *NILU, CEE, Health Effects Group, Kjeller, Norway*
GUANGZHAO MAO • *Department of Pharmaceutical Sciences, Wayne State University, Detroit, MI, USA*
HAI-QUAN MAO • *Department of Materials Science and Engineering, Whiting School of Engineering, Johns Hopkins University, Baltimore, MD, USA; Translational Tissue Engineering Center and Whitaker Biomedical Engineering Institute, Johns Hopkins School of Medicine, Baltimore, MD, USA*
ARNAUD MARQUETTE • *Université de Strasbourg/CNRS, UMR 7177, Institut de Chimie, Strasbourg, France*
LEIGH M. MARSH • *Institute of Laboratory Medicine and Pathobiochemistry, Molecular Diagnostics, Phillips-Univeristät Marburg, Marburg, Germany; Ludwig Boltzmann Institute for Lung Vasculature Research, Graz, Austria*

A. JAMES MASON • *Institute of Pharmaceutical Science, King's College London, London, UK*
OLIVIA M. MERKEL • *Department of Pharmaceutical Sciences, Wayne State University, Detroit, MI, USA*
NATHALIE MIGNET • *Unité de Pharmacologie Chimique et Génétique, CNRS, UMR 8151, Paris, France; Inserm, U 1022, Paris, France; Faculté des Sciences Pharmaceutiques et Biologiques, Université Paris Descartes, Paris, France; ENSCP, Paris, France*
S. MOEIN MOGHIMI • *Centre for Pharmaceutical Nanotechnology, University of Copenhagen, Copenhagen, Denmark*
MARIE MORILLE • *INSERM U646, Angers, France*
NAOTO MORIYOSHI • *Department of Pharmacokinetics and Biopharmaceutics, Subdivision of Biopharmaceutical Science, Institute of Health Biosciences, The University of Tokushima, Tokushima, Japan*
OLGA MYKHAYLYK • *Institute of Experimental Oncology and Therapy Research, Klinikum rechts der Isar, Technische Universität München, Munich, Germany*
YOSHIHISA NAMIKI • *Institute of Clinical Medicine and Research, The Jikei University School of Medicine, Kashiwa, Japan*
MANFRED OGRIS • *Center for System based Drug Research, Department of Pharmacy, Pharmaceutical Biotechnology, Ludwig-Maximilians-University, Munich, Germany; Center for NanoScience (CeNS), Ludwig-Maximilians-University, Munich, Germany*
DAVID OUPICKY • *Department of Pharmaceutical Sciences, Wayne State University, Detroit, MI, USA*
ANGELA K. PANNIER • *Department of Biological Systems Engineering, University of Nebraska-Lincoln, Lincoln, NE, USA*
LADAN PARHAMIFAR • *Centre for Pharmaceutical Nanotechnology, University of Copenhagen, Copenhagen Ø, Denmark*
CHRISTIAN PLANK • *Institute of Experimental Oncology and Therapy Research, Klinikum rechts der Isar, Technische Universität München, Munich, Germany*
YONG REN • *Department of Materials Science and Engineering, Whiting School of Engineering, Johns Hopkins University, Baltimore, MD, USA; Translational Tissue Engineering Center and Whitaker Biomedical Engineering Institute, Johns Hopkins School of Medicine, Baltimore, MD, USA*
WOLFGANG RÖDL • *Department of Pharmacy, Center for System Based Drug Research, Pharmaceutical Biotechnology, Ludwig-Maximilians-University, Munich, Germany*
CLAIRE ROME • *INSERM U823, Institut Albert Bonniot, Grenoble, France; Université Joseph Fourier, Grenoble, France*
NIEK N. SANDERS • *Laboratory for Gene Therapy, Faculty of Veterinary Medicine, Ghent University, Merelbeke, Belgium*
DAVID SCHAFFERT • *Department of Pharmacy, Center for System based Drug Research, Pharmaceutical Biotechnology, Ludwig-Maximilians-University, Munich, Germany; Department of Molecular Biology, Aarhus University, Aarhus C, Denmark*
TATIANA SEGURA • *Chemical and Biomolecular Engineering Department, University of California Los Angeles, Los Angeles, CA, USA*
DIALECHTI VLASKOU • *Institute of Experimental Oncology and Therapy Research, Klinikum rechts der Isar, Technische Universität München, Munich, Germany*
ERNST WAGNER • *Department of Pharmacy, Center for System based Drug Research, Pharmaceutical Biotechnology, Ludwig-Maximilians-University, Munich, Germany; Center for NanoScience (CeNS), Ludwig-Maximilians-University, Munich, Germany*

LEI WAN • *Department of Chemical Engineering and Material Sciences, Wayne State University, Detroit, MI, USA*
JOHN-MICHAEL WILLIFORD • *Department of Biomedical Engineering, Johns Hopkins University School of Medicine, Baltimore, MD, USA*
EVA-CHRISTINA WURSTER • *Department of Pharmaceutical Technology, University of Regensburg, Regensburg, Germany*
JUN-JIE YAN • *CAS Key Lab of Soft Matter Chemistry, Department of Polymer Science and Engineering, University of Science and Technology of China, Hefei, People's Republic of China*
YE-ZI YOU • *CAS Key Lab of Soft Matter Chemistry, Department of Polymer Science and Engineering, University of Science and Technology of China, Hefei, People's Republic of China*
ZHI-QIANG YU • *CAS Key Lab of Soft Matter Chemistry, Department of Polymer Science and Engineering, University of Science and Technology of China, Hefei, People's Republic of China*
MARK D. ZABEL • *Department of Microbiology, Immunology and Pathology, College of Veterinary Medicine and Biomedical Sciences, Colorado State University, Fort Collins, CO, USA*
YI ZOU • *Department of Chemical Engineering and Material Sciences, Wayne State University, Detroit, MI, USA*

Chapter 1

Toxicological Aspects for Nanomaterial in Humans

Maria Dusinska, Zuzana Magdolenova, and Lise Marie Fjellsbø

Abstract

Among beneficial applications of nanotechnology, nanomedicine offers perhaps the greatest potential for improving human conditions and quality of life. Engineered nanomaterials (ENMs), with their unique properties, have potential to improve therapy of many human disorders. The properties that make ENMs so useful could also lead to unintentional adverse health effects. Challenges arising from physicochemical properties of ENMs, their characterization, exposure, and hazard assessment and other key issues of ENM safety are discussed. There is still scant knowledge about ENM cellular uptake, transport across biological barriers, distribution within the body, and possible mechanisms of toxicity. The safety of ENMs should be tested to minimize possible risk before the application. However, existing toxicity tests need to be adapted to fit to the unique features related to the nanosized material and appropriate controls and reference material should be considered.

Key words: Nanomaterial safety, Mechanisms of toxicity, In vitro tests, In vivo tests, Uptake, Hazard assessment

1. Introduction

The area of nanomedicine brings humans into direct contact with engineered nanomaterials (ENMs). With increased use of ENMs in medicine, many improvements in disease treatment are expected. However, concern has been raised that the properties that make ENMs so useful could also cause unintended effects on human health. The same unique physical and chemical properties that may be beneficial in nanomedicine may be associated with potentially deleterious effects on human health. Despite a lack of knowledge of the side effects from ENMs, they are widely used in research, industry, and medicine. It is therefore important to minimize the risk related to their possible adverse health effects.

Manfred Ogris and David Oupicky (eds.), *Nanotechnology for Nucleic Acid Delivery: Methods and Protocols*, Methods in Molecular Biology, vol. 948, DOI 10.1007/978-1-62703-140-0_1,

Nanotoxicology is an emerging field, dealing with human and environmental hazard caused by structures smaller than 100 nm in at least one dimension (1–3). However, toxicity testing of ENMs is not straightforward (4) as data obtained for the bulk compound cannot be directly applied to the nanosize material (5). The underlying mechanisms, for example oxidative stress, immunotoxicity, and genotoxicity, need to be investigated and possible adverse effects should be considered. Some of the methods already developed into OECD guidelines for conventional chemicals can be also used for testing of NMs (6). However, there are methodological concerns about ENMs regarding dosimetry, dispersion, lack of washout, uptake, and interaction of ENMs with cells and tissues. Development of standardized and validated tests for ENMs is therefore needed.

Generally, chemicals are regulated under REACH (registration, evaluation, authorization and restriction of chemicals) and nano-pharmaceuticals and nanomedical products and devices under regulatory directives of EMEA in Europe (7), or FDA in the USA. Pharmaceuticals are tested at three levels. First, tests are performed in vitro in cell or tissue cultures. If these tests are promising, further testing is performed in vivo in animal models before clinical studies are performed on humans. Existing regulatory frameworks in principle cover all important aspects of production and products; however, different behavior of ENMs from the corresponding bulk material suggests the need for specific nano-policies or adaptation of existing frames (8).

2. Exposure Assessment and Characterization of ENMs

Properties of ENMs change depending on present environment, air, liquid, or solid compartments, such as food. Thus the whole life cycle of the ENMs should be taken into consideration from production through application and disposal when assessing exposure (5).

In nanomedicine exposed groups at risk are patients, medical staff applying ENMs, and biotechnologists exposed during ENM manufacturing. The main routes of possible exposure are intravenous injection and surgical implantation, and also inhalation and oral and dermal exposures are possible. Secondary exposure of ENMs is possible through wastewater after disposal/recycling and release into the environment.

When in liquid suspension, precipitation, aggregation, or agglomeration could occur (9, 10). Careful characterization of the ENMs should therefore be performed both in the stock solution and in the matrix used during application, in addition to investigating the kinetics of the ENMs when entering the body. However, dosimetry and metrics of ENMs are challenging as there is no consensus or standardization when it comes to mass, number, surface area, or other metrics, and this makes comparisons and objective analysis difficult (3). Unfortunately, there is still lack of validated analytical and biological methods as well as certified reference standards for exposure and hazard assessment for establishing safe doses of ENMs. Well-characterized and defined ENMs to be used as certified reference standards need to be developed as quality controls additionally to standardized test protocols.

3. Biokinetics, Uptake, and Transport

There is a pressing need to understand how ENMs interact with organs, tissues, and cells, and what is their bioavailability and biopersistence. How are ENMs transported into the body, transferred across biological barriers, and translocated to different tissues. ENM behavior is likely influenced by size, shape, reactivity, surface coating, and surface charge of the ENM (11). The intrinsic physicochemical properties of the ENMs will strongly influence the kinetics in the body and subsequently target organs and toxicity. The small size and the particle shape enable uptake into blood and lymph circulation and distribution to tissues in the body that normally are protected by biological barriers. Although ENMs in nanomedicine are designed with high target specificity, and consequently reduced toxicity in nontarget cells and tissues, it is important to explore the potential toxic properties before applying to humans.

ENMs can penetrate cell membranes and epithelial or endothelial barriers in the body. Cellular uptake mechanisms of ENMs include diffusion, phagocytosis, pinocytosis, and receptor-mediated endocytosis. Uptake is highly dependent on physicochemical properties and surface characteristics such as polarity. Nonpolar or lipophilic molecules more easily cross membranes than polar compounds (12). Surface area and composition strongly determine reactivity, dispersion, interaction with biological environments including cellular macromolecules, and thus toxicity of ENMs (1, 13–15).

ENMs designed for drug delivery deliberately overcome biological barriers or they may cross them unintentionally as a result of environmental exposure. Once they have been taken up, they can potentially be deposited in any region of the body (16, 17). Due to their small size or specific functionalization, ENMs may cross important biological barriers and cause toxicity to very sensitive systems such as the brain or the developing fetus.

ENMs can be transported through the blood and accumulate in secondary target tissues and organs such as liver, spleen, kidney, placenta, the cardiovascular system, and central nervous system (CNS) where they may cause adverse effects (17). Biodistribution and mechanisms underlying toxicity are still largely unknown.

4. Possible Cellular Mechanisms of Toxicity

The toxicity of ENMs depends on many properties such as dimensions, size, shape, chemical composition, surface chemistry, as well as coating (18–20). The main proposed modes of action are oxidative stress, inflammation, and genotoxicity.

The generation of reactive oxygen species (ROS) is the key mechanism by which ENMs exert their pro-inflammatory and pro-atherogenic effects on the respiratory and cardiovascular tracts, and this can account also for ENM toxicity (21). During ENM exposure, an excess of ROS production can occur and the antioxidant defenses available can be overwhelmed. Glutathione (GSH) is depleted and oxidized glutathione (GSSG) accumulates, resulting in a drop in the GSH/GSSG ratio. Oxidative stress may initiate an inflammatory response, modified cellular redox signaling, as well as perturbed mitochondrial function and cell death. Inflammation is induced by up-regulation of redox-sensitive transcription factors (such as NFkB, AP1, and Nrf2) and MAPKs (such as ERK, JNK, and p38) (22, 23). The potential for inflammatory and pro-oxidant activity is largely dependent on ENM surface chemistry and in vivo surface modifications (24).

Several studies show that ENMs may induce genotoxicity, but there are many conflicting results in the literature (25). Genotoxicity may be produced by direct interaction of ENMs with the genetic material, indirectly by ROS production, or by toxic ions released from soluble ENMs. Secondary genotoxicity may result from oxidative DNA attack by ROS via activated phagocytes (neutrophils, macrophages) during nanoparticle-elicited inflammation (26). ENMs that cross cellular membranes may reach the nucleus through diffusion across the nuclear membrane or transportation through the nuclear pore complexes, and interact directly with DNA. In dividing cells, access is also provided by dissolution of the nuclear envelope during mitosis. For instance, functionalized single-walled carbon nanotubes (SWCNT) or silver ENMs have been reported to enter the cell nucleus, and C60 nanoparticles have been found to deform nucleotides (27, 28).

ROS can induce oxidation of bases in DNA (e.g., 8-oxoG) and strand breaks. These lesions can give rise to mutations and thus be potentially carcinogenic. Transition metal ions, such as

Fe^{2+}, Ag^{+}, Cu^{+}, Mn^{2+}, Cr^{5+}, and Ni^{2+}, released from soluble ENMs may also contribute to DNA damage by ROS production via the Fenton reaction. Coating of ENMs with redox-cycling organic chemicals (such as quinones), and metal impurities in CNTs, can amplify chemical changes in the ENMs' environment. A reaction of H_2O_2 with silver ENMs is proposed to cause formation of Ag^{+} in vivo (29).

There is some evidence that additionally to oxidative stress, inflammation, and genotoxicity, other adverse effects such as cardiovascular, immunological, and neurological disorders and possibly other unintentional effects can be of concern when it comes to the ENM toxicity (30, 31).

5. In Vitro Toxicity Tests for Nanoparticles

Toxicity tests for ENMs should address key physicochemical properties, exposure, biopersistence, uptake, transport through biological barriers, and possible mechanisms of toxicity. Ideally, all possible toxicity endpoints, such as ROS generation, cell activation, inflammation, immunotoxicity, and genotoxicity, and other generic endpoints including carcinogenicity, cardiovascular toxicity, neurotoxicity, and reproductive and developmental toxicity should be covered.

Different ENM characteristics such as size, shape, surface properties, composition, solubility, aggregation/agglomeration, ENM uptake, and presence of toxins and transition metals affiliated with the ENMs have to be taken into consideration additionally to dose, exposure time, and cell type used as they could influence the results of the toxicity testing.

5.1. In Vitro Models

The biological effects associated with different ENMs should be studied using relevant model systems at cellular or tissue level derived from the target organs or in vivo directly in specific organs and tissues (28). Mechanisms of action can better be explored in cell culture systems, which can be more easily controlled and yield more reproducible factors than in vivo systems. Human model systems derived from human cells are recommended for in vitro assays, because they can be used to predict toxicity of ENMs in humans (32). Existing OECD guidelines for in vitro tests have limitations with respect to assessment of biological hazards of ENMs and do not include characterization. Human-related toxicity can be assessed using battery of in vitro assays covering a wide range of endpoints and species, using primary cell culture or transformed cell lines of different origin (e.g., human, monkey, mice, hamster, or rat). In comparison with stable cell lines primary cultures often diverge between different laboratories, and are much more susceptible to environmental changes. However, they have normal karyotype and thus offer an advantage by being closer to the in vivo situation.

Common cell cultures used for toxicity testing are based on a 2D cell layer, growing on a flat substrate that does not exactly recapitulate the structure, function, or physiology of living tissues. In vivo, cells are organized three dimensionally and very often the function of individual cells depends on interactions of proteins and neighboring cells within this 3D organization (33).

5.2. Standards

One of the problems of toxicity testing of ENMs is lack of validated reference standards. Both positive and negative controls are needed when toxicity of conventional chemicals is assessed; this applies even more for ENM testing. When it comes to nanomedicine, it is suggested that ENMs already available on the market should be chosen as benchmark/negative control or reference standards. Benchmark ENMs are under development by OECD, while positive controls are still not available.

The extent to which ENMs are dispersed will influence the outcome of cytotoxicity testing, and is affected by physicochemical properties of ENMs, dispersion medium, temperature, ultrasound energy used for dispersion, dispersing agents, and sequence of dispersion preparation steps (34). ENMs are prone to agglomerate at physiological conditions due to large surface area. Selection of an appropriate vehicle is therefore crucial (15).

ENM dispersion/agglomeration is a concentration-dependent phenomenon (35); thus agglomeration state, uptake mechanisms, and toxic effects may vary non-linearly with dose. Additionally, ENMs often interact with the cell growth medium, eliciting in itself a toxic response that may have little or no physiological relevance in vivo.

5.3. Uptake and Transport Tests

Uptake, biodistribution, and concentration are important parameters besides toxicity testing of ENMs. The concentration of ENMs in biological samples and uptake should be screened before any toxicity studies. Several approaches for uptake can be used in vitro. Cells can be grown to confluence on semipermeable membranes in specialized inserts, e.g., Transwells®, placed in tissue culture wells to provide access to both apical and basal compartments. Integrity of the cell barrier can be determined by measuring marker transfer such as Lucifer Yellow dye or radioactive mannitol, combined with measurement of trans-epithelial/endothelial electrical resistance (TEER) (36, 37). Any adherent cell line that forms an intact monolayer and is representative of the organ of origin can be used as a barrier model (38, 39). The transport mechanisms and cellular uptake can be determined by energy inhibition studies under varying metabolic conditions by using appropriate pathway inhibitors such as sucrose or chlorpromazine (clathrin-dependent), filipin or nystatin (caveolin-dependent), 5-ethylisopropyl amiloride (EIPA) or cytochalasin D (macropinocytosis) in combination with imaging by confocal or electron microscopy (40).

5.4. Cytotoxicity Tests

Cell lines of different origins such as intestine or lung cells can be used to predict acute toxicity by estimation of LC_{50} values, the lethal concentration for 50% of the cells. Most cytotoxicity assays are based on colorimetric assays involving optical activity of organic dyes, and are prone to interference due to interaction between the dye and ENMs. Therefore careful consideration and several controls should be included when performing these tests on ENMs to avoid false positive or false negative results. The most used assays are the Lactate Dehydrogenase (LDH) assay; tetrazolium assays such as MTT, XTT, and WST; or the Alamar Blue assay. OECD guidelines have been developed for the Neutral Red assay based on neutral red uptake. Neutral red dye is weakly cationic and can diffuse across the membrane and accumulate in cellular lysosomes. Increased cell permeability and lysosome fragility are associated with the latter stages of cell death; thus the level of cellular dye is indicative of the level of viability. As internalized ENMs can be engulfed in lysosomes, increased lysosomal activity does not always reflect decreased cell viability.

Different endpoints measured by the above-mentioned assays represent distinct metabolic (dys-)functions and the assays can differ significantly in terms of their sensitivity. Probably the most reliable assays for cytotoxicity testing of ENMs are the proliferation assay and clonogenic assay (plating efficiency assays). In the clonogenic assay single cells are seeded at low density, and colonies formed after several cell divisions are counted representing both viability (colony number) and proliferative capacity (colony size) (41).

5.5. Genotoxicity Tests

There are numerous genotoxicity assays that cover different endpoints, such as measurement of single- and double-strand breaks, point mutations, deletions, chromosomal aberrations, micronuclei formation, DNA repair, and cell-cycle interactions (25, 42). These endpoints can be measured in vitro (human and mammalian cells) as well as in vivo (human biomonitoring and animal studies). Use of the same endpoints in various biological systems in different species enables comparison of interspecies effects. In vitro genotoxicity assays represent simple, robust, and time- and cost-effective testing of targeted toxicity and underlying mechanisms. The most commonly applied methods for detecting genotoxicity include the bacterial Ames test, the comet assay, and cytogenetic assays (micronucleus and chromosomal aberration assays, including the use of fluorescence in situ hybridization and chromosome paintings).

Ames test (bacterial reverse mutation) (43) is the most commonly used genotoxicity assay in regulatory toxicology based on induction of reverse mutation in the histidine gene, but there are a number of concerns regarding the suitability of the test for ENM testing due to the bacterial cell wall and the small size of bacteria. Thus it is not surprising that tested ENMs have been mostly negative (44–52).

Comet assay or single cell gel electrophoresis is one of the most common tests used for ENM genotoxicity testing. It can detect both strand breaks and DNA oxidation (53, 54) at the level of a single cell, and can be applied both in vitro as well as in vivo. The assay is currently under validation by ECVAM/JaCVAM, and a draft guideline protocol for regulatory purposes is being developed. Additionally, photogenotoxic effects of ENMs can be measured in combination with ultraviolet radiation (55).

Micronucleus (MN) assay both in vitro and in vivo has been recommended for assessing mutagenic and clastogenic effects of compounds (56, 57) and, like comet assay, can easily be applied in different target cells and tissues (58, 59). Micronuclei are formed during anaphase from chromosomal fragments or whole chromosomes that are left behind when the nucleus divides (56). The MN test detects both clastogenic and aneugenic effects (60) and is suitable for detection of genotoxicity of a wide range of compounds, including ENMs (44, 61).

Chromosomal aberration test is often applied in mammalian systems, both in vitro and in vivo, and has been used also for testing of genotoxicity of ENMs (51, 62). The in vitro chromosomal aberration test identifies agents that cause structural chromosomal or chromatid breaks in cultured mammalian cell lines or primary cultures (63, 64). Chromosomal aberrations are the cause of many human genetic diseases, and there is substantial evidence that chromosomal damage and related events causing alterations in oncogenes and tumor suppressor genes of somatic cells are involved in cancer induction in humans and experimental animals.

The most recent review on nanogenotoxicity (25) shows that the most frequently used genotoxicity tests are comet assay (42 studies), followed by micronucleus test (27 studies), chromosome aberration, and Ames test (both 10 studies). For genotoxicity/mutagenicity testing of chemicals which can be adopted also for ENMs, a battery of in vitro tests covering a wide range of genotoxic endpoints and species is recommended; usually three different in vitro tests are recommended. If at least two out of three tests are positive, the compound is considered genotoxic, and is also likely to be carcinogenic.

5.6. Carcinogenotoxicity In Vitro

Recent recommendation for testing conventional chemicals suggests inclusion of in vitro transformation assay as part of first screening (65). The three main in vitro transformation assays are the Syrian Hamster Embryo cell (SHE), the BALB/c 3T3, and the C3H10T1/2 assays. The SHE test is a promising assay since it is sensitive to both genotoxic and non-genotoxic chemicals and able to identify up to 90% of carcinogens, e.g., only 5–10% of the initial number of carcinogens tested were not detected. The in vitro transformation assays are under validation and draft OECD guidelines are available. Thus the inclusion of one of these assays in tiered ENM testing strategy together with mammalian genotoxicity tests can contribute to safety assessment of ENMs.

6. In Vivo Toxicity Testing

In addition to validated in vitro test methods, validated in vivo tests are highly needed in the field of ENMs. Interactions of ENMs with macromolecules and other biological components can only be explored properly in in vivo models (11, 66, 67). In vivo testing is expensive and time-consuming and also encompasses an ethical issue. Nonetheless, in vivo studies have to be performed at least for absorption, distribution, metabolism, and excretion (ADME); interactions with cells, tissues, and proteins; hormonal effects; and long-term chronic effects. It is also important to ensure that the vehicle is suitable for good dispersion of the ENMs to avoid intrinsic toxicity.

Pharmacokinetic studies exploring ADME of ENMs are needed as a basis to determine toxicity as well as for designing ENMs for diagnostics and therapy. Only kinetics studies in vivo can provide data about concentrations in target organs and cells, uptake into blood, and distribution in the body.

Toxicity is generally dose dependent, although that might not be completely true for ENMs. Additionally, substances such as mutagenic and carcinogenic compounds are known to induce non-threshold effects. The need to perform in vivo chronic toxicity and carcinogenicity studies should be considered based upon results from in vivo ADME and in vitro genotoxicity studies. Positive results from in vitro genotoxicity tests normally require further testing in vivo. The genotoxicity assays on human or mammalian cells in vitro can be also applied in vivo. If in vivo genotoxicity and carcinogenicity testing is considered necessary, the tests should be designed according to appropriate target organs or tissues as well as in vitro genotoxicity endpoints. ENMs could induce toxic effects which are not caught by the endpoints under the current protocols (4) and thus might also be a need for additional endpoints, as cardiovascular, immunological, and neurological (30, 31).

7. Conclusion

The unique features of ENMs may be coupled with toxicity that needs to be assessed. The leading mechanisms of possible toxicity are oxidative stress, inflammation, genotoxicity, and carcinogenicity but other generic endpoints which may lead to cardiovascular, neurological diseases or developmental and reproductive toxicity are not excluded. Existing validated and standardized tests for toxicity testing of ENMs both in vitro and in vivo need to be adopted for ENM testing and new methods need to be developed. Additionally to appropriate battery of tests, testing strategy needs to take into consideration physicochemical characterization of ENMs, exposure assessment, and uptake and transport of ENMs and these need to be integral part of the testing strategy. Validated reference material and appropriate controls relevant to nanosized material need to be also included.

Acknowledgment

Supported by European Commission Seventh Framework Programme [Health-2007-1.3-4], Contract no: 201335, www.nanotest-fp7.eu

References

1. Donaldson K, Stone V, Tran CL et al (2004) Nanotoxicology. Occup Environ Med 61: 727–728
2. Haynes CL (2010) The emerging field of nanotoxicology. Anal Bioanal Chem 398:587–588
3. Maynard AD, Warheit DB, Philbert MA (2011) The new toxicology of sophisticated materials: nanotoxicology and beyond. Toxicol Sci 120(Suppl 1):S109–S129
4. Kroll A, Pillukat MH, Hahn D et al (2009) Current in vitro methods in nanoparticle risk assessment: limitations and challenges. Eur J Pharm Biopharm 72:370–377
5. Feliu N, Fadeel B (2010) Nanotoxicology: no small matter. Nanoscale 2:2514–2520
6. Dusinska M, Fjellsbo L, Magdolenova Z et al (2009) Testing strategies for the safety of nanoparticles used in medical applications. Nanomedicine (Lond) 4:605–607
7. EMEA (2006) Reflection Paper on Nanotechnology-Based Medicinal Products for Human Use, EMEA/CHMP/79769/2006. www.emea.europa.eu/pdfs/human/genetherapy/7976906en.pdf
8. Gwinn MR, Tran L (2010) Risk management of nanomaterials. Wiley Interdiscip Rev Nanomed Nanobiotechnol 2:130–137
9. Stone V, Nowack B, Baun A et al (2010) Nanomaterials for environmental studies: classification, reference material issues, and strategies for physico-chemical characterization. Sci Total Environ 408:1745–17454
10. Bouwmeester H, Lynch I, Marvin HJP et al (2011) Minimal analytical characterisation of engineered nanomaterials needed for hazard assessment in biological matrices. Nanotoxicology 5:1–11
11. Commitee ES (2011) Guidance on the risk assessment of the application of nanoscience and nanotechnologies in the food and feed chain. EFSA Journal 9:2140–2176
12. Faraji AH, Wipf P (2009) Nanoparticles in cellular drug delivery. Bioorg Med Chem 17:2950–2962
13. Warheit DB, Sayes CM, Reed KL et al (2008) Health effects related to nanoparticle exposures: environmental, health and safety considerations for assessing hazards and risks. Pharmacol Ther 120:35–42

14. Kunzmann A, Andersson B, Thurnherr T et al (2011) Toxicology of engineered nanomaterials: focus on biocompatibility, biodistribution and biodegradation. Biochim Biophys Acta 1810:361–373
15. Dhawan A, Sharma V (2010) Toxicity assessment of nanomaterials: methods and challenges. Anal Bioanal Chem 398:589–605
16. Borm PJ, Muller-Schulte D (2006) Nanoparticles in drug delivery and environmental exposure: same size, same risks? Nanomedicine (Lond) 1:235–249
17. Oberdorster G, Oberdorster E, Oberdorster J (2005) Nanotoxicology: an emerging discipline evolving from studies of ultrafine particles. Environ Health Perspect 113:823–839
18. Adiseshaiah PP, Hall JB, McNeil SE (2009) Nanomaterial standards for efficacy and toxicity assessment. Nanotechnology 2:99–112
19. Goodman CM, McCusker CD, Yilmaz T et al (2004) Toxicity of gold nanoparticles functionalized with cationic and anionic side chains. Bioconjug Chem 15:897–900
20. Kayat J, Gajbhiye V, Tekade RK et al (2011) Pulmonary toxicity of carbon nanotubes: a systematic report. Nanomedicine 7:40–49
21. Nel A (2005) Atmosphere air pollution-related illness: effects of particles. Science 308:804
22. Eom HJ, Choi J (2010) p38 MAPK activation, DNA damage, cell cycle arrest and apoptosis as mechanisms of toxicity of silver nanoparticles in Jurkat T cells. Environ Sci Technol 44:8337–8342
23. Hsin YH, Chen CF, Huang S et al (2008) The apoptotic effect of nanosilver is mediated by a ROS- and JNK-dependent mechanism involving the mitochondrial pathway in NIH3T3 cells. Toxicol Lett 179:130–139
24. Dailey LA, Jekel N, Fink L et al (2006) Investigation of proinflammatory potential of biodegradable nanoparticle drug delivery systems in the lung. Toxicol Appl Pharmacol 215:100–108
25. Magdolenova Z, Dhawan A, Collins A, Stone V, Dusinska M Mechanisms of Genotoxicity. A Review of Recent in vitro and in vivo Studies with Engineered Nanoparticles. Nanotoxicology. Submitted
26. Stone V, Johnston H, Schins RP (2009) Development of in vitro systems for nanotoxicology: methodological considerations. Crit Rev Toxicol 39:613–626
27. Singh N, Manshian B, Jenkins GJ et al (2009) NanoGenotoxicology: the DNA damaging potential of engineered nanomaterials. Biomaterials 30:3891–3914
28. Liang XJ, Chen C, Zhao Y et al (2008) Biopharmaceutics and therapeutic potential of engineered nanomaterials. Curr Drug Metab 9:697–709
29. Asharani PV, Low Kah Mun G, Hande MP et al (2009) Cytotoxicity and genotoxicity of silver nanoparticles in human cells. ACS Nano 3:279–290
30. Seaton A, Tran L, Aitken R et al (2010) Nanoparticles, human health hazard and regulation. J R Soc Interface 7(Suppl 1): S119–S129
31. Simko M, Mattsson MO (2010) Risks from accidental exposures to engineered nanoparticles and neurological health effects: a critical review. Part Fibre Toxicol 7:42
32. Oberdörster G, Maynard A, Donaldson K et al (2005) Principles for characterizing the potential human health effects from exposure to nanomaterials: elements of a screening strategy. Part Fibre Toxicol 2:8
33. Abbott A (2003) Cell culture: biology's new dimension. Nature 424:870–872
34. Byrne H. J., Lynch I., de Jong W. H. et al. (2010) Protocols for assessment of biological hazards of engineered nanoparticles. European network on the health and environmental impact of nanomaterials, pp 1–30, http://www.nanoimpactnet.eu/uploads/file/Reports_Publications/D1.7%20Protocols%20for%20Assessment%20Bio-Hazards%20in%20ENMs.pdf
35. Priya BR, Byrne HJ (2008) Investigation of sodium dodecyl benzene sulphonate assisted dispersion and debundling of single wall carbon nanotubes. J Phys Chem C 112:332–337
36. Liu F, Soares MJ, Audus KL (1997) Permeability properties of monolayers of the human trophoblast cell line BeWo. Am J Physiol 273: C1596–C1604
37. Ampasavate C, Chandorkar GA, Vande Velde DG et al (2002) Transport and metabolism of opioid peptides across BeWo cells, an in vitro model of the placental barrier. Int J Pharm 233:85–98
38. Saunders M (2009) Transplacental transport of nanomaterials. Wiley Interdiscip Rev Nanomed Nanobiotechnol 1:671–684
39. Brown J, Reading SJ, Jones S et al (2000) Critical evaluation of ECV304 as a human endothelial cell model defined by genetic analysis and functional responses: a comparison with the human bladder cancer derived epithelial cell line T24/83. Lab Invest 80:37–45
40. Saovapakhiran A, D'Emanuele A, Attwood D et al (2009) Surface modification of PAMAM dendrimers modulates the mechanism of cellular internalization. Bioconjug Chem 20:693–701
41. Herzog E, Casey A, Lyng FM et al (2007) A new approach to the toxicity testing of carbon-based nanomaterials-the clonogenic assay. Toxicol Lett 174:49–60

42. Ng CT, Li JJ, Bay BH et al (2010) Current studies into the genotoxic effects of nanomaterials. J Nucleic Acids pii:947859
43. Mortelmans K, Zeiger E (2000) The Ames Salmonella/microsome mutagenicity assay. Mutat Res 455:29–60
44. Landsiedel R, Kapp MD, Schulz M et al (2009) Genotoxicity investigations on nanomaterials: methods, preparation and characterization of test material, potential artifacts and limitations-many questions, some answers. Mutat Res 681:241–258
45. Shinohara N, Matsumoto K, Endoh S et al (2009) In vitro and in vivo genotoxicity tests on fullerene C60 nanoparticles. Toxicol Lett 191:289–296
46. Mori T, Takada H, Ito S et al (2006) Preclinical studies on safety of fullerene upon acute oral administration and evaluation for no mutagenesis. Toxicology 225:48–54
47. Wirnitzer U, Herbold B, Voetz M et al (2009) Studies on the in vitro genotoxicity of baytubes, agglomerates of engineered multi-walled carbon-nanotubes (MWCNT). Toxicol Lett 186:160–165
48. Di Sotto A, Chiaretti M, Carru GA et al (2009) Multi-walled carbon nanotubes: lack of mutagenic activity in the bacterial reverse mutation assay. Toxicol Lett 184:192–197
49. Balasubramanyam A, Sailaja N, Mahboob M et al (2010) In vitro mutagenicity assessment of aluminium oxide nanomaterials using the Salmonella/microsome assay. Toxicol In Vitro 24:1871–1876
50. Maenosono S, Suzuki T, Saita S (2007) Mutagenicity of water-soluble FePt nanoparticles in Ames test. J Toxicol Sci 32:575–579
51. Maenosono S, Yoshida R, Saita S (2009) Evaluation of genotoxicity of amine-terminated water-dispersible FePt nanoparticles in the Ames test and in vitro chromosomal aberration test. J Toxicol Sci 34:349–354
52. Yoshida R, Kitamura D, Maenosono S (2009) Mutagenicity of water-soluble ZnO nanoparticles in Ames test. J Toxicol Sci 34:119–122
53. Collins AR, Dusinska M, Gedik CM et al (1996) Oxidative damage to DNA: do we have a reliable biomarker? Environ Health Perspect 104(Suppl 3):465–469
54. Dusinska M, Collins AR (1996) Detection of oxidised purines and UV-induced photoproducts in DNA, by inclusion of lesion-specific enzymes in the comet assay (single cell gel electrophoresis). ATLA 24:405–411
55. Jha AN (2008) Ecotoxicological applications and significance of the comet assay. Mutagenesis 23:207–221
56. O.E.C.D. (1997) Guidelines for testing chemicals. Mammalian erythrocyte micronucleus test. Vol. Guideline 474, Adopted: 21st July 1997, pp 1–10, http://www.oecd.org/chemicalsafety/assessmentofchemicals/1948442.pdf
57. Fenech M (2007) Cytokinesis-block micronucleus cytome assay. Nat Protoc 2: 1084–1104
58. Laingam S, Froscio SM, Humpage AR (2008) Flow-cytometric analysis of in vitro micronucleus formation: comparative studies with WIL2-NS human lymphoblastoid and L5178Y mouse lymphoma cell lines. Mutat Res 656:19–26
59. Fenech M, Kirsch-Volders M, Natarajan AT et al (2011) Molecular mechanisms of micronucleus, nucleoplasmic bridge and nuclear bud formation in mammalian and human cells. Mutagenesis 26:125–132
60. Gonzalez L, Sanderson BJ, Kirsch-Volders M (2011) Adaptations of the in vitro MN assay for the genotoxicity assessment of nanomaterials. Mutagenesis 26:185–191
61. Doak SH, Griffiths SM, Manshian B et al (2009) Confounding experimental considerations in nanogenotoxicology. Mutagenesis 24:285–293
62. Hackenberg S, Scherzed A, Kessler M et al (2011) Silver nanoparticles: evaluation of DNA damage, toxicity and functional impairment in human mesenchymal stem cells. Toxicol Lett 201:27–33
63. Galloway SM, Aardema MJ, Ishidate M Jr et al (1994) Report from working group on in vitro tests for chromosomal aberrations. Mutat Res 312:241–261
64. Galloway SM, Armstrong MJ, Reuben C et al (1987) Chromosome aberrations and sister chromatid exchanges in Chinese hamster ovary cells: evaluations of 108 chemicals. Environ Mol Mutagen 10(Suppl 10):1–175
65. Benigni R, Bossa C (2011) Alternative strategies for carcinogenicity assessment: an efficient and simplified approach based on in vitro mutagenicity and cell transformation assays. Mutagenesis 26:455–460
66. Clift MJ, Gehr P, Rothen-Rutishauser B (2011) Nanotoxicology: a perspective and discussion of whether or not in vitro testing is a valid alternative. Arch Toxicol 85: 723–731
67. Fischer HC, Chan WC (2007) Nanotoxicity: the growing need for in vivo study. Curr Opin Biotechnol 18:565–571

Chapter 2

Lactate Dehydrogenase Assay for Assessment of Polycation Cytotoxicity

Ladan Parhamifar, Helene Andersen, and S. Moein Moghimi

Abstract

Cellular toxicity and/or cell death entail complex mechanisms that require detailed evaluation for proper characterization. A detailed mechanistic assessment of cytotoxicity is essential for design and construction of more effective polycations for nucleic acid delivery. A single toxicity assay cannot stand alone in determining the type and extent of damage or cell death mechanism. In this chapter we describe a lactate dehydrogenase (LDH) assay for high-throughput screening that can be used as a starting point for further detailed cytotoxicity determination. LDH release is considered an early event in necrosis but a late event in apoptosis. An accurate temporal assessment of the toxic responses is crucial as late apoptosis may convert into necrosis as well as in situations where cell death is initiated without any visible cell morphological changes or responses in assays measuring late events, resulting in early ongoing toxicity being overlooked.

Key words: Polycations, Toxicity, Lactate dehydrogenase, Plasma membrane damage, Apoptosis, Necrosis

1. Introduction

Several methods are available for determining cellular toxicity such as the lactate dehydrogenase (LDH) assay, annexinV/PI, nuclear fragmentation assays, dye exclusion assays, morphological assessments, mitochondrial based assays, and analysis of various biochemical markers involved in apoptotic (1), necroptotic (2) (programmed necrosis), or necrotic pathways (3). LDH release is considered an early event in necrosis but a late event in apoptosis. The assay can be used for high-throughput analysis in a 96-well format and is easy to perform and informative if interpreted

Manfred Ogris and David Oupicky (eds.), *Nanotechnology for Nucleic Acid Delivery: Methods and Protocols*, Methods in Molecular Biology, vol. 948, DOI 10.1007/978-1-62703-140-0_2,

correctly. This chapter provides a detailed introduction, protocol, and useful considerations specific to this method.

LDH is a stable cytosolic enzyme present in mammalian cells that leaks out of the cells following plasma membrane damage or rupture. The LDH assay is based on quantifying the enzyme activity in the cell medium and can be used in a variety of ways. Firstly, the assay may be used to directly assess cytotoxicity by measuring the release of LDH from dead or severely damaged cells either subjected to a drug/particle or to effector cells (cytotoxic T cells, natural killer cells, lymphokine-activated killer cells, and monocytes) (Promega, Clontech). Secondly, the assay may be used to count the total number of cells (by complete lysis) or to count the amount of surviving cells after removal of the dead cells (viability test). Only the first assay will be discussed in relation to polycation cytotoxicity; however it should be mentioned that the viability test can also be used in assessing the effect of polycation safety. The essential principle behind the assay is that LDH released into the growth medium (from damaged/dead cells) catalyzes the conversion of lactate to pyruvate and converts NAD^+ into NADH. NADH in the presence of diaphorase transforms the yellow tetrazolium salt 2-(*p*-iodophenyl)-3-(*p*-nitrophenyl)-5-phenyltetrazolium chloride (INT) into red formazan. The level of color change is proportional to the amount of LDH released and thus the extent of cell death and membrane damage (Fig. 1). This color change can be measured at the absorbance bandwidth of 490–520 nm. The assay is quick and takes under 1 h to perform rendering it excellent for preliminary screening purposes.

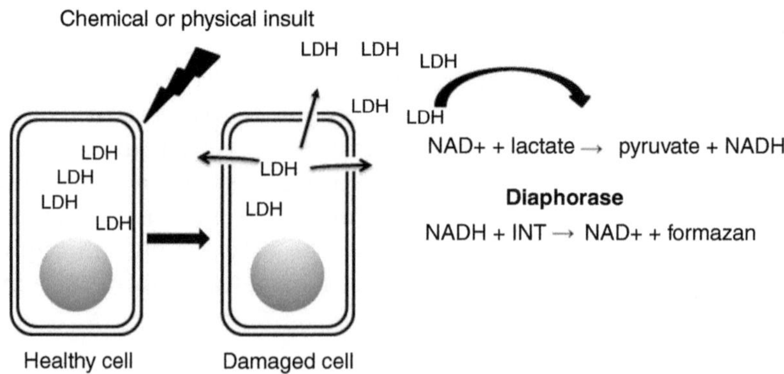

Fig. 1. LDH assay chemical reaction.

2. Materials

LDH kits are available from a variety of companies and are all based on the same principle; however, they may vary slightly in their procedures. In this chapter the various companies are sometimes given as references, as the information described is provided by their protocols. The assay can be performed both in a colorimetric and flourometric fashions. The principle is the same (see Subheadings 1 and 3) but the substrate converted is different. All reagents, except 1% BSA/PBS, are usually included in the kit but the components may be named, composed, or provided differently. Some kits provide the components in such a fashion that all reagents are mixed together and added to the assay plate and measurement is performed directly after 30 min (Cayman), whereas others include or suggest as an optional step (Clontech) addition of stop solution before measurement (Promega). Abcam and BioVision also offer the (4-[3-(4-iodophenyl)-2-(4-nitrophenyl)-2*H*-5-tetrazolio]-1,3-benzene disulfonate) WST substrate which produces a bright yellow color resulting in less growth medium being required for measurement, thus significantly reducing the interference of medium color and allowing the use of 10% FBS. Examples of companies that carry LDH kits: Abcam, Clontech, Promega, Sigma Aldrich, Cayman Chemicals, BioVision, and GE Biosciences.

2.1. Media, Materials, and Equipment

1. Growth medium (if possible phenol red free) with appropriate supplements.
2. Distilled water.
3. Growth plate, 96-well flat bottom for adherent cells or round bottom for suspension cells (other well size plates can be used).
4. Assay plate, 96-well clear flat-bottom.
5. Plate reader with correct filter (varies between kits).
6. Plate centrifuge.
7. Hemocytometer.
8. Light microscope.
9. PBS + 1% BSA (0.2 g/L KCl, 8.0 g/L NaCl, 0.2 g/L KH_2PO_4, 1.15 g/L Na_2HPO_4, 1% (w/v) bovine serum albumin, dissolved in deionized water and sterile filtered).
10. Kit components:
    (a) Lysis buffer (Triton X-100) working concentration is 0.9–1% v/v.
    (b) Stop solution: 1 M acetic acid solution (Promega) or 1 N hydrochloric acid (Sigma Aldrich).

(c) Substrate mix (can also be supplied as separate components depending on the kit).

(d) Assay buffer.

(e) LDH positive control/standard (diluted in 1% BSA/PBS).

3. Methods

As mentioned in Subheading 1 there are many commercial kits available for measuring LDH release and they are based on absorbance or fluorescence measurements. The assay protocols are very similar for both types of measurements, although differences will be highlighted in this section. This protocol is based mainly on the Promega kit and therefore certain references are made to Promega; however similar points might also be relevant in other kit descriptions.

3.1. Standard Controls to Be Included in LDH Cytotoxicity Assays (CytoTox 96, Promega, LDH Detection Kit, Clontech)

1. Spontaneous LDH release: Transfer growth medium from a triplicate set of wells containing untreated cells in the growth plate to the assay plate (see Notes 1 and 2).
2. Maximum release control: Transfer growth medium from a triplicate set of wells with lysed cells in the growth plate to the assay plate. Add lysis buffer from the kit or lyse the cells in the growth plate with Triton X-100 (final concentration 0.9%, v/v). Add the lysis buffer for 45 min before assay start.
3. Medium background control: Growth medium added to triplicate set of wells in assay plate. This control is used to correct for the absorbance of phenol red and LDH present in the serum.
4. Volume correction control: Add the lysis buffer and growth medium to a triplicate set of wells in the assay plate. This corrects for the change in volume when adding lysis buffer to the maximum release control cells.
5. Substance control: Add medium with the cytotoxic agent, in the concentration used with the target cells, to a triplicate set of wells in the assay plate.
6. Positive control: Dilute LDH positive control/standards diluted in 1% BSA/PBS and add it to a triplicate set of wells in the assay plate.

3.2. Optimization of Target Cell Number

The amount of LDH released from cells varies among different cell types (4). Therefore, the optimal cell density should be determined for each cell type. The optimal cell density is where the ratio/difference between maximum release and spontaneous release is the

Table 1
Layout for optimization of cell number in 96-well plate format

	1	2	3	4	5	6	7	8	9	10	11	12
A	Background											
B	Cell density 1			Cell density 8			Cell density 1			Cell density 8		
C	Cell density 2			Cell density 9			Cell density 2			Cell density 9		
D	Cell density 3			Cell density 10			Cell density 3			Cell density 10		
E	Cell density 4			Cell density 11			Cell density 4			Cell density 11		
F	Cell density 5			Cell density 12			Cell density 5			Cell density 12		
G	Cell density 6			Cell density 13			Cell density 6			Cell density 13		
H	Cell density 7			Cell density 14			Cell density 7			Cell density 14		
	Spontaneous LDH release						Maximum release					

highest. For most cell lines the optimal concentration is between 2.5×10^4 and 10^5 cells/mL (see Notes 3–6).

1. Prepare a serial dilution of your target cells (see Note 5).
2. Divide the plate into two, where one half is used to measure spontaneous release and the other half for measuring maximum release. An example of a plate layout is shown in Table 1.
3. Add the cells to a 96-well tissue culture plate in triplicates.
4. Adherent cells should adhere overnight (or longer) at 37°C and 5% CO_2.
5. Suspension cells can be treated just after seeding.
6. Add lysis buffer or Triton X-100 (0.9% v/v, final concentration) to the wells used for maximum release and incubate at 37°C. This step should be started 45 min before the final time point.
7. Add medium to the spontaneous LDH release wells (low control), so the final volume of medium is the same in all the wells (treated and untreated). The final volume should be the same when performing the cytotoxicity assay.
8. Prepare the controls (see above) and continue with the LDH release assay.

3.3. Preparation of Reagent

1. Prepare the assay buffer as described in the kit. In some of the available kits, the assay buffer is included and ready to use (CytoTox 96, Promega) (see Notes 7 and 8).
2. Prepare the reaction solution/substrate mix before use.

3.4. LDH Release Assay

1. Seed out the test cells in the optimal concentration. Let the cells attach overnight at 37°C and 5% CO_2, or until they have reach the appropriate density (see Notes 2, 3, and 9).
2. Aspirate the medium to remove any released LDH during culturing.
3. Add fresh medium to all wells. Keeping the volume so it just covers the bottom of the well can help reduce the possibility of the polycation to bind to the well walls.
4. Add the polycation or polyplexes/lipoplexes (diluted in growth medium) in the concentration to be assayed to the designated wells.
5. Incubate at 37°C and 5% CO_2 for the time required to assay the polycation.
6. Pellet the cells (250 × *g* for 4 min). This step may be skipped when working with adherent cells (see Note 10).
7. Transfer 50 μL of the supernatant to a new flat-bottomed 96-well plate (the volume varies between the kits). For this step and in the rest of the assay a Biomek 3000 robot (Beckman Coulter) can be a valuable tool. In some kits this step can be left out and the reaction solution can be added directly to the culture medium (e.g., The Cytotoxicity Detection Kit^{PLUS}, Roche Applied Science, and CytoScan—*Fluoro* Assay, GE Biosciences) (see Notes 11 and 12).
8. Add 50 μL reaction solution to each well (the volume varies between the kits) (see Note 11).
9. Incubate the plate for 30 min at ambient temperature protected from the light. The plate can be placed on an orbital shaker during incubation (see Notes 13 and 14).
10. If recommended in the kit, add stop solution to each well (CytoTox 96, Promega, and CytoScan—*Fluoro* Assay, GE Biosciences). The absorbance/fluorescence can now be measured directly (LDH Cytotoxicity Kit, Cayman Chemical) or the stop solution is an option (LDH Cytotoxicity Detection Kit, Clontech), after which the measurement should be performed within 1 h.
11. Make sure that there are no bubbles in the wells or remove them before measuring absorbance (see Note 15).
12. Measure the absorbance at the recommended wavelength using a Multiskan MS ELISA reader (Labsystems) or a similar instrument. The absorbance/fluorescence should be measured within an hour after addition of the stop solution (see Notes 1, 4, 8, 13, 16–21).

3.5. Calculations

The percentage of cytotoxicity can be calculated and this gives the result in percentage of maximum release. Calculate the average of the triplicate of the experimental value, spontaneous release, and maximum release. If several concentrations of the polycation and several time points have been tested, the cytotoxic profile can be mapped (5) (see Notes 22, 23).

4. Notes

1. Serum and phenol red, both found in cell growth medium, can give rise to high background in the assay (depending on which kit is used); serum because of its LDH activity and phenol red due to its color. Different sera contain different amounts of LDH; for example calf-serum has high LDH activity whilst human AB serum has low LDH activity. Eliminating phenol red from the media or decreasing the percentage of serum from 10 to 5% can reduce this effect (6). However, some kits offer substrates and buffers where serum and phenol red do not interfere with the background (Abcam, BioVision). For troubleshooting regarding underestimation see Notes 16–18.
2. Some protocols suggest using 1% BSA in the culture medium instead of FBS (Clontech) to reduce background LDH activity; however lack of serum whether it is replaced by BSA or not can affect cell viability and therefore is discouraged (Promega).
3. Even though the kits are normally set up for high-throughput readout in 96-well format, the assay can be performed using other well sizes by adjusting cell number and volumes accordingly. This may be important when cell confluency matters for drug delivery by polycations or in case the cells need to grow for a certain number of days before the assay is performed. Accurate cell confluency can be difficult to achieve in 96-well format due to cell size and growth rate.
4. High absorbance values in the untreated or spontaneous release cells may be due to cells already having a leaky plasma membrane. This can stem from suboptimal growth conditions or disturbance of cells due to rough pipetting or high centrifugation rates ($\geq 250 \times g$). To avoid these high values refrain from the above-mentioned handling of the cells and keep the cell density lower than 1.5×10^6/mL (Promega).
5. The kits usually specify the sensitivity of the assay by supplying information regarding the lowest amount of cells that should be used.

6. In a study by Wolterbeek and van der Meer the total cellular content of LDH was related to the cell volume as opposed to cell number. Therefore they suggest that it is more accurate to measure cell volume to obtain a ratio, which is independent of cell numbers and allows more accurate comparison of independent experiments on the same cell line (7).
7. If there is a precipitate in the assay buffer this may be removed by centrifugation with 300 × *g* for 5 min. However the precipitate should not interfere with the assay (Promega).
8. Make sure that all reagents are stored properly (at correct temperatures and protected from light) and not frozen and thawed too many times. This can reduce the sensitivity of the assay.
9. Optimize the cell number so that the values in the readout are not too low or too high (Promega). Values should be within an accurate range of absorbance. This can be tested during cell number optimization assay and can vary between plate readers. But in general absorbance values above 1.5–2 are not considered accurate.
10. If centrifugation is not performed one must be very careful to make sure that there is no cell-debris or particle disturbing the readout. The disturbance can manifest as inconsistencies in the readout.
11. Inconsistent pipetting can also give variations in the assay (normally it is recommended that each treatment should be performed in triplicate or quadruplicate). Therefore correct pipetting with as few pipetting steps as possible by using multichannel pipettes or using kits where the reagents are added directly to the cells can decrease variation.
12. A downside with kits where the reagents are added directly to the cells is that the cells might then not be useful for further assaying. However, in kits where the supernatant is transferred to an assay plate and assayed there, the cells can be used for further assaying.
13. If the toxicity is high leading to high values (above 1.5–2) the reaction time with the substrate-mix buffer can be shortened (Promega).
14. Make sure that the reaction is performed in the dark as the light can degrade the substrate giving rise to values lower than expected (Promega).
15. Bubbles/foaming should be avoided as these can interfere with the readout (7).
16. The LDH assay can under some circumstances underestimate the amount of toxicity. If the LDH is released into the media at an early stage and the assay is performed too late the LDH might be degraded. To avoid this, cells can be lysed in the

beginning and at the end of the assay and compared. This will estimate any potential degradation during the course of the assay.

17. A second concern of underestimation can arise if the test polycation/particles (polyplexes and lipoplexes) interfere with the LDH assay, thus affecting the chemical reaction. This can be tested by adding the polycation/particles together with the lysis buffer to check whether interference occurs. If it does, the values will be lowered compared with the lysed cells without the addition of polycation/particles (8).
18. A third concern of underestimation is if the polycation/particles in question stops cell proliferation. Smaller number of cells will show lower LDH content. The correct control for this would be to add the polycation/particles for the duration of the experiment and then lyse the cells and compare them to the untreated lysed cells (8).
19. If the readout values are high and a spectrophotometer is used that has a low sensitivity the readout will be unreliable. Depending on the instrument used the absorbance readout should preferably not exceed 1.5–2.
20. Some particles may interfere with the readout by interfering with the enzyme. Therefore a morphological estimation of the cells must also be performed to ensure that the toxicity measured accurately corresponds to the observed toxicity. To test if the polycation/test particles could interfere with the LDH activity, the polycation/particles (diluted in medium) is added to the assay plate (in triplicate) followed by addition of LDH and the reaction mixture provided in the kit (Substance control II, Roche). The readout is then compared to the control sample without the addition of the polycation/particles.
21. LDH release is an indication of plasma membrane damage which is a late stage in apoptosis (but early in necrosis), so a lack of increase does not necessarily mean that there is no toxicity. A time- and concentration-dependent assay should be performed and the assay should be used in conjunction with other toxicity assays for accurate interpretation of the toxicity.
22. The kits usually provide means of calculating the cytotoxicity (see Subheading 3) which can be presented in various ways. A few examples include percent of LDH release compared to untreated cells, percent of LDH release compared to total LDH content (obtained from lysed cells), and amount of LDH released in units/mL.
23. A standard curve of LDH activity can be made if instead of percentage of toxicity, the amount of LDH released (units/mL) needs to be calculated (Clontech, BioVision).

References

1. Degterev A, Yuan J (2008) Expansion and evolution of cell death programmes. Nat Rev Mol Cell Biol 9:378–390
2. Galluzzi L, Kroemer G (2008) Necroptosis: a specialized pathway of programmed necrosis. Cell 135:1161–1163
3. Hetz CA, Hunn M, Rojas P, Torres V, Leyton L, Quest AF (2002) Caspase-dependent initiation of apoptosis and necrosis by the Fas receptor in lymphoid cells: onset of necrosis is associated with delayed ceramide increase. J Cell Sci 115:4671–4683
4. Korzeniewski C, Callewaert DM (1983) An enzyme-release assay for natural cytotoxicity. J Immunol Methods 64:313–320
5. Moghimi SM, Symonds P, Murray JC, Hunter AC, Debska G, Szewczyk A (2005) A two-stage poly(ethylenimine)-mediated cytotoxicity: implications for gene transfer/therapy. Mol Ther 11:990–995
6. Decker T, Lohmann-Matthes ML (1988) A quick and simple method for the quantitation of lactate dehydrogenase release in measurements of cellular cytotoxicity and tumor necrosis factor (TNF) activity. J Immunol Methods 115:61–69
7. Wolterbeek HT, van der Meer AJ (2005) Optimization, application, and interpretation of lactate dehydrogenase measurements in microwell determination of cell number and toxicity. Assay Drug Dev Technol 3: 675–682
8. Stone V, Johnston H, Schins RP (2009) Development of in vitro systems for nanotoxicology: methodological considerations. Crit Rev Toxicol 39:613–626

Chapter 3

Combined Fluorimetric Caspase 3/7 Assay and Bradford Protein Determination for Assessment of Polycation-Mediated Cytotoxicity

Anna K. Larsen, Arnaldur Hall, Henrik Lundsgart, and S. Moein Moghimi

Abstract

Cationic polyplexes and lipoplexes are widely used as artificial systems for nucleic acid delivery into the cells, but they can also induce cell death. Mechanistic understanding of cell toxicity and biological side effects of these cationic entities is essential for optimization strategies and design of safe and efficient nucleic acid delivery systems. Numerous methods are presently available to detect and delineate cytotoxicity and cell death-mediated signals in cell cultures. Activation of caspases is part of the classical apoptosis program and increased caspase activity is therefore a well-established hallmark of programmed cell death. Additional methods to monitor cell death-related signals must, however, also be carried out to fully define the type of cell toxicity in play. These may include methods that detect plasma membrane damage, loss of mitochondrial membrane potential, phosphatidylserine exposure, and cell morphological changes (e.g., membrane blebbing, nuclear changes, cytoplasmic swelling, cell rounding). Here we describe a 96-well format protocol for detection of capsase-3/7 activity in cell lysates, based on a fluorescent caspase-3 assay, combined with a method to simultaneously determine relative protein contents in the individual wells.

Key words: Caspase activity, Cell-death assays, Apoptosis, Necrosis, Gene therapy, Cationic drug carriers

1. Introduction

Cationic polyplexes have long been used for nucleic acid delivery in many transfection and silencing protocols, both in vitro and in vivo, and promising results have emerged from such attempts (1). Nevertheless, safety concerns and technical problems, mainly associated with the cellular delivery systems, are still hampering the production of successful nucleic acid-based drug candidates. Accumulating evidence from studies of cationic carrier systems,

Manfred Ogris and David Oupicky (eds.), *Nanotechnology for Nucleic Acid Delivery: Methods and Protocols*, Methods in Molecular Biology, vol. 948, DOI 10.1007/978-1-62703-140-0_3, © Springer Science+Business Media, LLC 2013

comprising cationic polymers, cationic lipids, and cell-penetrating peptides, suggests that diverse biological responses or adverse side effects, including apoptotic and necrotic cell death, may be associated with the use of these carriers. Polyethylenimines (PEIs) are among the most prominent polycations available, and are widely used as transfection agents, ascribed to their high transfection efficiency. Polycations can condensate RNA and DNA into nanostructures (polyplexes), which protect the nucleic acid from degradation; an essential prerequisite for efficient delivery into cells. A major challenge in the design of polycation-based gene therapeutics is to overcome immediate or later phase-response cytotoxicity of synthetic carriers without significantly reducing their delivery efficacy, as well as ensuring target cell specificity. In order to rationally design new improved polycation-based delivery systems, and to properly monitor the performance of existing ones, detailed mechanistic understanding of their complex cellular toxicity profiles is essential. Characterization of biological side effects and evaluation of cytotoxicity should be based on several different types of assays to determine the extent and type of cell death-response(s) involved.

Caspases are a well-known family of cysteine proteases with essential regulatory properties in both apoptotic and necrotic cell death (2, 3). Caspases cleave their substrates after aspartic acid residues, and are synthesized as proenzymes that require proteolytic processing to become active enzymes. Initiator caspases (caspases-1, -2, -4, -5, -8, -9, -10, -11, and -12) process downstream executioner or effector caspases (caspases-3, -6, -7, and -14), which in turn cleave further targets and have the capacity to trigger complex cellular cascades that may lead to cell death. Although an intimate connection between caspases-activation and cell death mechanisms (classically apoptosis, but also necrosis) has been well established (2–4), the existence of caspase-independent forms of programmed cell-death has also been reported (5, 6) as well as studies showing vital functions of caspases in proliferation and differentiation (7, 8). Caspase 3 is a prominent cell death marker and a point of convergence as effector enzyme in the caspase cascade activation that typically occurs during apoptosis.

A large array of methods, based on different techniques, have been developed to detect the protein levels, or enzyme activity, of various initiator- and effector-caspases and a vast number of assays and kits are commercially available for this purpose. Many activity-assays are based on caspase-mediated proteolysis of a short (often tetrameric) peptide sequence, coupled to a chemical group (a dye or a fluorophore) that following proteolytic cleavage of the peptide either (1) changes its spectral properties leading to changes in absorbance intensity (colorimetric assays) or (2) changes the fluorescent properties of a substrate due to the release of a fluorescent group (fluorescent assays). The change in fluorescence or dye intensity directly reflects the level of active caspase enzymes capable of hydrolyzing the specific peptide substrate. Depending on the caspase enzyme of interest, the fluorescently labeled substrate can be exchanged to address the activity of other caspases. The readout of activity-based caspases assays can be monitored using diverse technological platforms, including fluorescence plate readers or fluorescence spectrophotometers. The procedures described in this chapter are based on a combination of a caspase 3 fluorimetric assay, CASP3F (Sigma), and the Bradford Protein Determination procedure (BioRad), and were developed to be carried out in 96-well format for high-throughput assays on a robotic platform (BIOMEK 3000), but the procedures may well be carried out manually using appropriate plate reader(s).

2. Materials

This fluorimetric caspases-3 assay is a modified protocol based on the CASP3F kit available from Sigma-Aldrich. Reagents and buffers for this assay may also be prepared in-house with chemicals and reagents from other suppliers (see Note 1).

1. Growth media, PBS (D8587, Sigma), trypsin–EDTA 10× solution (T4174, Sigma), PenStrep (P0781, Sigma) solution, and other standard reagents such as Fetal Calf Serum (FCS, e.g., Fischer Scientific; PAA-A15-101) and equipment for cultivation of the cell line(s). The assay may be carried out with either suspension or adherent cells in biological replicates. If suspension cells are used and biological replicates desired (in contrast to technical replicates) a plate centrifuge is also needed.
2. Apoptosis inducers: Staurosporine: A stock solution (0.2 mg/mL in DMSO) is diluted into growth medium at 1 μg/mL (see Note 2). Doxorubicin: A stock solution of 50 mM (aq) is diluted into serum-free medium at 100 μM and allowed to incubate with the cells for a defined period (see Note 3). Medium containing the inducers are allowed to incubate with the cells for a defined period of time (e.g., 30–60 min, or longer), typically in parallel with the same exposure time as for the polycations.
3. Protein Assay Dye Reagent Concentrate (BioRad #500-0006) based on the method by Bradford (9). The Protein Dye Reagent is diluted (1:5) with ultrapure water before use.
4. CASP3F Kit components:
   (a) 5× Lysis Buffer: 250 mM HEPES, pH 7.4, 25 mM CHAPS, 25 mM DTT. Working concentration is 1×.

Ultrapure (17 MΩ) water for dilution is supplied in the CASP3F kit.

(b) 10× Assay Buffer: 200 mM HEPES, pH 7.4, 1% CHAPS, 50 mM DTT, 20 mM EDTA. Working concentration is 1× (see Note 4).

(c) Caspase-3 substrate solution, Acetyl-Asp-Glu-Val-Asp-7-amido-4-methylcoumarin (Ac-DEVD-AMC) (A1086, Sigma). Keep the substrate protected from light (see Note 5).

(d) Caspase-3 recombinant lyophilized enzyme (C5974, Sigma). A stock solution of 100 μg/mL recombinant caspase-3 is prepared by reconstituting the vial content (5 μg) with 50 μL ultrapure water. Working concentration is 0.5 μg/mL (1:200 dilution in 1× Assay Buffer or 1× Lysis Buffer).

(e) Caspase-3 inhibitor, Acetyl-Asp-Glu-Val-Asp-CHO (Ac-DEVD-CHO) (A0835, Sigma). A stock solution (2 mM) of the caspases-3 inhibitor is made by dissolving the vial content (5 μg) by addition of 500 μL DMSO (see Note 6).

5. Assay plates and tissue culture plates: Flat-bottom 96-well tissue culture growth plates (e.g., Costar #3595 from Corning Inc.) are used to seed out the cells to be investigated. The optimal cell density in the 96-well growth plate as well as the exact timing of the assay should be carefully determined for each cell line of interest, as the caspase-3 response is highly cell type dependent (see Note 7).

   Two types of flat-bottom assay plates are required for each growth plate: one 96-well clear (e.g., Sterilin) and one 96-well black flat-bottom (e.g., black Greiner 96F FIA Polystyrene plate).

6. Plate reader(s) with top reading setup, and appropriate filter sets for measuring fluorescence (Excitation: 360 nm/Emission: 460 nm) and absorbance maximum at 595 nm (see Note 8). Preferably, the plate reader should be capable of performing/carrying out kinetic reads using/with time intervals set by the user (see Note 9).

3. Methods

All procedures are carried out at room temperature unless otherwise specified. The procedure described here is a modified version of the Caspase-3 fluorimetric assay found in Sigma's CASP3F kit protocol. Methods to measure the level of cytotoxicity (cell death signal) markers in a cell culture should always be adjusted and optimized to the cell line(s) of interest, particularly if (when) the mechanistic profile of cell death is not fully clear. Several types of polycationic materials, with the capacity to mediate cellular co-uptake of nucleic acids in the form of nanosize particles, also induce apoptotic- and/or necrosis-like cellular responses (10–12). As a consequence, subpopulations of the cultured cells might gradually retract and detach from the tissue-culture growth surface. Therefore, this protocol describes a modified version of the Caspase-3 fluorimetric assay, so it also includes a single-well measurement of the protein content in the cell lysates, in order to correct for loss of cellular material. This is achieved by normalization of single-well fluorescent measurements with the total protein content in the individual wells.

3.1. Standard Controls to be Included in the Caspase-3 Assays (CASP3F, Sigma)

All controls are included in biological quadruplicate or triplicate wells. There are two (optional) non-induced, negative cell controls: (1) with or (2) without caspase-3 inhibitor present in the cell lysates.

1. Reagent Blank (background): 5 μL 1× Assay buffer mixed with 200 μL Reaction Mixture.
2. Induced (positive control) cells: Cells treated with an apoptosis inducer in the growth medium for an appropriate amount of time before performing the assay (5 μL of cell lysates mixed with 200 μL Reaction Mixture) (see Note 10).
3. Non-induced cells (negative control 1): Cells that have not been treated with any drug in the growth medium but have had the medium changed in the same manner as the cells in the experimental wells and positive control/induced cells (5 μL of cell lysates mixed with 200 μL Reaction Mixture).
4. Non-induced cells + caspase-3 inhibitor (negative control 2): Negative control cells that have not been treated with any drug in the growth medium, but where Caspase-3 inhibitor has been added to the cell lysates before mixing an aliquot (5 μL) with 200 μL Reaction Mixture (see Note 11).
5. Caspase-3 enzyme positive control: An aliquot of reconstituted recombinant Caspase-3 (100 μg/mL) is diluted 1:200 in 1× Assay or 1× Lysis buffer (0.5 μg/mL final). A 5 μL aliquot of this is added to 200 μL Reaction Mixture (see Note 12).

3.2. Preparation of Cells

The combined fluorimetric Caspase-3/protein determination assay may be carried out using either adherent cells or suspension cells. The protocol described below refers to adherent cells only, but can be adapted to fit suspension cells such as Jurkat T-cells (see Note 13). However, seeding densities of 5,000–10,000 cells per well in a 96-well tissue culture plate (corresponding to approx.

15,000–30,000 cells/cm^2) should fit well with many adherent cell lines. The cells are grown for 24–48 h prior to the assay, depending on the nature of the experiment. The optimal cell density must also be adjusted to the timing and the objective of the experiment (see Note 14).

1. Seed the cells in the appropriate density and grow for 24–48 h (37°C and 5% CO_2 in a humid atmosphere). Serial dilution of the target cells may be used for optimization of the cell density.
2. Treatment of cells in the experimental wells, with the cytotoxic substance in the preferred concentration range, is carried out in biological triplicate or quadruplicates. The potential cytotoxic substance may either be added to the existing growth medium or be added by replacement of the growth medium with medium/buffer containing the desired dilutions of the agent (see Note 15).
3. Incubate or expose cells in the experimental wells to the cytotoxic substance for the defined periods of time (37°C and 5% CO_2 in a humid atmosphere) (see Note 16).
4. Optional: Replace the medium/buffer containing the cytotoxic substance(s) with normal growth medium after incubation or exposure of cells to the cytotoxic substance.
5. Parallel to treatment/exposure of the experimental cells, the wells designated "Induced (positive control) cells" are treated with the apoptotic inducer (e.g., doxorubicin, staurosporine), in triplicate or quadruplicate wells.
6. Incubate 96-well growth plates (37°C and 5% CO_2 in a humid atmosphere) with the experimental and induced (positive control cell) wells for the specified amount of time. Typically a 6–24-h incubation is sufficient for cells treated with an apoptotic inducer (depending on the cell type) before Caspase-3 activity can be detected.
7. Perform the Combined Caspase-3/Protein determination assay according to the procedure described here below.

3.3. Preparation of Reagents

1. Thaw and prepare the amount of Lysis buffer (1×) required for the planned number of wells by dilution with water (use 25 μL Lysis buffer/well).
2. Thaw and prepare the amount of Assay buffer (1×) needed for the planned number of wells (use 200 μL/well). (A small aliquot of 1× Assay buffer might be stored separately for dilution of the Caspase enzyme control and Caspase-3 inhibitor).
3. Prepare the Reaction Mixture by adding an aliquot of the Caspase-substrate (10 mM DMSO stock) to the required amount of 1× Assay Buffer, as specified in the kit (1:600-fold

dilution is used in the CASP3F kit). The Reaction Mixture should be protected against light.

4. Prepare (dilute 1:5) the amount of BioRad Protein Dye Reagent required for the planned number of wells in the assay (use 200 μL/well).
5. Prepare the Caspase-3 enzyme positive control, by dilution of (1 μL) enzyme stock in (200 μL) 1× Assay or 1× Lysis buffer (i.e., 0.5 μg/mL final concentration).
6. Prepare a 200 μM working solution of Caspase-3 inhibitor (Ac-DEVD-CHO) by tenfold dilution in 1× Assay Buffer.

3.4. Combined Fluorimetric Caspase-3/ Bradford Protein Determination Assay

This part of the protocol may be carried out using a programmable robotic instrument (BIOMEK 3000). If carried out by hand, we recommend using a 1–200 μL 8-channel pipette whenever possible. Three 96-well plates are used: (1) the cell growth plate, (2) a black caspases-3 assay plate, and (3) a (clear) protein determination assay plate. A plate layout is presented in Fig. 1.

1. Remove growth medium from the cells by very gentle aspiration, or by pouring off the medium (see Note 17).
2. Immediately dispense 25 μL Lysis buffer to all wells in use in the growth plate, using a multichannel pipette, and incubate on ice for 15–20 min.

A1 Blank	A2 Induced	A3 1	A4 3	A5 5	A6 7	A7 9	A8 11	A9 13	A10 15	A11 17	A12 19
B1 Blank	B2 Induced	B3 1	B4 3	B5 5	B6 7	B7 9	B8 11	B9 13	B10 15	B11 17	B12 19
C1 Blank	C2 Induced	C3 1	C4 3	C5 5	C6 7	C7 9	C8 11	C9 13	C10 15	C11 17	C12 19
D1 Blank	D2 Induced	D3 1	D4 3	D5 5	D6 7	D7 9	D8 11	D9 13	D10 15	D11 17	D12 19
E1 Caspase	E2 Inhibit	E3 2	E4 4	E5 6	E6 8	E7 10	E8 12	E9 14	E10 16	E11 18	E12 20
F1 Caspase	F2 Inhibit	F3 2	F4 4	F5 6	F6 8	F7 10	F8 12	F9 14	F10 16	F11 18	F12 20
G1 Caspase	G2 Inhibit	G3 2	G 4 4	G 5 6	G6 8	G 7 10	G 8 12	G9 14	G 10 16	G 11 18	G 12 20
H Caspase	H2 Inhibit	H 3 2	H 4 4	H 5 6	H 6 8	H 7 10	H 8 12	H 9 14	H 10 16	H 11 18	H 12 20

Negative control: Background control / Growth medium blank:

Positive control: Caspase -3 enzyme:

Positive control cells: Apoptotic Induced cells:

Non-Induced cells or Non induced cells + Inhibitor:

Experimental wells:

Fig. 1. Suggested plate layout for combined caspases 3 and Bradford protein determination assays.

3. During the incubation, add 2 μL Caspase-3 inhibitor (Ac-DEVD-CHO) working solution (200 μM) to the cells in each wells designated "Non-induced cells + caspase-3 inhibitor" (negative control) in the growth plate.
4. During the incubation, dispense 200 μL diluted BioRad Protein Dye Reagent to all the wells in use, in the Protein Assay plate.
5. After incubation on ice, 200 μL Reaction Mixture is dispensed, column by column, to all wells in use in the growth plate. For each column (each addition-step), the Reaction Mixture is mixed well with the cell lysates by pipetting three times up and down, and 200 μL of the mixed liquids are transferred directly to the (black) Capsase-3 assay plate. When all cell lysate suspensions have been transferred to the Capsase-3 assay plate, cover the wells with foil to protect the Caspase substrate from light exposure.
6. Next, in the same way (column by column) transfer 5–10 μL of the remaining cell lysates in the growth plate to the corresponding wells in the Protein Assay plate, and mix the liquids by pipetting up/down just after each transfer step.
7. Incubate the Protein Assay plate at room temperature for at least 5 min (and no longer than 60 min), and then measure the absorbance at 595 nm (560–610 nm) with a plate reader. Make sure that there are no bubbles in the wells or remove them before measuring of absorbance.
8. Allow the enzyme reaction in the Caspase-3 assay plate to occur for 30–120 min at room temperature or 37°C, or until the fluorescent signal that stems from the 'positive control (apoptotically induced) cells' is well above that of the background (reagent blank) and/or non-induced (negative control) cells.
9. Make sure that there are no bubbles in the wells or remove them before measuring fluorescence.
10. If possible, read fluorescence in a kinetic mode every 10 min for, e.g., 60 min, using 360 nm excitation and 460 nm emission filters (see Note 18).

3.5. Calculations

The Caspase-3 activity is calculated by normalization (well to well) of the blank-subtracted fluorescent values obtained from the Caspase-3 Assay plate, with the blank-subtracted absorbance (A_{595}) values from the Protein Assay plate. Calculate the average of the triplicate or quadruplicate values of the normalized data, and plot the data, e.g., as a function of polycation concentration, exposure time, cell type, or type of polycation being examined. Normalization of the fluorescence data results in better signal-to-noise ratios in the calculated data, compared to the non-normalized form, as normalization takes into account the cell loss that may have occurred before running the assay, e.g., due to the exposure times or concentrations of the polycation used. Furthermore, normalization of the fluorescence values (caspase-activity data) with the total protein content values also gives much smoother curves in plots of kinetic data, with fluorescence intensity (per total protein content), as a function of time.

4. Notes

1. DTT 1 M stock should be stored at –20°C, and repeated freeze-thawing cycles must be avoided.
2. Staurosporine DMSO stock is stored at –20°C. When using staurosporine as an apoptosis inducer, it is important to include relevant DMSO/medium control cells.
3. When using Doxorubicin as an apoptosis inducer in SFM, it is important to include relevant serum starvation control cells, if the cells are left without serum for longer/extended periods of time. Doxorubicin is stored at 4°C.
4. If preparing any of these reagents in-house, always use 17 MΩ ultrapure water and molecular biology grade chemicals of a high quality.
5. Store aliquots of the reconstituted enzyme at –70°C and avoid repeating freeze-thawing cycles. BSA might be added as a stabilizer to the dilute enzyme solutions.
6. Store inhibitor stock solutions at –20°C.
7. Higher cell densities are likely to be more resistant to cytotoxic substances, and a clear cytotoxic response is thus more difficult or more slowly obtained from more densely seeded cell cultures. On the other hand, the cell density must be sufficiently high to produce enough activated caspase enzymes to be detected in the assay. The timing of the assay is also an important factor that requires careful optimization. In order to obtain a correct mechanistic cytotoxicity profile, it is desirable to know which type(s) of cell death that is in play. Therefore a time span *post* treatment is allowed before an appropriate signal-to-background ratio of caspases-3/7 activity is obtained.
8. Filter settings between 560 and 610 nm to measure protein dye are acceptable.
9. If more high-put-through conditions are desirable, this assay can be programmed onto and carried out using a semiautomatic system, such as the BIOMEK 3000 (Beckman Coulter) robotic platform, coupled with a DTX800 Multimode Reader.
10. Various types of apoptosis-inducing compounds are commercially available. The amount of time that the cells should be exposed to selected inducer(s) should be determined experimentally,

and according to recommendations in the literature. In our hands adherent cell cultures (e.g. NCI-H1299 from ATCC and HepG2 #85011430 from ECACC) exposed to staurosporine (1 μg/mL final concentration in the growth medium for 20 min) readily induce caspase-3/7 activity after only 4 h of exposure, whereas in cells exposed to doxorubicin (typically 100 μM in serum-free medium for 20 min) caspases-3/7 activity is detectable later and typically after 8–24 h.

11. This cell control is included to see if nonspecific substrate hydrolysis occurs during the assay. Significant difference between the fluorescence (emission) values obtained for negative control 1 and 2 indicates that the substrate, Ac-DEVD-AMC, is subject to elevated background hydrolysis, mediated either by caspase-3 or by other proteases with similar substrate specificity that recognize the Asp-Glu-Val-Asp sequence motif.
12. This control provides a good test of whether the assay components are working. Comparison of caspase-3 activities from whole-cell lysates with those of the pure (recombinant) caspase-3 enzyme is however not very biologically relevant, unless a standard curve for 7-amino-4-methylcoumarin (AMC) that is included in the CASP3F kit from Sigma is also made.
13. To perform the assay using suspension cells in biological replicates, as described in this protocol for adherent cells, a centrifuge equipped with a plate rotor is needed. The optimal cell density to be used in the assay varies among different cell types, and must be determined experimentally.
14. If cells are seeded 1 day before the experiment, e.g., to mimic transfection conditions, a slightly higher cell density should be considered, compared to conditions where cells are allowed to grow for 2 days prior to the assay.
15. If the treatment is carried out by medium replacement, also remember to change the growth medium in the wells with non-treated (negative control) cells, in parallel.
16. The duration of cell exposure to the cytotoxic substance, as well as the time after cell exposure/treatment and until the assay is performed, should be carefully optimized. Depending on the expected cytotoxic mechanism of the potentially cytotoxic substance, and the selected concentration range being tested, exposure times may vary, e.g., between 30 min and 4 h. The timescale after the cell exposure (to polycations) and until the assay is performed is however much longer, from a few up to 24 h.
17. If pouring off the medium, very quickly allow excess medium to drip off by placing the cell growth plate upside down on a piece of tissue paper, before proceeding to the next step, where

the lysis buffer is added. Make sure that the cells are kept wet at all times.

18. The kinetic reading mode is well suited when relatively short incubation times (30 min at 37°C) are used before measuring fluorescence. A linear relationship between the elapsed incubation time and the 460 nm fluorescent emission values is expected and desired, and linearity may serve as a quality assessment of the assay. On the other hand, if very long incubation times are selected before measuring fluorescence, kinetic readings might reveal that the signal is bending off and becomes nonlinear.

Acknowledgments

This work was supported by the Danish Agency for Science, Technology and Innovation (Det Frie Forskningsråd for Teknologi og Produktion), ref. 274-08-0534, and Det Strategiske Forskningsråd, ref. 09-065746/DSF.

References

1. Parhamifar L, Larsen AK, Hunter AC, Andresen TL, Moghimi SM (2010) Polycation cytotoxicity: a delicate matter for nucleic acid therapy-focus on polyethylenimine. Soft Matter 6:4001–4009
2. Wilson KP, Black JA, Thomson JA, Kim EE, Griffith JP, Navia MA, Murcko MA, Chambers SP, Aldape RA, Raybuck SA, Livingston DJ (1994) Structure and mechanism of interleukin-1 beta converting enzyme. Nature 370:270–275
3. Yuan J, Shaham S, Ledoux S, Ellis HM, Horvitz HR (1993) The C. elegans cell death gene ced-3 encodes a protein similar to mammalian interleukin-1 beta-converting enzyme. Cell 75:641–652
4. Lakhani SA, Masud A, Kuida K, Porter GA Jr, Booth CJ, Mehal WZ, Inayat I, Flavell RA (2006) Caspases 3 and 7: key mediators of mitochondrial events of apoptosis. Science 311:847–851
5. Kroemer G, Martin SJ (2005) Caspase-independent cell death. Nat Med 11:725–730
6. Tait SW, Green DR (2008) Caspase-independent cell death: leaving the set without the final cut. Oncogene 27:6452–6461
7. Lamkanfi M, Festjens N, Declercq W, Vanden BT, Vandenabeele P (2007) Caspases in cell survival, proliferation and differentiation. Cell Death Differ 14:44–55
8. Yi CH, Yuan J (2009) The Jekyll and Hyde functions of caspases. Dev Cell 16:21–34
9. Bradford MM (1976) A rapid and sensitive method for the quantitation of microgram quantities of protein utilizing the principle of protein-dye binding. Anal Biochem 72: 248–254
10. Moghimi SM, Symonds P, Murray JC, Hunter AC, Debska G, Szewczyk A (2005) A two-stage poly(ethylenimine)-mediated cytotoxicity: implications for gene transfer/therapy. Mol Ther 11:990–995
11. Symonds P, Murray JC, Hunter AC, Debska G, Szewczyk A, Moghimi SM (2005) Low and high molecular weight poly(L-lysine)s/poly(L-lysine)-DNA complexes initiate mitochondrial-mediated apoptosis differently. FEBS Lett 579:6191–6198
12. Verdurmen WP, Brock R (2011) Biological responses towards cationic peptides and drug carriers. Trends Pharmacol Sci 32:116–124

Chapter 4

Anti-PEG IgM Production via a PEGylated Nano-Carrier System for Nucleic Acid Delivery

Masako Ichihara, Naoto Moriyoshi, Amr S. Abu Lila, Tatsuhiro Ishida, and Hiroshi Kiwada

Abstract

For the systemic application of nucleic acids such as plasmid DNA and small interfering RNA, safe and efficient carriers that overcome the poor pharmacokinetic properties of nucleic acids are required. A cationic liposome that can formulate lipoplexes with nucleic acids has significant promise as an efficient delivery system in gene therapy. To achieve in vivo stability and long circulation, most lipoplexes are modified with PEG (PEGylation). However, we reported that PEGylated liposomes lose their long-circulating properties when they are injected repeatedly at certain intervals in the same animal. This unexpected and undesirable phenomenon is referred to as the accelerated blood clearance (ABC) phenomenon. Anti-PEG IgM produced in response to the first dose of PEGylated liposomes has proven to be a major cause of the ABC phenomenon. Therefore, in a repeated dosing schedule, the detection of anti-PEG IgM in an animal treated with PEGylated lipoplex could be essential to predict the occurrence of the ABC phenomenon. This chapter introduces a method for the evaluation of serum anti-PEG IgM by a simple ELISA procedure, and describes some precautions associated with this method.

Key words: Polyethylene glycol, PEGylated nano-carrier, Anti-PEG IgM, Accelerated blood clearance phenomenon, Enzyme-linked immunosorbent assay

1. Introduction

Surface modification of liposomes with polyethylene glycol (PEG) improves the pharmacokinetics of liposomes after intravenous injection (1). PEG, a hydrophilic polymer, provides a steric barrier for liposomes to avoid interaction with opsonins and subsequent phagocytosis by the cells of the mononuclear phagocyte system, which results in prolonged duration of liposome circulation (1–3). PEGylated liposomes with a mean size of around 100 nm are

Manfred Ogris and David Oupicky (eds.), *Nanotechnology for Nucleic Acid Delivery: Methods and Protocols*, Methods in Molecular Biology, vol. 948, DOI 10.1007/978-1-62703-140-0_4,

attractive for tumor targeting due to the enhanced permeability and retention (EPR) effect (4), in which the liposomes accumulate in tissue with leaky blood vessels after intravenous injection. PEGylated cationic liposome is a potent nonviral vector for the systemic delivery of nucleic acids. Lipid aggregation of nucleic acids provides protection from nuclease degradation, passive targeting to disease sites, and enhancement of the intracellular delivery of nucleic acids (5).

PEGylated cationic liposomes are considered to be an ideal carrier of nucleic acids. However, it is well documented that in rats, mice, and Rhesus monkeys, intravenous injection of PEGylated liposomes triggers the rapid clearance of a subsequent dose of the same type of liposomes, injected a few days later (6–9). This unexpected phenomenon is referred to as the accelerated blood clearance (ABC) phenomenon. We have shown that anti-PEG IgM induced by the first dose of PEGylated liposomes is responsible for the rapid clearance of the second dose (10–12). A similar phenomenon was also observed after repeated injections of PEG-modified (PEGylated) lipoplex (5, 13), and the presence of nucleic acids (plasmid DNA, pDNA, and small interfering RNA, siRNA) in the PEGylated cationic liposomes strongly enhanced the production of anti-PEG IgM, and consequently the ABC phenomenon.

If the circulation of PEGylated cationic liposomes containing nucleic acids is shortened upon repeated injection, then their therapeutic efficacy is lower. From a series of studies, we found that the serum level of anti-PEG IgM in animals treated with PEGylated liposomes is a reliable marker that can predict the ABC phenomenon. We developed a simple ELISA method to determine serum anti-PEG IgM (14). The serum level evaluated with this method is well correlated with the hepatic clearance (CLh) of the second dose, which is a good indicator of the magnitude of the induced ABC phenomenon (14). In this section, we present an ELISA method for the evaluation of serum anti-PEG IgM, and illustrate some points of caution for an appropriate assay of anti-PEG IgM.

2. Materials

2.1. siRNA and pDNA Stocks and Dilutions

1. pDNA and siRNA of interest.
2. TE buffer, pH 8.0: 10 mM Tris–HCl, 1 mM EDTA. The final pH is adjusted to 8.0. Store at room temperature (see Note 1).
3. Phosphate-buffered saline (PBS), pH 7.4: 137 mM NaCl, 2.7 mM KCl, 4.3 mM Na_2HPO_4, and 1.47 mM KH_2PO_4. The final pH is adjusted to 7.4 (see Note 2).
4. Double-distilled water (see Note 1).

5. Culture medium (see Note 1).
6. Flat-top microtubes (see Note 2).
7. Conical tubes (BD Falcon™, NJ, USA) (see Note 2).

2.2. PEGylated Cationic Liposome Stock and Dilution

1. Cationic liposome of interest.
2. 1,2-Distearoyl-*sn*-glycero-3-phosphoethanolamine-*n*-[methoxy (polyethylene glycol)-2000] ($mPEG_{2000}$-DSPE) (NOF, Tokyo, Japan).
3. Double-distilled water (see Note 1).
4. 150 mM NaCl solution (see Note 1).
5. Culture medium (see Note 1).
6. Flat-top microtubes (see Note 2).
7. Conical tubes (BD Falcon™, NJ, USA) (see Note 2).

2.3. Complex Formation

1. Culture medium (see Note 1).
2. 9% Sucrose solution (see Note 1).
3. Flat-top microtubes (see Note 2).
4. Conical tubes (BD Falcon™, NJ, USA) (see Note 2).

2.4. Gel Electrophoresis

1. Powdered agarose for routine use, stored at room temperature.
2. Running Tris–Borate–EDTA (TBE) buffer (1×): 45 nM Tris–Borate, 1 mM Na_2EDTA. The final pH was adjusted to 8.2, and it was stored at room temperature.
3. Ethidium bromide (EtBr) solution (10 mg/L in water), stored in a refrigerator at 2–8°C. EtBr is very sensitive to light, so aluminum foil was used to protect the brown glass flask.
4. Double-distilled water (see Note 1).

2.5. Preparation of Stock Solution for ELISA (See Note 3)

1. Blocking buffer: 50 mM Tris, 140 mM NaCl, 1% bovine serum albumin (BSA). The final pH is adjusted to 8.0, and it is stored in a refrigerator at 2–8°C.
2. Washing buffer: 50 mM Tris, 140 mM NaCl, 0.05% Tween 20. The final pH is adjusted to 8.0, and it is stored at room temperature.
3. Sample/antibody diluents: 50 mM Tris, 140 mM NaCl, 1% BSA, 0.05% Tween 20. The final pH is adjusted to 8.0, and it is stored in a refrigerator at 2–8°C.
4. Substrate development solution: 36 mM anhydrous citric acid, 67 mM Na_2HPO_4, 0.05% hydrogen peroxide, stored in a refrigerator at 2–8°C.
5. Stop solution: 2 M H_2SO_4, stored at room temperature.

2.6. Preparation of Target Plate (Coating Plate)

1. mPEG$_{2000}$-DSPE (NOF, Tokyo, Japan).
2. Plate coating solution: 0.2 mM mPEG$_{2000}$-DSPE in 100% ethanol, prepared at the time of use.
3. 96-Well plate: Polystyrene EIA/RIA plate (Corning, NY, USA).
4. 8-Channel adjustable pipettes and pipette tips (see Notes 4 and 5).

2.7. Blocking the Plate

1. Blocking buffer.
2. Washing buffer.
3. 8-Channel adjustable pipettes and pipette tips (see Notes 4 and 5).

2.8. Sample Dilution

1. Sample/antibody diluents (see Note 6).
2. Samples (plasma or serum).
3. Negative control (blank) (see Note 7).
4. Positive control (see Note 8).
5. Pipettes (~1,000 and ~20 μL) and pipette tips (see Notes 4 and 5).
6. Flat-top microtubes.

2.9. Preparation of Horseradish Peroxidase-Conjugated Anti-IgM Antibody Solution

1. Horseradish peroxidase (HRP)-conjugated antibody: HRP-conjugated anti-mouse IgM antibody (Millipore, Billerica, MA, USA) (see Note 9).
2. Sample/antibody diluents (see Note 6).
3. Conical tubes (BD Falcon™, NJ, USA).
4. Pipettes (~1,000 and ~20 μL) and pipette tips (see Notes 4 and 5).
5. 8-Channel adjustable pipettes and pipette tips (see Notes 4 and 5).

2.10. Enzyme Reaction

1. *o*-Phenylene diamine (OPD): 20 mg/tablet (Sigma, MO, USA).
2. Substrate development solution.
3. Enzyme substrate: 1 mg/mL OPD solution prepared from substrate development solution.
4. Stop solution.
5. Conical tubes (BD Falcon™, NJ, USA).
6. 8-Channel adjustable pipettes and pipette tips (see Notes 4 and 5).

2.11. Reading Absorbance

1. Microplate reader: Sunrise (TECAN Japan, Kanagawa, Japan), Magellan™ (software).

2.12. Special Equipment

1. Automatic plate washer (if necessary): Auto Mini Washer AMW-8 (BioTec, Tokyo, Japan).

2. Shaker: Double shaker (TAITEC, Saitama, Japan).
3. Thermostat shaker: Bioshaker BR-15LF (TAITEC, Saitama, Japan).
4. –20°C Freezer.

3. Methods

All steps are conducted at room temperature (see Note 10). Gentle agitation of the plate, using a plate shaker, is recommended during long-term incubation (see Note 11). A flowchart for the ELISA method is illustrated in Fig. 1.

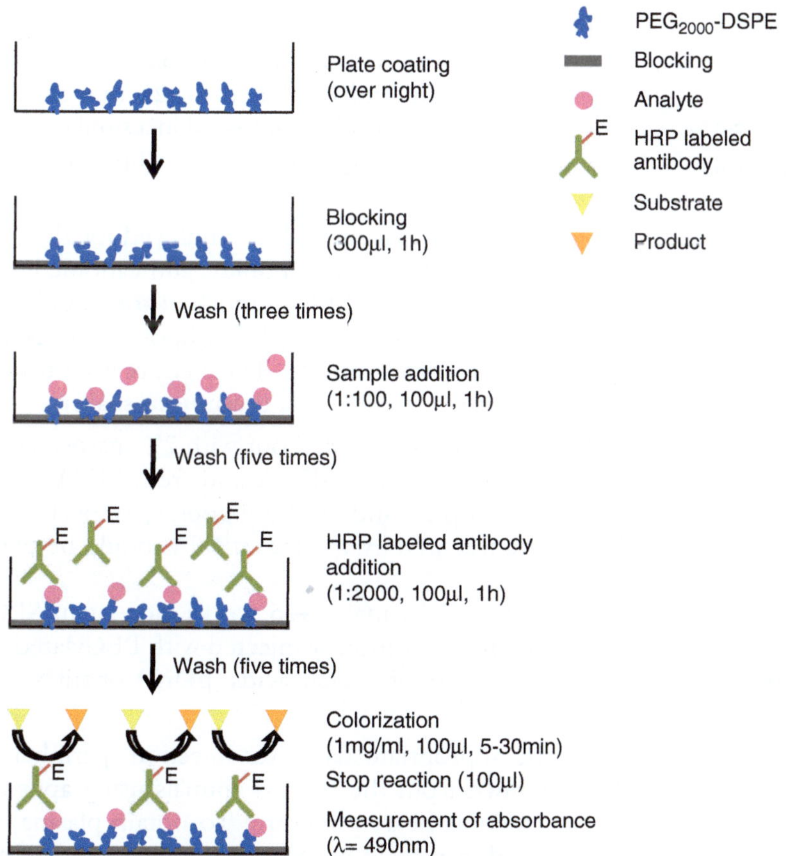

Fig. 1. Schematic diagram for the ELISA method. The principle of detection of serum anti-PEG IgM with ELISA is illustrated. $mPEG_{2000}$-DSPE is immobilized in the wells of a 96-well plate. Samples and controls are added into wells. Anti-PEG IgM in the samples is captured by the immobilized $mPEG_{2000}$-DSPE. After washing away any unbound anti-PEG IgM, HRP-conjugated antibody specific for IgM is added to the wells. After washing to remove any unbound HRP-conjugated antibody, an enzyme substrate solution is added to the wells. The enzyme reaction yields a *yellow* product that turns *orange* when the stop solution is added. Anti-PEG IgM in samples is evaluated from the absorbance of the samples.

3.1. Preparation of Cationic Liposomes

1. Cationic liposomes, composed, for example, of DC-6-14:POPC:CHOL:DOPE (10:30:30:30, molar ratio), are prepared as previously described (13, 15, 16). Briefly, the lipids are dissolved in chloroform. After evaporation of the organic solvent, the resulting thin lipid film is hydrated in 9% sucrose to produce multilamellar vesicles (MLV).
2. The MLV are sized by repeated extrusion through polycarbonate membrane filters (Nuclepore, CA, USA) with consecutive pore sizes of 400, 200, 100, and 80 nm.
3. The mean diameters and zeta potentials of the resulting liposomes are determined using a NICOMP 370 HL submicron particle analyzer (Particle Sizing System, CA, USA). In the sample liposome, the mean diameter and zeta potential of the cationic liposome were 90.7 ± 2.3 nm and 18.5 ± 0.5 mV, respectively.

3.2. Preparation of PEGylated Lipoplexes Formulated with Nucleic Acids (pDNA or siRNA) and PEGylated Cationic Liposomes

1. For the formulation of pDNA-lipoplex, pDNA (10 μg) and cationic liposomes (1 μmol phospholipids) are mixed at a 3.82 (±) charge ratio and incubated for 20 min at room temperature. siRNA-lipoplex is formulated in a similar manner, by mixing siRNA (12.5 μg) and cationic liposomes (0.625 μmol phospholipids).
2. A post-insertion technique is employed for the PEGylation of pDNA-lipoplex, siRNA-lipoplex, and cationic liposomes (17). Briefly, $mPEG_{2000}$-DSPE (5 mol% of total lipid) in 9% sucrose solution is added to the pDNA-lipoplex, siRNA-lipoplex, or cationic liposome solution. The mixture is vortexed and shaken gently for 1 h at 37°C in a thermostat shaker.
3. Electrophoresis is carried out with 2% agarose gel to check for the presence of free pDNA and free siRNA in the prepared pDNA-lipoplex and siRNA-lipoplex, respectively. No bands relating free pDNA and free siRNA should be observed.

3.3. Animal Experiment and Sample Collection

1. Mice (BALB/c, male, 4–5 weeks old, Japan SLC, Shizuoka, Japan) are intravenously injected with PEGylated lipoplexes, at a defined dose of nucleic acids (pDNA or siRNA, 0.1–10 μg/mouse).
2. At the appropriate day after injection, peripheral blood was withdrawn from the treated animals using appropriate methods, such as heart puncture. To obtain plasma, the blood is collected directly in heparin-coated tubes and allowed to stand for 30 min at room temperature to sediment the blood cells.
3. Plasma samples are obtained by separation with a centrifuge at $1407 \times g$ at 4°C for 15 min.
4. For the serum samples, the blood is placed for 30 min at room temperature until the blood cells precipitated in the bottom of

tubes, and then centrifuged at 1407 × *g* for 15 min at 4°C. After centrifugation, both aliquots were collected and frozen at −20°C until further use.

3.4. Preparation of Target Plate (Coating Plate)

The wells of the 96-well polystyrene plate are coated with 50 μL of the coating solution (10 nmol of $mPEG_{2000}$-DSPE). The plate was covered and incubated overnight at room temperature to allow the $mPEG_{2000}$-DSPE to stick to the plate. If time is short, the plate may be incubated in a thermostat shaker set at 37°C for about 2 h, and then checked to assure that the organic reagent is fully evaporated.

3.5. Blocking the Plate

1. To block nonspecific binding sites for anti-mouse IgM antibody, 300 μL of blocking buffer is added to the target plate, which is then incubated for 1 h.
2. After incubation, the plate is washed three times with a washing buffer, either by an automatic washer or manually using 8-channel adjustable pipettes. When using 8-channel adjustable pipettes, wells are filled with washing buffer and the plate is inverted between each washing step to dispose of the washing buffer. After the final wash, any remaining solution left in the well is removed by blotting against clean paper towels (see Note 12).

3.6. Preparation of Samples

1. When frozen samples are used, it was necessary to thaw the samples at room temperature during the blocking step.
2. The samples are mixed thoroughly using a vortex, and diluted with sample/antibody diluents in microtubes. For example, 10 μL of serum sample was added to 990 μL of sample/antibody diluents and mixed well.

3.7. Addition of Samples to Pre-coated Plate

1. The plate layout is prepared as a record of the samples, positive and negative controls. An example is illustrated in Fig. 2.
2. The diluted serum samples, negative control serum and positive control serum, are added at room temperature (100 μL) to duplicate or multiple wells (see Notes 13 and 14).
3. The plate is then incubated for 1 h, and then washed five times with washing buffer as described above.

3.8. Addition of HRP-Conjugated Antibody

1. HRP-conjugated anti-mouse IgM antibody is diluted with sample/antibody diluents at 1:2,000 in a conical tube (15 or 50 mL). At least 10 mL of HRP-conjugated antibody per plate is required.
2. Using a conical tube (15 or 50 mL), 6 μL of HRP-conjugated antibody is diluted with 12 mL of sample/antibody diluents and mixed vigorously with a vortex.

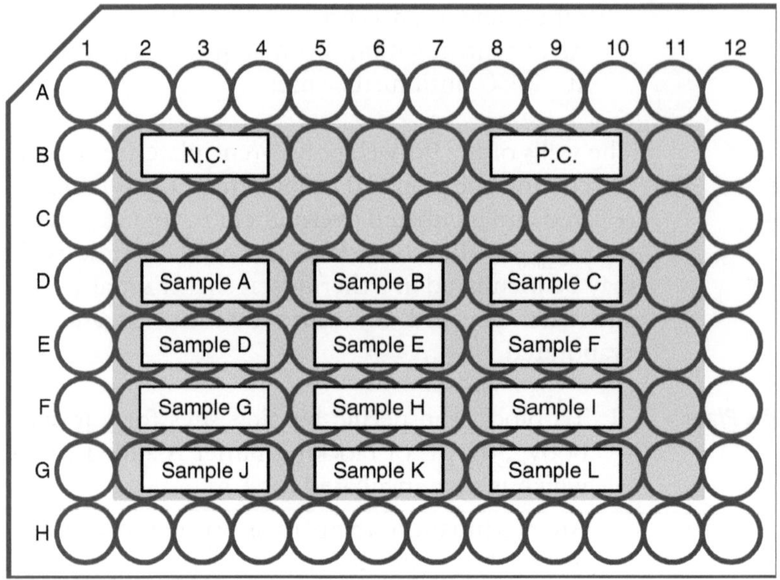

Fig. 2. Plate layout (example). If possible, sample wells were set apart from positive control cells to avoid contamination.

3. HRP-conjugated antibody solution (100 μL) is added to each well.
4. The plate is incubated for 1 h and then washed five times with washing buffer, as described above.

3.9. Colorization

1. As an example, 20 mL of substrate solution is required for two plates. Using a 50 mL conical tube, one OPD tablet (20 mg) is dissolved with 20 mL of enzyme buffer and mixed vigorously with a vortex, resulting in a uniform solution.
2. To each well is added 100 μL of prepared enzyme substrate solution (1 mg/mL) followed by incubation at room temperature for a predefined interval (5–20 min) until suitable color intensity developed (see Note 15).
3. The reaction is stopped by the addition of stop solution (100 μL) to each well, and the plate was then shaken gently to ensure that the color is uniformly distributed in the well (see Note 16).

3.10. Measurement of Absorbance

The plate was read using a microplate reader set at 490 nm within 30 min (see Note 17).

3.11. Calculation of Results

The absorbance in the sample (OD_{sample}) and positive control ($OD_{p.c.}$) is subtracted from that in the negative control ($OD_{n.c.}$, blank), and these values are averaged for the same samples. The results are expressed as the ratio ($OD_{sample}/OD_{p.c.}$) of the absorbance in the sample (OD_{sample}) to absorbance in the positive control ($OD_{p.c.}$).

3.12. An Example of the Detection of Serum Anti-PEG IgM Using the ELISA Method

1. Two dosage levels (low: 0.1 μmol phospholipid (PL)/kg, and high: 100 μmol PL/kg) of PEGylated neutral liposome, cationic liposome, and pDNA-lipoplex are intravenously injected into mice (BALB/c, male, 4 weeks old, Japan SLC, Shizuoka, Japan).
2. At day 5 after injection, the blood is withdrawn and serum is collected. The serum sample from the non-treated mice (naïve mice) is used as a blank.
3. An ELISA plate is prepared, and serum samples and both controls are diluted to 1:100 with sample/antibody diluents.
4. The ELISA procedure is carried out and the absorbance of the samples is measured.
5. The average absorbance of the blank is then subtracted from the raw data, and the resulting values for each sample are averaged. The results are shown in Fig. 3.
6. At a low dose, a single injection of both PEGylated neutral and cationic liposomes without nucleic acids causes a significant induction of anti-PEG IgM, which is consistent with our

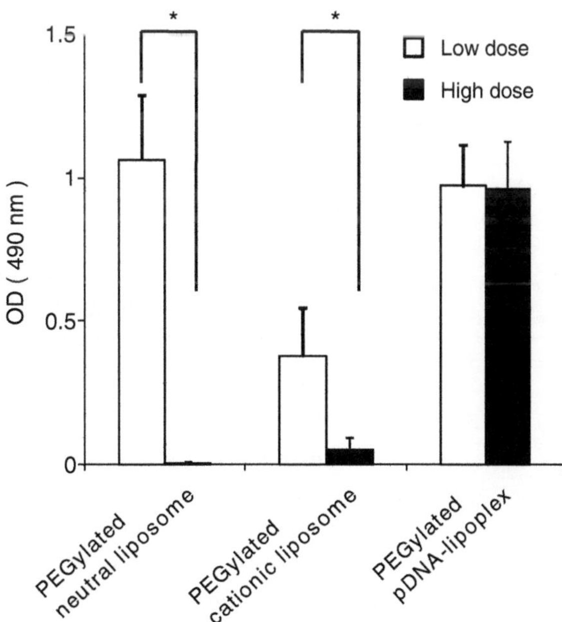

Fig. 3. Serum anti-PEG IgM induced by a single injection of PEGylated neutral, cationic, or pDNA-lipoplex. Two dose levels (low: 0.1 μmol phospholipid (PL)/kg, and high: 100 μmol PL/kg) of PEGylated neutral charged, cationic, and pDNA-lipoplex were intravenously injected into mice. At day 5 after injection, the blood was withdrawn and serum was collected. The serum samples collected from naïve (untreated) mice were used as a blank. Serum anti-PEG IgM was detected with ELISA. Each value represents the mean ± SD, ($n=3$), $^{*}p<0.05$.

earlier results (18). In similar fashion, pDNA-lipoplex strongly induces anti-PEG IgM at a low dose. At a high dose, although no induction of anti-PEG IgM is observed in mice treated with empty PEGylated neutral and cationic liposomes, anti-PEG IgM is strongly induced in mice treated with PEGylated pDNA-lipoplex. This result suggests that the encapsulation of nucleic acids such as pDNA within PEGylated cationic liposomes can activate the immune system, and the nucleic acids make these PEGylated liposomes more immunogenic.

4. Notes

1. Solutions used in the preparation of lipoplex (double-distilled water and buffer) should be of sterile DNase- and RNase-free grade and stored in a refrigerator at 2–8°C. All solutions should be prepared with water that has a resistivity of 18.2 MΩ cm and total organic content of less than five parts per billion.
2. Ensure that all glassware and plasticware are of a DNase- and RNase-free grade. It is recommended that DNase- and RNase-free plastic tubes are used throughout the preparation of pDNA-lipoplex and siRNA-lipoplex.
3. Bring all reagents to room temperature before use. Gently mix reagents prior to use. When chilled reagents are used, the occurrence of edge effects in a 96-well plate, manifested in increased signals in the outer wells, can be caused by temperature differences across the plate. Because of the insulating properties of the polystyrene, the outer wells reach ambient temperature faster than those in the middle of the plate. This problem can be eliminated by ensuring that all reagents are at room temperature prior to their addition to the wells.
4. Ensure that pipettes are working correctly and are calibrated routinely.
5. Ensure that pipette tips are pushed on far enough to create a good seal. Make sure that the pipette tips are all picking up and releasing the correct amount of reagents. This will greatly affect the consistency of results between duplicates or triplicates.
6. Sodium azide can be an HRP inhibitor, so it is not generally used as a preservative in this buffer.
7. The samples obtained from non-treated animals served as the negative control (blank).

8. If necessary, samples obtained from animals that had been injected with empty PEGylated liposome can serve as the positive control.
9. If the antibody is lyophilized, it is necessary to reconstitute with an appropriate solvent, such as sterile distilled water. After reconstitution, the solution is stable for several weeks in a refrigerator at 2–8°C. However, after dilution, it should be used within 24 h. For extended storage after reconstitution, the antibody solution could be divided and frozen in a deep freezer (–80°C).
10. To detect IgM other than mouse IgM, an appropriate anti-IgM antibody must be selected.
11. If the incubation temperature is lower than the ambient temperature (15–25°C), it is recommended that a thermostat shaker set at ambient temperature be used.
12. During a long incubation time, it is necessary to shake the plates gently. The use of a plate shaker is considered essential in the development of enzyme reactions, to ensure that the substrate is distributed evenly in the well and reliable results are obtained.
13. Inadequate plate washing easily leads to high background signals and significant sample-to-sample variation. Therefore, the wash step is essential for reproducible results.
14. When a 96-well polystyrene plate is used, an increased signal in the outer wells, i.e., the edge effect, sometimes occurs.
15. To ensure accurate results, it is recommended that the samples be added to the plate in duplicate or triplicate.
16. The color intensity in the positive control wells provides an indication of enzyme reaction development. The incubation time should be adjusted to no more than the maximum absorbance of the plate reader in the positive control.
17. To ensure accurate results, the stop solution should be added to the plate in the same order as the addition of the enzyme substrate.
18. The bottom of the plate directly affects the reading of the absorbance. It is recommended that the bottom of the plate be cleaned carefully with a clean paper towel wet with 70% ethanol before it is read.

5. Troubleshooting

5.1. High Background Across the Entire Plate

1. The concentration of the HRP-conjugated antibody is too high or the enzyme reaction is left for too long. Check the dilution ratio of HRP-conjugated antibody to sample/antibody

diluents. Try to dilute the conjugate again. Otherwise, stop the chromogenic reaction as soon as the plate has developed enough to read the absorbance.

2. The plate is left too long before reading the absorbance. A chromogenic reaction could keep developing at a slower rate, even though the stop solution was added. Read the plate immediately after the enzymatic reaction is stopped.
3. Nonspecific binding of the antibody: Confirm that the plate-blocking step was fully carried out and a suitable blocking buffer was used. A variety of agents such as gelatin, casein BSA, and powdered milk have been employed. These agents are usually included at 0.1–1.0%. If the problem persists, it is recommended that the percentage of blocking reagents be increased to 5.0%.

5.2. Low Absorbance Values

1. Insufficient incubation time during the development of the enzyme reaction. A longer incubation time is recommended.
2. Inappropriate incubation temperature: HRP-conjugated antibody will have optimum binding activity at the correct temperature. Ensure that incubation is carried out at the correct temperature in the development of the chromogenic reaction.
3. The concentration of samples or HRP-conjugated antibody is too low. It is recommended that the concentrations of samples and HRP-conjugated antibody be reassessed.
4. The blocking step is carried out under inappropriate conditions. It is recommended that the blocking conditions, such as incubation time or concentration, should be changed.

5.3. Inconsistent Absorbance Across the Plate

1. Incorrect pipetting: Ensure that pipettes work correctly and are constantly calibrated. Ensure that pipette tips are pushed on far enough to create a good seal. When diluting samples and reagents are added, watch to make sure that the pipette tips are accurately picking up and releasing the correct amounts of samples and reagents.
2. Washing buffer is left in the wells. Any remaining liquid in the well can affect the assay, adversely affecting the accuracy of the results. Make sure that no liquid remains after the washing step.

5.4. Chromogenic Reaction is Slow

Degradation of enzyme substrate solution: Prepare the enzyme substrate (OPD solution) immediately before use and consume it immediately.

Acknowledgments

The authors thank Dr. James L. McDonald for his helpful advice in developing the English manuscript. This research was supported by a Grant-in-Aid for Young Scientists (A) (21689002) and a Grant-in-Aid for Scientific Research (B) (23390012) from the Ministry of Education, Culture, Sports, Science and Technology (MEXT), Japan.

References

1. Klibanov AL, Maruyama K, Torchilin VP, Huang L (1990) Amphipathic polyethyleneglycols effectively prolong the circulation time of liposomes. FEBS Lett 268:235–237
2. Allen TM, Hansen C (1991) Pharmacokinetics of stealth versus conventional liposomes: effect of dose. Biochim Biophys Acta 1068:133–141
3. Lasic DD (1996) Doxorubicin in sterically stabilized liposomes. Nature 380:561–562
4. Maeda H, Bharate GY, Daruwalla J (2009) Polymeric drugs for efficient tumor-targeted drug delivery based on EPR-effect. Eur J Pharm Biopharm 71:409–419
5. Judge A, McClintock K, Phelps JR, Maclachlan I (2006) Hypersensitivity and loss of disease site targeting caused by antibody responses to PEGylated liposomes. Mol Ther 13:328–337
6. Dams ET, Laverman P, Oyen WJ, Storm G, Scherphof GL, van Der Meer JW, Corstens FH, Boerman OC (2000) Accelerated blood clearance and altered biodistribution of repeated injections of sterically stabilized liposomes. J Pharmacol Exp Ther 292:1071–1079
7. Laverman P, Carstens MG, Boerman OC, Dams ET, Oyen WJ, van Rooijen N, Corstens FH, Storm G (2001) Factors affecting the accelerated blood clearance of polyethylene glycol-liposomes upon repeated injection. J Pharmacol Exp Ther 298:607–612
8. Ishida T, Maeda R, Ichihara M, Irimura K, Kiwada H (2003) Accelerated clearance of PEGylated liposomes in rats after repeated injections. J Control Release 88:35–42
9. Ishida T, Masuda K, Ichikawa T, Ichihara M, Irimura K, Kiwada H (2003) Accelerated clearance of a second injection of PEGylated liposomes in mice. Int J Pharm 255:167–174
10. Ishida T, Ichihara M, Wang X, Kiwada H (2006) Spleen plays an important role in the induction of accelerated blood clearance of PEGylated liposomes. J Control Release 115: 243–250
11. Ishida T, Ichihara M, Wang X, Yamamoto K, Kimura J, Majima E, Kiwada H (2006) Injection of PEGylated liposomes in rats elicits PEG-specific IgM, which is responsible for rapid elimination of a second dose of PEGylated liposomes. J Control Release 112:15–25
12. Wang X, Ishida T, Kiwada H (2007) Anti-PEG IgM elicited by injection of liposomes is involved in the enhanced blood clearance of a subsequent dose of PEGylated liposomes. J Control Release 119:236–244
13. Tagami T, Nakamura K, Shimizu T, Yamazaki N, Ishida T, Kiwada H (2010) CpG motifs in pDNA-sequences increase anti-PEG IgM production induced by PEG-coated pDNA-lipoplexes. J Control Release 142:160–166
14. Ishida T, Wang X, Shimizu T, Nawata K, Kiwada H (2007) PEGylated liposomes elicit an anti-PEG IgM response in a T cell-independent manner. J Control Release 122:349–355
15. Tagami T, Nakamura K, Shimizu T, Ishida T, Kiwada H (2009) Effect of siRNA in PEG-coated siRNA-lipoplex on anti-PEG IgM production. J Control Release 137:234–240
16. Tagami T, Uehara Y, Moriyoshi N, Ishida T, Kiwada H (2011) Anti-PEG IgM production by siRNA encapsulated in a PEGylated lipid nanocarrier is dependent on the sequence of the siRNA. J Control Release 151:149–154
17. Allen TM, Sapra P, Moase E (2002) Use of the post-insertion method for the formation of ligand-coupled liposomes. Cell Mol Biol Lett 7:889–894
18. Ichihara M, Shimizu T, Imoto A, Hashiguchi Y, Uehara Y, Ishida T, KIwada H (2011) Anti-PEG IgM response against PEGylated liposomes in mice and rats. Pharmaceutics 3:1–11

Acknowledgements

The authors thank Dr. [illegible] for his helpful advice in [illegible] development [illegible] This research was supported [illegible] a Grant-in-Aid for Young Scientists [illegible] and a [illegible] Grant-in-Aid for Scientific Research [illegible] from the Ministry of Education, Culture, Sports, Science and Technology [illegible] (MEXT), Japan.

References

[illegible]

Chapter 5

Near-Infrared Optical Imaging of Nucleic Acid Nanocarriers In Vivo

Claire Rome, Julien Gravier, Marie Morille, Gilles Divita, Anne-Laure Bolcato-Bellemin, Véronique Josserand, and Jean-Luc Coll

Abstract

Noninvasive, real-time optical imaging methods are well suited to follow the in vivo distribution of nucleic acid nanocarriers, their dissociation, and the resulting gene expression or inhibition. Indeed, most small animal imaging devices perform bioluminescence and fluorescence measurements without moving the animal, allowing a simple, rapid, and cost-effective method of investigation of several parameters at a time, in longitudinal experiments that can last for days or weeks.

Here we help the reader in choosing adapted near-infrared (NIR) fluorophores or pairs of fluorophores for Förster resonance energy transfer assays, imaging of reporter genes, as well as nanocarriers for in vivo gene and siRNA delivery. In addition, we present the labeling methods of these macromolecules and of their payload and the protocols to detect them using bioluminescence and NIR fluorescence imaging in mice.

Key words: Nucleic acid delivery, In vivo imaging, NIR fluorescence, Nanocarriers

1. Introduction

Optical imaging based on 2D Fluorescence reflectance (FRI) is a noninvasive method that allows to follow molecules injected into a mouse in real time, at a good spatial (nm to mm range) and temporal resolution (μs to ms) and a high sensibility (fM to pM range) and from whole-body down to the subcellular scale. However, the information that can be obtained through absorbing biological tissues is strongly depth-weighted and depends on the thickness and optical properties of the tissues to be imaged. This method is thus only semiquantitative. The tissues are less absorbing in a near-infrared

Manfred Ogris and David Oupicky (eds.), *Nanotechnology for Nucleic Acid Delivery: Methods and Protocols*, Methods in Molecular Biology, vol. 948, DOI 10.1007/978-1-62703-140-0_5, © Springer Science+Business Media, LLC 2013

(NIR) spectral window ranging from 650 to 900 nm and it is thus important to use adapted fluorophores whenever possible.

The recent description of increasing amounts of organic or inorganic NIR fluorophores contributed to the development of optical imaging methods for the investigation of deep tissues (1, 2). These fluorophores can be used for labeling molecules of interest and will accelerate significantly the preclinical studies. By separately labeling the vector and the nucleic acid using an adapted pair of fluorophores, we can obtain direct information on the distribution of the intact particle, its capacity to release the nucleic acid, the subcellular distribution of each component, and finally a direct measure of the reporter gene expression.

Several reporter genes of the luciferase family were adapted to bioluminescence imaging (BLI) in vivo. Commonly used fluorescent proteins like green fluorescent proteins (GFP) and red fluorescent proteins can be used in FRI but are more adapted for in vitro applications because they emit in the visible spectrum. β-Galactosidase, an enzyme encoded by the *lacZ* gene of *Escherichia coli*, can also be followed in vivo using NIR substrates like DDAOG (3). By combining FRI and BLI, we can obtain multiple information especially well suited for the evaluation of the biodistribution of lipid nanocapsules (LNCs), polyethylenimine (PEI) polyplexes, or cell-penetrating peptides (CPPs), three nonviral systems well suited for the delivery of DNA and siRNA in mice (4–6).

2. Materials

1. To visualize the efficiency and biodistribution of in vivo gene delivery, different reporter genes can be used. For in vivo bioluminescence, plasmids encoding for *firefly luciferase* are very commonly used (pgWIZ-*luciferase* (GENLANTIS, San Diego, USA) for example). The bacterial β-*galactosidase* gene expression can be detected in vivo using NIR fluorescent imaging. For plasmid DNA, $OD_{260/280}$ ratio should be greater than 1.8. It is best to use DNA prepared in water, free of salt, RNAs, proteins, or endotoxins.
2. Concerning siRNA (21–23 nucleotides duplex RNA with symmetric 2 nucleotides 3′ overhangs), we use high-quality desalted siRNA (PAGE or HPLC purified). As for primer/probe design, a G/C content around 50% is usually selected and oligonucleotides with three or more of any purine nucleotide in a row are avoided. Prediction algorithms provide some guidance for siRNA sequence selection (see Note 1).

3. Cationic lipids 1,2-dioleoyl-3-trimethylammonium-propane (DOTAP) and 1,2-dioleyl-*sn*-glycero-3-phosphoethanolamine (DOPE) (Avanti Polar Lipids, Inc, Alabaster, USA).
4. Labrafac® WL 1349 (caprylic–capric acid triglycerides, European Pharmacopoeia, IV, 2002).
5. Oleic Plurol® (polyglyceryl-6 dioleate) (Gattefosse S.A., Saint-Priest, France).
6. Solutol® HS-15 (30% of free polyethylene glycol (PEG) 660 and 70% of PEG 660 hydroxystearate (HS-PEG) European Pharmacopeia, IV, 2002) (BASF, Ludwigshafen, Germany) (7).
7. DiD fluorochrome [1,10-dioctadecyl-3,3,30,30-tetramethylindodicarbocyanine perchlorate (DiD, em. = 644 nm; exc. = 665 nm)] (Invitrogen, Carlsbad, CA, USA).
8. PD10 Sephadex columns (Amersham Biosciences Europe, Orsay, France).
9. Millipore Amicon® Ultra-15 centrifugal filter devices (Millipore, St Quentin-Yvelines, France).
10. 1,2-Distearoyl-*sn*-glycero-3-phosphoethanolamine-*N*-[methoxy(polyethyleneglycol)-2000] (DSPE-mPEG2000) (mean molecular weight (MMW) = 2,805 g/mol) was provided from Avanti Polar Lipids (Inc, Alabaster, USA).
11. Malvern Zetasizer® (Nano Series DTS 1060, Malvern Instruments S.A., Worcestershire, UK).
12. C18 column Interchrom UP5 WOD/25M Uptisphere 300 5 ODB, 250 × 21.2 mm.
13. Maleimide dyes (maleimide-FITC or Maleimide-Alexa) from Molecular Probes. Inc (Invitrogen, Carlsbad, CA, USA).
14. Female 6–8-week-old nude mice. All animal experiments are conducted in agreement with the "Principles of laboratory animal care" (NIH publication No. 86–23, revised 1985) and approved by the local ethical Committee.
15. d-Luciferin. Powder (Promega, Charbonnières, France) is dissolved in DPBS without $CaCl_2$ and $MgCl_2$ at 100 mg/ml. The solution is sterilized by filtration across a 0.2 μm membrane and stored at −80°C sheltered from light.
16. DDAOG. 9*H*-(1,3-dichloro-9,9-dimethylacridin-2-one-7-yl) β-d-galactopyranoside (Molecular Probes, Invitrogen, Carlsbad, CA, USA) is dissolved in dimethyl sulfoxide (DMSO)/PBS v/v to get 5 mg/ml.
17. Light-emitting diodes (LEDs) 633 or 660 nm equipped with interference filters.
18. Fluorescence images are acquired with a back-thinned CCD cooled (−70°C) Camera (ORCAII-BT-512G, Hamamatsu Photonics, Massy, France) filtered either by a colored glass

long pass filter RG 665 plus a band pass filter 680 (MellesGriot, Voisins le Bretonneux, France) when illuminating at 633 nm or a high pass filter RG 9 (Schott, Clichy, France) when illuminating at 660 nm.

19. PBS pH 7.4 (Gibco, Invitrogen, Carlsbad, CA, USA).
20. Ultra-Pure DNAse–RNAse-free Sterile Water (GIBCO; Invitrogen, Carlsbad, CA, USA).
21. DMSO (Sigma; Saint-Quentin Fallavier, France).
22. Chloroform (Sigma, Saint-Quentin Fallavier, France).

3. Methods

3.1. General Considerations for Fluorescent Dye Selection

Commercially available fluorophores are available in a wide variety of detection wavelengths, hydrophilicity, and reactive moieties. Since the choice of absorption and emission wavelength mainly depends on the available apparatus, we here focus on selection criteria related to hydrophobicity/hydrophilicity and chemical reactivity (Table 1). Even with the right fluorophore, fluorescence imaging will only reveal the molecule location independently of the nanoparticle. Förster resonance energy transfer (FRET) occurs between a donor and a receptor dye within the range of the Förster radius, typically ranging from 5 to 10 nm depending on the pair considered. This property makes it a useful phenomenon to gather information in the nanometer range through spectroscopic methods. When donor and acceptor dyes are in close vicinity, excitation of the former can

Table 1
Selection criteria for the choice of a fluorescent probe

Target	Fluorophore type	Grafting group	Reactive group	pH range	Examples
Specific molecule	Reactive	Primary amine ($-NH_2$)	Isothiocyanate	~9	FITC, RITC
			Succinimidyl ester	~8	Alexafluor®, Cydyes
		Thiol (–SH)	Maleimide	~7	Alexafluor®, Fluoprobes®
Whole nanoparticle	Hydrophobic	None—non-covalent interaction		Any pH compatible with nano-particle and fluorophore	DiI, DiD

be passed to the latter via non-radiative pathways. Macroscopically, FRET is attested by a decrease of the donor fluorescence and an increase of the acceptor emission upon irradiation of the donor. It is therefore possible to know whether or not dyes are still encapsulated within a nanoparticle by studying its emission spectrum. Upon excitation of the donor, free dyes will display normal donor fluorescence while nanoparticles will display strong acceptor emission.

Dark dye quenching is also based on FRET, except that the acceptor dye is nonfluorescent. In this case, the energy transfer results in a decrease of the donor emission while the acceptor remains nonfluorescent. This results in a simple on/off switch in fluorescence, where emission is recovered when the dyes are released from the nanoparticle. Though this method provide a more straightforward mean to track where and when the dyes escape from the nanoparticles, it also means that in contrast to classical FRET circulating nanoparticles are nonfluorescent and cannot be tracked through spectroscopic method unless a third dye is introduced.

To occur, FRET requires that spectral absorption of the acceptor overlaps sufficiently with the donor emission. Though precise calculations are required, this generally means that absorption peaks of the donor and acceptor should be spaced by approximately 50–100 nm. On the one hand, with less than 50 nm, spectral overlap between donor and acceptor emissions might hinder FRET measurements. On the other hand, spectral differences of more than 100 nm will often result in a low spectral overlap, therefore very short Förster radii. The following table provides a few examples of NIR dyes (fluorescent or dark dyes) along with possible FRET pairing dyes (Tables 2 and 3).

Table 2
Fluorescent dyes grouped depending on their spectral properties and possible dark dye quenchers

Maximum absorption (emission) wavelength (nm)	Fluorescent dye	Dark dye
633 (647)	AF 633	
650 (670)	AF 647, DiD, Cy5, IRdye 650	
660 (690)	AF 660	QSY 21
680 (700)	AF 680, Cy5.5, IRdye 680	BHQ-3
700 (720)	AF 700, IRdye 700	
750 (775)	AF750, DiR, Cy7, IRdye 750	IRdye QC-1

Table 3
Possible near-infrared FRET pairs with absorptions and emissions maxima, donor fluorescence quantum yield, acceptor molar extinction coefficient (providers' data), and estimated Förster radius (using providers' data in water or PBS)

Donor/acceptor pair	Donor absorbance (nm)	Donor emission (nm)	Acceptor absorbance (nm)	Acceptor emission (nm)	QY	l/mol/cm	Förster radius (nm)
AF633/AF660	632	647	663	690	0.3	132,000	6.5
AF633/AF680	632	647	679	702	0.3	183,000	7
AF650/AF700	650	668	702	723	0.33	192,000	7
AF650/AF750	650	668	749	775	0.33	240,000	6.5
AF660/AF750	663	690	749	775	0.37	240,000	7
AF680/AF750	679	702	749	775	0.36	240,000	7.5
AF680/IRdye800	679	702	774	789	0.36	240,000	7
AF700/AF750	702	723	749	775	0.25	240,000	7.5
AR700/IRdye800	702	723	774	789	0.25	240,000	7
AF700/AF790	702	723	785	810	0.25	260,000	7

3.2. Preparation of DiD-Labeled DNA Lipid Nanocapsules

LNCs were developed according to a solvent-free process based on an emulsion phase inversion. DNA complexed with cationic lipids, i.e., DOTAP/DOPE, was encapsulated into LNCs leading to the formation of stable nanocarriers (DNA LNCs) with a size less than 130 nm. Amphiphilic PEG coating [PEG lipid derivative (DSPE-mPEG2000)] at different concentrations was selected to make DNA LNCs stealthy.

3.2.1. Preparation of the Aqueous Phase Containing DNA and Cationic Liposome Complexes

1. Dissolve DOTAP and DOPE in chloroform at a concentration of 20 mg/ml. To obtain a final concentration of positive charge of 25 mM (+) in 3 ml, mix DOTAP/DOPE at a molar ration of 1:1, i.e., add 2.617 ml of DOTAP solution (20 mg/ml) to 2.790 ml of DOPE (20 mg/ml) in a 25 ml round-bottom flask and assure a homogeneous mixture of lipids (see Note 2).
2. Once the lipids are thoroughly mixed in the organic solvent, remove the organic solvent by rotary evaporator at room temperature under vacuum to yield a thin lipid film without solvent residues.
3. Hydrate the dry lipid film by adding 3 ml of purified deionized water (18 MΩ cm, MilliQ), and incubate overnight at 4°C.
4. After this incubation step, vigorously mix the lipid suspension by vortexing. The thus-formed multilamellar vesicles are sonicated for 20 min in a bath sonicator to allow the formation of small unilamellar vesicles.
5. To form lipoplexes, mix DNA and cationic liposomes at a charge ratio (±) of 5 in 150 mM NaCl. Briefly, add 660 μg of pDNA in a solution containing 300 mM NaCl to obtain a final concentration of 660 μg of pDNA in 400 μl NaCl. Then add the same volume of previously prepared cationic liposomes (400 μl). Mix thoroughly with a pipette, and let incubate for 30 min before use.

3.2.2. Preparation of the Lipid Phase Containing DiD Stained Labrafac®

1. Dissolve 5 mg of DiD in 1 ml of acetone to obtain a final solution at 5 mg/ml in a scintillation vial (see Note 3). Gently mix for 2 h.
2. Mix 100 mg of 5 mg/ml DiD solution with 900 mg of Labrafac®. Mix for 15 min, and let the bottle open under a cell culture hood for 2–3 h to allow the acetone evaporate. Control the whole acetone evaporation by weighting before and after evaporation.

3.2.3. Preparation of DiD-Labeled DNA LNCs

1. Weigh all the LNC components in a scintillation vial. First, weigh 58 mg of Solutol HS-15®(see Note 4), and then add 38.7 mg of oleic Plurol®, 96.16 mg of DiD stained lipophilic Labrafac® (Subheading 3.2.2), 14.1 mg of NaCl, and 800 μl of aqueous solution containing lipoplexes (Subheading 3.2.1).

2. Mix all the components together under magnetic stirring and heat to 60°C using a hot plate stirrer.
3. Apply six temperature heat-cooling cycles between 20 and 60°C under magnetic stirring, to obtain a phase inversion from an oil-in-water to a water-in-oil emulsion. Change in color can be observed in a zone between the two emulsions. In this region, the system appears translucent with blue glints, which is representative of microemulsions; this zone was called the phase inversion area (7). After these temperature cycles, in the phase inversion zone, the mixture should be rapidly cooled by dilution with 500 μl of cold water (4°C), leading to the formation of DiD-labeled DNA LNCs in water (see Note 5).
4. Follow the size, polydispersity, and surface charge (zeta potential) of DiD-labeled DNA LNCs using dynamic light scattering (DLS, size) and electrophoretic mobility (zeta potential) (Malvern Zetasizer®, Nano Series DTS 1060, Malvern Instruments S.A., Worcestershire, UK).
5. Confirm the encapsulation DNA using electrophoresis migration of broken (Triton X-100) or intact DNA LNCs in an agarose gel (1%) containing ethidium bromide (method described in (5, 8, 9)).

3.2.4. Purification of DiD-Labeled DNA LNCs

1. Equilibrate a PD10 Sephadex® column with approximately 25 ml elution buffer.
2. Add DiD-labeled DNA LNC suspension of a total volume of 2.5 ml per column. If the volume is less than 2.5 ml, then add buffer until the total volume of 2.5 ml is achieved.
3. Elute with buffer and collect the flow-through. Collect the fractions containing DiD-labeled DNA LNCs. If after this purification step, the LNC concentration is low, a concentration step may be required.
4. Add 12 ml maximum of collected fraction of DiD-labeled DNA LNCs in an Amicon® Ultra-15 centrifugal filter device.
5. Centrifuge for 30 min at 4,000 rpm at room temperature.
6. The thus concentrated DiD-Labeled DNA LNCs are collected from the filter unit sample reservoir, while the ultrafiltrate is collected in the provided centrifuge tube.
7. The DiD-labeled DNA LNC concentration is fixed at 150 mg/ml. Place the DiD-labeled DNA LNC suspension collected in the upper part to this reference concentration and to obtain a final NaCl concentration of 150 mM.
8. Control the size, polydispersity, and surface charge (zeta potential) of DiD-labeled DNA LNCs using DLS (Malvern Zetasizer®, Nano Series DTS 1060, Malvern Instruments S.A., Worcestershire, UK).

9. Control the encapsulation properties of DiD-labeled DNA LNCs using electrophoresis migration of broken (Triton X-100) or intact DNA LNCs in an agarose gel (1%) containing ethidium bromide (method developed in (5, 9)).

3.2.5. Formulation of PEGylated DiD-Labeled DNA LNCs

1. Weigh DSPE-mPEG2000 as a function of the desired concentration in a vial. As an example, add 28.5 mg of DSPE-mPEG2000 for a final molar concentration of 10 mM.
2. Add 1 ml of DiD-labeled DNA LNCs and mix by vortexing.
3. Co-incubate polymers and purified DiD-labeled DNA LNCs in a vial placed in a water bath for 4 h at 30°C.
4. After this co-incubation step, mix by vortexing and put the tube in an ice bath.

3.3. Fluorescent Labeling of Cell-Penetrating Peptides

CPPs constitute promising tools for noninvasive cellular import of siRNA. CPP-based strategies have been successfully applied for ex vivo and in vivo delivery of therapeutic siRNA. Recently, a new peptide-based system, CADY, has been described for efficient delivery of siRNA. CADY is a secondary amphiphatic peptide able to form stable non-covalent complexes with siRNA and to improve their cellular uptake. It is possible to modify this peptide for following CADY/siRNA complexes in vivo. The protocol described below outlines a general procedure suitable for conjugation of most thiol-reactive probes (iodoacetamides, maleimides) to peptides and procedures for peptide purification.

1. CADY (20-residues: Ac-GLWRALWRLLRSLWRLLWRA-cya; MW: 2,653 Da) is synthesized by solid-phase peptide synthesis using AEDI-expensin resin with (fluorenylmethoxy)-carbonyl (Fmoc) continuous (Pionner, Applied Biosystems, Foster City, CA) as described previously (10). CADY is purified by semi-preparative reverse-phase high-performance liquid chromatography (RP-HPLC) and identified by electrospray mass spectrometry and amino acid analysis (10). The peptide is acetylated at its N-terminus and has a cysteamide group at its C-terminus.
2. Take the vial containing the peptide powder out of the freezer and equilibrate for 30 min at room temperature without opening the vial. Resuspend CADY at a final concentration of 2 mg/ml (774.5 μM) in ultrapure RNAse- and DNAse-free water containing 2% DMSO. CADY powder should be first solubilized directly in DMSO, and then add calculated volume of water to reach the 2 mg/ml final CADY concentration and 2% DMSO.
3. Mix gently by tapping the tube and sonicate the CADY solution for 10 min in the water bath sonicator. Sonication is essential to prevent peptide aggregation and finalize solubilization.

4. CADY (1 mM) is dissolved in phosphate buffer saline (pH 7.5) at room temperature. At this pH, the C-terminal thiol group is nucleophilic and reacts exclusively with the fluorescent reagent. In contrast peptide amines are protonated and therefore not reactive.
5. Prepare a 10 mM stock solution of the reactive dye (Maleimide-FITC or Maleimide-Alexa from Molecular Probes. Inc., Invitrogen, Carlsbad, CA, USA) in DMSO and protect stock solutions from light by wrapping the tubes with an aluminum foil. Maleimide stock solutions can be stored at –20°C for a month.
6. A step of reduction of disulfide bonds in the peptide can be added to the protocol using a tenfold molar excess of TCEP reducing agent. For labeling with iodoacetamides or maleimides, TCEP can be kept in the solution during conjugation.
7. Add 20-fold molar excess of the reactive dye dropwise to the peptide solution.
8. Allow the reaction to proceed for 2 h at room temperature or overnight at 4°C. For more detail *see The Handbook: A Guide to Fluorescent Probes and Labeling Technologies* (probes.invitrogen.com).
9. Upon completion of the reaction with the protein, an excess of glutathione, and 2-β-mercaptoethanol, is present, ensuring that no reactive species are formed during the purification step.
10. Fluorescently labeled peptide can be separated on a Sephadex G-25 gel filtration column equilibrated in phosphate buffer saline (NAP-10 column). However, for in cellular or in vivo imaging, fluorescently labeled peptides are further purified by RP-HPLC using a C18 reverse-phase HPLC column (Interchrom UP5 HDO/25M Modulo-cart Uptisphere, 250×10 mm) and identified by electrospray mass spectrometry.
11. The degree of labeling can be calculated using the following formula:

$$\frac{A_x}{\varepsilon} \times \frac{\text{MW of protein}}{\text{mg protein / mL}} = \frac{\text{moles of dye}}{\text{moles of protein}},$$

where A_x = the absorbance value of the dye at the absorption maximum wavelength and ε = molar extinction coefficient of the dye or the reagent at the absorption maximum wavelength (according to *The Handbook: A Guide to Fluorescent Probes and Labeling Technologies*).

The procedure for CADY–siRNA complex formation constitutes a major factor in the success and efficiency of CADY technology and should be followed carefully. It was well described previously in (11).

3.4. Synthesis of PEI FluoR Complexes with Nucleic Acids (Fig. 1)

1. A solution (1.5 ml) of borate buffer (0.2 M) is added into a vial containing 50 mg of PEI and the pH is adjusted to 8 by adding a solution of NaOH (10 M).
2. 7.0 mg of tetramethylrhodamine isothiocyanate (RITC) is dissolved in 200 μL of DMSO and the resulting solution is slowly added to the prior solution under agitation.
3. The reaction mixture is stirred overnight at room temperature with exclusion of light.
4. Thin-layer chromatography (TLC) (CH2Cl2/MeOH; 80/20) of the reaction mixture is then carried out to evaluate the presence of starting material RITC. If required, the product may be purified by column chromatography (Sephadex G25 PD10) in water.
5. 34 mg of PEI FluoR is typically isolated (75% yield) and is then dissolved in 1.8 ml of water. The exact concentration of the PEI FluoR solution may be determined by proton NMR and colorimetric measurement at 493 nm (ca. 150 mM).
6. The preparation of the PEI FluoR/nucleic acid complexes should be performed in a laminar flow hood using sterile 10% glucose solution. Dilute the nucleic acid (plasmid DNA or siRNA; Table 4) using the 10% glucose stock solution in double-distilled sterile water to prepare a solution of half the injection volume of 5% glucose. Vortex gently or mix by pipetting up and down (see Note 6).
7. Dilute the PE-FluoR reagent using the 10% glucose stock solution and sterile water to prepare a solution of half the injection volume of 5% glucose. Vortex gently and spin down.
8. Add the diluted PEI FluoR to the diluted nucleic acid all at once, vortex gently, and spin down.
9. Incubate for 15 min at room temperature. From this time point, the complexes are stable for 2 h at room temperature and for 24 h if stored at 4°C.

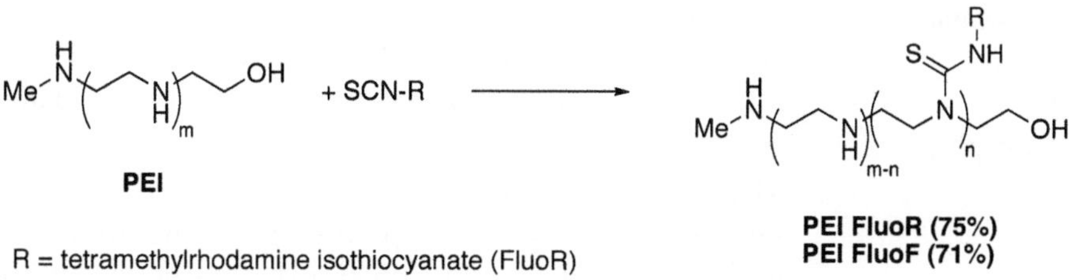

PEI + SCN-R → PEI FluoR (75%) / PEI FluoF (71%)

R = tetramethylrhodamine isothiocyanate (FluoR)

Fig. 1. Schematic representation of the synthesis of PEI FluoR.

Table 4
Recommended conditions for common administration routes in mice (http://www.polyplus-transfection.com/in-vivo-reagents-therapeutics/dna-sirna-delivery-in-vivo-jetpei/)

Animal	Site of injection	Starting conditions	Nucleic acid optimization range	Injection volume optimization range (5% glucose)
Mouse	IV Tail vein/retro-orbital	40 μg nucleic acid 6.4 μl reagent 200 μl of 5% glucose	40–60 μg (1.6–2.4 mg/kg)	200–400 μl
	Intraperitoneal IP	100 μg nucleic acid 16 μl reagent 1 ml 5%glucose	100–200 μg (4–8 mg/kg)	1 ml
	Subcutaneous (s.c.)	5 μg nucleic acid 0.6 μl reagent 10 μl of 5% glucose	3–5 μg	5–15 μl

10. Perform injections into animals using complexes equilibrated at room temperature (Table 4).
11. The biodistribution of PEI-FluoR can be followed as early as 1–4 h following injection, using fluorescence of the whole animal or slices of fixed tissues depending on the target organ.
12. Monitor gene expression as required at the appropriate time point (1–96 h after the last injection) depending on the mode of injection and the targeted organ.
13. With the current formulation and delivery approaches employed today, silencing can be achieved in several tissues and cell types, most notably through direct delivery of siRNA to the back of the eye, to the central nervous system, and to lung epithelial cells, and through systemic delivery of siRNA to hepatocytes and tumor cells.

3.5. Pharmacokinetic Study by In Vivo Fluorescence Reflectance Imaging

1. Mice are anesthetized (isoflurane/oxygen 3.5% for induction and 1.5% thereafter) and placed on a warm holder (37°C) for intravenous injection of the fluorescent mixture (100–200 μL) via the tail vein.
2. Mice are immediately placed in dorsal position in the fluorescence imaging setup and a real-time imaging sequence is acquired over 15 min. Afterwards anesthesia is suspended till the next imaging time point.

3. Fluorescence and bright field images are acquired 1, 2, 3, 5, and 24 h after injection in four mouse-body positions (ventral, dorsal, and laterals).
4. At 24 h post injection, mice are anesthetized and blood is sampled by heart puncture. Mice are then sacrificed and the principal organs are removed (brain, heart, lungs, liver, spleen, pancreas, stomach, guts, kidneys, suprarenal glands, ovaries, uterus, muscle, fat, skin, lymph nodes).
5. Blood is centrifuged (20 min, 2,300×*g*, 4°C) and 200 μL of plasma is sampled for fluorescence imaging with the isolated organs.
6. Each fluorescence image can be superimposed on the bright field image for better location of the fluorescent signal. Semiquantitative data are obtained from the fluorescence images by drawing regions of interest (ROI) on the area to be quantified. The results are expressed as a number of Relative Light Units per pixel for a 100-ms exposure time (RLU/pix/100 ms).

3.6. Luc Gene Transfer Follow-Up by In Vivo Bioluminescence Imaging

1. In vivo luciferase expression is monitored by bioluminescence 24 h after *luciferase* gene transfer in vivo. On the day of experimentation the solution is diluted to 10 mg/ml in PBS without $CaCl_2$ and $MgCl_2$ and is kept protected from light.
2. Vigil mice are injected intraperitoneally with d-Luciferin (150 mg/kg) just before anesthesia (isoflurane 4% for induction and 1.5% thereafter).
3. Five minutes after Luciferin injection, bioluminescence images are acquired with the IVIS Kinetic (Caliper) (Fig. 2).
4. Semiquantitative data are obtained from the bioluminescence images by using the Living image software (Caliper). ROI are drawn around the areas to be quantified. The results are expressed as a number of photons per second (ph/s).
5. After imaging, mice are sacrificed and the main organs are removed for in vitro Luciferase activity measurement.

3.7. LacZ Gene Transfer Follow-Up by In Vivo Fluorescence Reflectance Imaging

1. In vivo β-Gal expression is monitored by NIR fluorescence 24 h after *LacZ* gene transfer in vivo. Mice are anesthetized (isoflurane/oxygen 3.5% for induction and 1.5% thereafter) and placed on a warm mat (37°C) for intracardiac injection of DDAOG (100 μL).
2. Twenty minutes after injection, fluorescence and bright field images are acquired for four mouse-body positions (ventral, dorsal, and laterals) (Fig. 3).

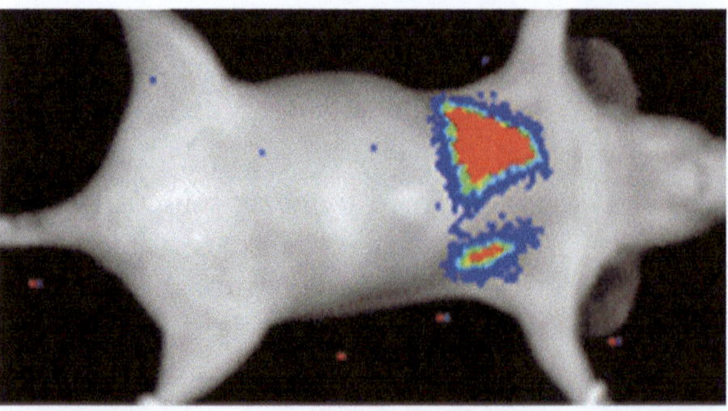

Fig. 2. In vivo lung bioluminescence imaging (BLI). Mice were injected intravenously with 50 μg (200 μl) of *luciferase* plasmid DNA combined with "In vivo-PEI." Whole-body BLI was performed 24 h after DNA–PEI injection and 10 min after Luciferine ip injection.

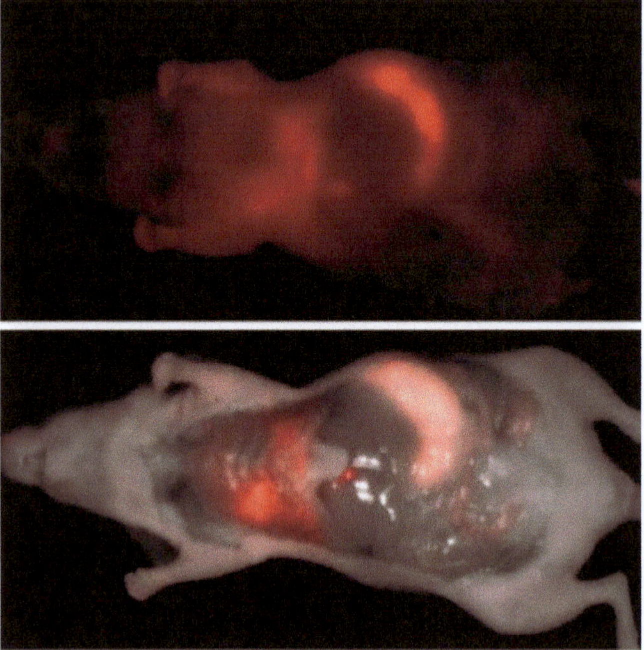

Fig. 3. In vivo lung NIR fluorescence imaging. Mice were injected with 50 μg (200 μl) of ß-*galactosidase* plasmid DNA combined with "In vivo-PEI." Whole-body fluorescence imaging was performed 24 h after complex DNA–PEI injection and after i.c. injection of DDAOG.

3. Just after imaging, mice are sacrificed and the main organs are removed (brain, heart, lungs, liver, spleen, pancreas, stomach, guts, kidneys, suprarenal glands, ovaries, uterus, muscle, fat, skin, lymph nodes) for ex vivo fluorescence imaging.
4. Each fluorescence image can be superimposed on the bright field image for better location of the fluorescent signal.

Semiquantitative data are obtained from the fluorescence images by drawing ROI on the area to be quantified. The results are expressed as a number of Relative Light Units per pixel for a 100-ms exposure time (RLU/pix/100 ms).

3.8. In Vitro Enzymatic Assays on Isolated Organs

1. Luciferase or β-Gal activities are also assayed in vitro using the Luciferase Assay System or the β-Gal Assay System (Promega) as recommended by the manufacturer.
2. Tissues are extracted and cut into small pieces with a razor blade. Fragments are mixed thoroughly in 1 ml lysis buffer.
3. After a 15-min incubation, the samples are frozen at −20°C.
4. The following day, the samples are thawed and brought to room temperature before centrifugation (2 min, 15,600 × *g*).
5. The supernatants are sampled for protein content measurements using the DC-comp Bio-Rad assay (Bio-Rad, Marnes-la-Coquette, France) and luciferase or β-Gal activity measurements.
6. To quantify luciferase, 10 μL of the extract are mixed with 100 μL of luciferase assay substrate and the luciferase activity is measured for 10 s on a photoluminometer (Berthold, Thoiry,France). The luciferase activities are calculated as a number of Relative Light Unit per 10 s per mg of protein (RLU/10s/mg). As a reference, using purified firefly luciferase under the same experimental conditions, we determined that 2 ng of this enzyme produced 10^8 RLU/10 s.
7. To quantify β-Gal, 50 μL of the extract are added to 50 μL of 2× Assay buffer and incubated for 30 min at 37°C. The reaction is stopped by adding 150 μL of sodium carbonate 1 M and the absorbance is read at 420 nm on a spectrophotometer (Bio-Rad, Marnes-la-Coquette, France). The β-Gal activities are expressed as a number of milliUnits (mU) per mg of protein (1 mU corresponding to the hydrolyze of 1 nmol of substrate per minute at pH 7.5 and 37°C). As a reference, using purified β-Gal under the same experimental conditions, we determined that 2 ng of this enzyme produced 0.654 mU.

4. Notes

1. Several sequence prediction Web sites exist for the determination of the sequence of an siRNA. They are often associated with companies that sell siRNA. *Whitehead siRNA Selection Web Server*. http://jura.wi.mit.edu/bioc/siRNAext—Developed and hosted by the Whitehead Institute, this Web site is

somewhat more complex, giving a large number of possible duplexes along with their thermodynamic properties. An off-target search can be done for each duplex, within the site but in a separate step (siRNA Selection Server: an automated siRNA oligonucleotide prediction server Bingbing Yuan, Robert Latek, Markus Hossbach, Thomas Tuschl, and Fran Lewitter). *Rosetta siRNA Design Algorithm.* Sigma-Aldrich has entered into an exclusive partnership with Rosetta Inpharmatics Sequence design: http://www.sigmaaldrich.com/life-science/your-favorite-gene-search.html. The Rosetta siRNA Design Algorithm utilizes Position-Specific Scoring Matrices (PSSM) and knowledge of the all-important siRNA seed region to predict the most effective and specific siRNA sequences for your target gene of interest. *BIOPREDsi algorithm.* Developed by the Novartis Institutes for BioMedical Research, this site features a very simple input and output, with only the essential information given (input is a gene Accession number or gene sequence, output is a user-defined number of optimized siRNA sequences: http://www.qiagen.com/geneglobe/default.aspx). siRNA targeting gene, and a derived siRNA harboring two mismatches, must be used for each target gene as control. A BLAST search is carried out to ensure that the chosen sequence does not target unwanted genes.

Fluorescent dyes are widely used to label at the 3′ end or 5′ end of the sense strand siRNA for tracking of siRNA. Fluorescently labeled siRNA could be synthesized by different companies according to the chosen sequences: Sigma (http://www.sigmaaldrich.com/life-science/), Dharmacon (http://www.dharmacon.com), Qiagen (http://www.qiagen.com), Ambion (http://www.invitrogen.com/site/us/en/home/brands/ambion.html), and Eurogentec (http://secure.eurogentec.com).

2. Add a plastic paraffin film at the bottom of round-bottom flask, and gently shake avoiding the solvent to evaporate or to enter in contact with the plastic paraffin film.
3. The stock solution of DiD in acetone (5 mg/ml) can be stored at −20°C, in a well-closed scintillation bottle.
4. As Solutol HS-15® is viscous, this compound is weighted first, and could be added to the bottle with a syringe.
5. This protocol leads to a final volume of 1.479 ml of DiD-labeled DNA LNCs. When a higher volume is needed, the proportions of the different LNC compounds have to be multiplied.
6. The final concentration of nucleic acid in the injection volume should not exceed 0.5 μg/μl.

Acknowledgments

The authors thank C Passirani, JP Benoit (INSERM Angers), I Texier-Nogues (CEA-LETI, Grenoble), and P Erbacher (Polyplus transfection, Illkirch, France) for their help and advices. This work was funded by the Agence Nationale pour la Recherche (ANR pNANO, CALIF, ANR BiotecS GLIOTHERAP, ANR CES NANOBIOTOX) and the INCA (PLbio Poro-Combo; PLBio Biosensimag).

References

1. Weissleder R (2001) A clearer vision for in vivo imaging. Nat Biotechnol 19:316–317
2. Weissleder R, Ntziachristos V (2003) Shedding light onto live molecular targets. Nat Med 9:123–128
3. Josserand V, Texier-Nogues I, Huber P, Favrot MC, Coll JL (2007) Non-invasive in vivo optical imaging of the lacZ and luc gene expression in mice. Gene Ther 14:1587–1593
4. Lin EH, Keramidas M, Rome C, Chiu WT, Wu CW, Coll JL, Deng WP (2011) Lifelong reporter gene imaging in the lungs of mice following polyethyleneimine-mediated sleeping-beauty transposon delivery. Biomaterials 32:1978–1985
5. Morille M, Montier T, Legras P, Carmoy N, Brodin P, Pitard B, Benoit JP, Passirani C (2010) Long-circulating DNA lipid nanocapsules as new vector for passive tumor targeting. Biomaterials 31:321–329
6. Coll JL, Chollet P, Brambilla E, Desplanques D, Behr JP, Favrot M (1999) In vivo delivery to tumors of DNA complexed with linear polyethylenimine. Hum Gene Ther 10:1659–1666
7. Heurtault B, Saulnier P, Pech B, Proust JE, Benoit JP (2002) A novel phase inversion-based process for the preparation of lipid nanocarriers. Pharm Res 19:875–880
8. Morille M, Passirani C, Letrou-Bonneval E, Benoit JP, Pitard B (2009) Galactosylated DNA lipid nanocapsules for efficient hepatocyte targeting. Int J Pharm 379:293–300
9. Morille M, Passirani C, Dufort S, Bastiat G, Pitard B, Coll JL, Benoit JP (2011) Tumor transfection after systemic injection of DNA lipid nanocapsules. Biomaterials 32:2327–2333
10. Crombez L, Aldrian-Herrada G, Konate K, Nguyen QN, McMaster GK, Brasseur R, Heitz F, Divita G (2009) A new potent secondary amphipathic cell-penetrating peptide for siRNA delivery into mammalian cells. Mol Ther 17:95–103
11. Crombez L, Divita G (2011) A non-covalent peptide-based strategy for siRNA delivery. Methods Mol Biol 683:349–360

Acknowledgments

The authors thank [illegible] [illegible] [illegible] [illegible] [illegible] was funded by the [illegible] [illegible] NANO [illegible] CARNOT [illegible] Biosciences.

References

[illegible]

Chapter 6

Lipids for Nucleic Acid Delivery: Synthesis and Particle Formation

Michel Bessodes and Nathalie Mignet

Abstract

Lipidic vesicles have been extensively studied for their capacity to condensate and deliver nucleic acids to the cells. Many different amphiphilic lipidic structures have been proposed, each of them bringing some advances in nonviral gene transfection. The ionic or neutral nature of the lipids induces tremendous differences in the behavior of the corresponding liposomes, from the complexation of nucleic acid to the delivery to the cell. An efficient delivery in vitro or in vivo also depends closely on the structure of the lipids and very often, efficient liposomes in vitro have been found useless for in vivo administration.

We wish to describe in this chapter the chemical synthesis of two different lipids, one cationic and the other essentially neutral, and the formulation to obtain liposomes and DNA/liposome complexes. The different ways and tricks for the formulation of the two different structures are especially highlighted.

Key words: Cationic lipid, Non-cationic lipid, Lipid synthesis, Ethanolic injection, Liposome formulation, Lipoplex characterization, DNA–lipid complex stabilization

1. Introduction

Amphiphilic lipids are able to auto-assemble to form spherical vesicles such as liposomes in aqueous solution (1). Synthetic lipids, like phospholipids, are amphiphiles which tend to organize themselves in an aqueous medium in order to maximize the hydrophobic interactions, repulsing the water content from the lipidic part, therefore reaching the more thermodynamically stable state. Their structure should allow the lipids to expose the cationic charges of their hydrophilic head in order for them to interact with DNA. The lipoplexes formed between the lipids and DNA protect DNA from enzymatic degradation and increase its cellular uptake leading to high levels of transfection in vitro (2). We have earlier described

Manfred Ogris and David Oupicky (eds.), *Nanotechnology for Nucleic Acid Delivery: Methods and Protocols*, Methods in Molecular Biology, vol. 948, DOI 10.1007/978-1-62703-140-0_6, © Springer Science+Business Media, LLC 2013

two families of such lipids, the first family contains cationic head able to strongly condense plasmidic DNA into liposomes through electrostatic interactions (3, 4), and the second family bears thiourea groups known to promote hydrogen bonds with phosphate-containing macromolecules (5–7).

A thorough structure to physicochemical and biological properties relationship has been established, thus allowing to highlight two structures (DMAPAP and DDSTU, Scheme 1) in each family with particularly interesting properties. The cationic liposomal formulation (DMAPAP/DOPE 1/1) is nowadays used as a reference in our laboratory as it transfects as efficiently as lipofectamine adherent cultured cells in vitro. However, for in vivo applications, we developed non-cationic lipids that we called thiourea lipids in order to increase blood circulation of the lipoplexes and reduce nonspecific interactions. We showed in vitro that cationic lipoplexes were internalized more efficiently than thiourea lipoplexes by the cells, and that thiourea lipoplexes were able to release DNA into the cells more efficiently than cationic complexes leading to gene transfection (Scheme 1) (7). When administered i.v. to mice, thiourea complexes were also shown to remain longer in the circulation (6).

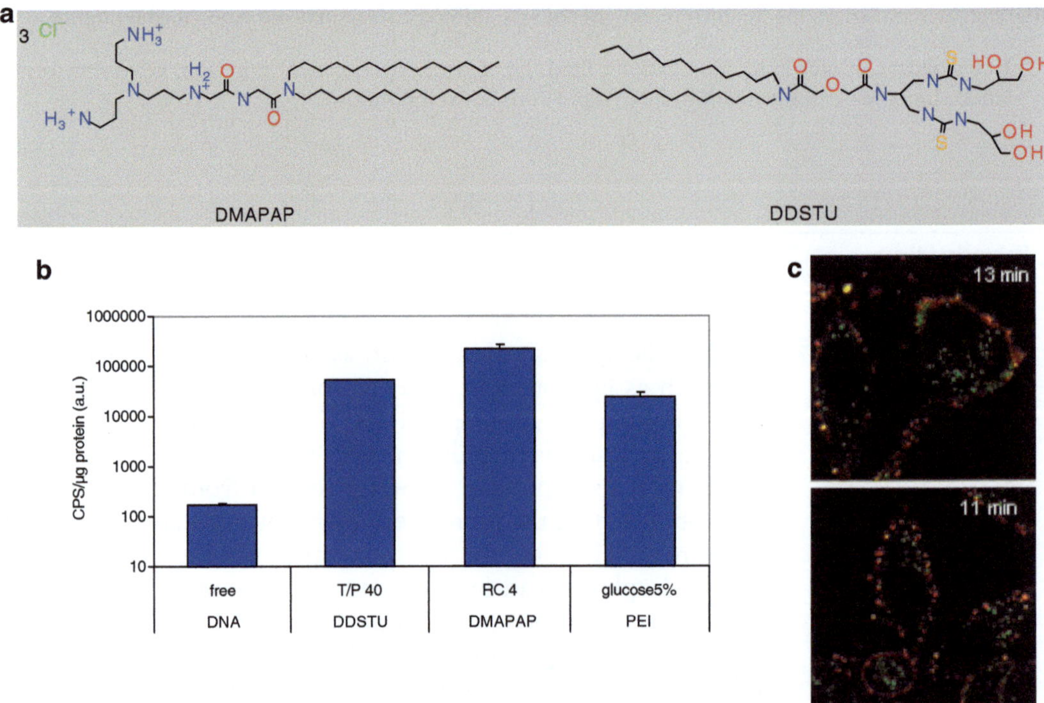

Scheme 1. (**a**) Chemical structure of our lipopolyamine (DMAPAP) and lipopolythiourea lead compounds (DDSTU). (**b**) Their efficacy in terms of luciferase expression by a luciferase encoding gene is shown on B16 cell lines in the presence of 10% serum (SVF). (**c**) DNA release from the complexes was shown to be faster with DDSTU as compared to DMAPAP on Hela cells (7). Lipids are labeled with rhodamine (*red*), the gene is labeled with fluorescein (*green*).

The practical syntheses, auto-assembly of the lipids, DNA association, characterization of the lipoplexes, and their stability assessment are described in this chapter.

2. Materials

2.1. Abbreviations of the Lipids Used

DOPE: L-α-Dioleoyl Phosphatidylethanolamine; the chemical names were generated with Symyx Draw 4.0® software, based on IUPAC rules. The cationic lipid whose name according to the nomenclature is 2-{3-[Bis-(3-amino-propyl)-amino]-propylamino}-*N*-ditetradecylcarbamoyl methyl-acetamide or RPR209120 that we called DMAPAP was previously described, using a different chemical route, in the supporting information of Thompson et al. (8). Synthesis of 2-[2-(didecylamino)-2-oxo-ethoxy]-*N*-[2-(2,3-dihydroxypropylcarbamothioylamino)-1-[(2,3-dihydroxypropylcarbamothioylamino)methyl]ethyl]acetamide, called DDSTU, was described slightly differently in reference (9).

2.2. Chemicals and DNA Provided or Synthesized

The chemicals were purchased from Sigma-Aldrich Company. The solvents were from Carlo Erba-SDS (analytical grade). TLC was performed on silica gel $60F_{254}$ aluminum sheets from Merck. L-α-DOPE was purchased from Avanti Polar Lipids, Picogreen® from Molecular Probes. pDNA was obtained as described in reference (7).

2.3. Equipment

Size and zeta potentials measurements were performed on a Zeta Sizer NanoSeries from Malvern Instruments equipped with an MPT2 autotitrator. Fluorescence was measured on a multilabel plate reader Wallac Victor2 1420 Multilabel Counter, Perkin Elmer, France, equipped with excitation and emission filters (350 ± 10, 450 ± 10 nm).

3. Methods

3.1. Synthesis of DMAPAP Cationic Lipid (see Note 1)

tert-Butyl 2-(3-hydroxypropylamino)acetate (1) (Scheme 2)

To a cooled solution of 3-aminopropanol (196 ml, 2.56 mol, 25 eq.) in DCM (250 ml), *tert*-butyl bromoacetate (20 g, 102.5 mmol, 1 eq.) in DCM (200 ml) was added dropwise. After 2 h the reaction medium was warmed to room temperature and kept for three more hours. The solution was then washed with $NaHCO_3$ sat (3 × 150 ml), sat. NaCl (150 ml). After drying over magnesium sulfate, filtration, and evaporation, a colorless oil was obtained (17.8 g, 92%).

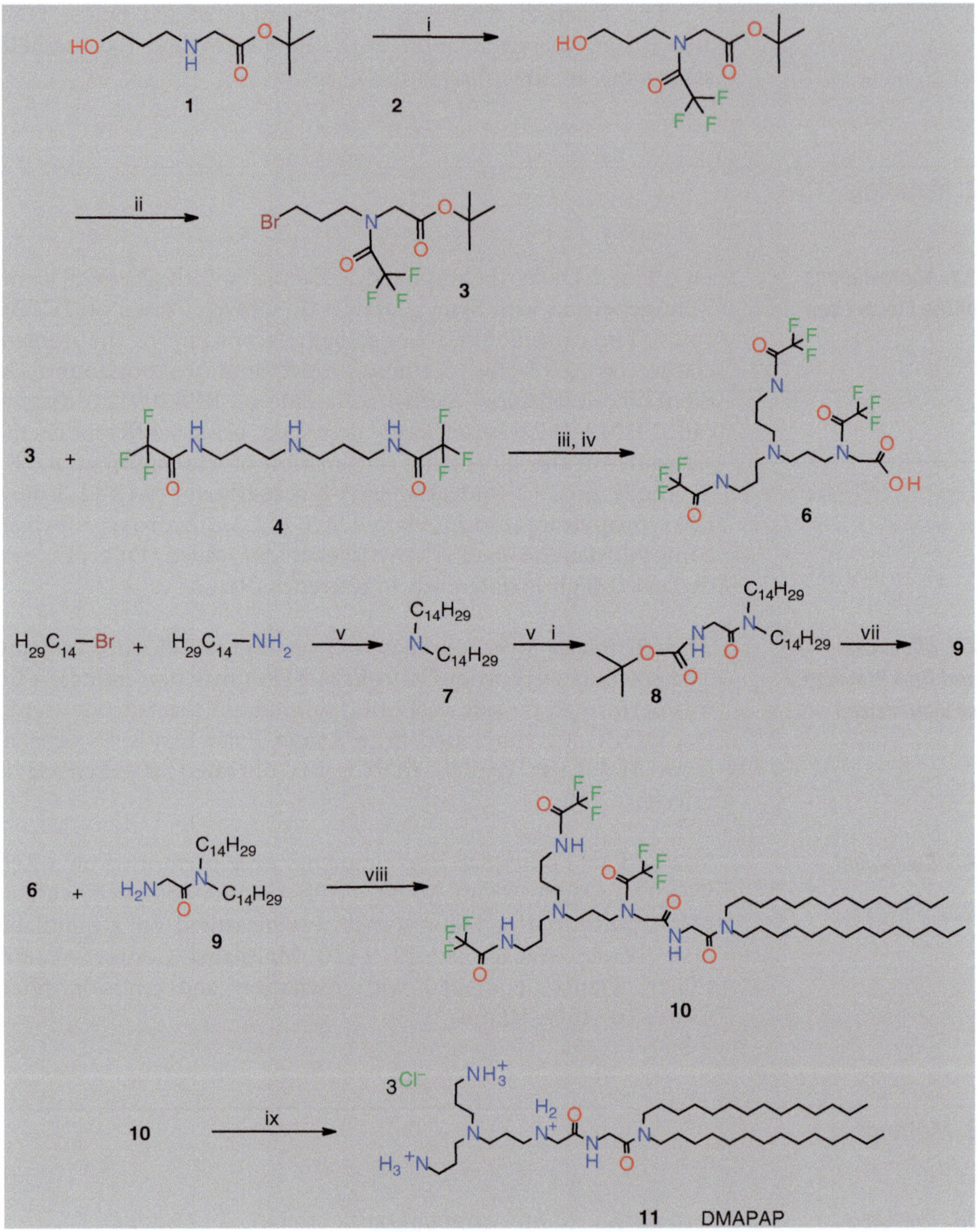

Scheme 2. *i*, $(TFA)_2O$, Et_3N, CH_2Cl_2; *ii*, CBr_4, Ph_3P, CH_3CN; *iii*, K_2CO_3, CH_3CN; *iv*, **5**+TFA, CH_2Cl_2; *v*, Na_2CO_3, EtOH, reflux; *vi*, Boc-Gly, BOP, DIEA, CH_2Cl_2; *vii*, TFA, H_2O; *viii*, BOP, DIEA, CH_2Cl_2; *ix*, 1N NaOH, THF.

TLC (Rf=0.55; DCM/MeOH 8/2; ninhydrin or I_2/H_2SO_4). *^1H NMR* (300 MHz, $(CD_3)_2SO$, δ ppm): 1.44 (s, 9H), 1.54 (m, *J*=6.5 Hz, 2H), 2.56 (t, *J*=6.5 Hz, 2H), 3.19 (s, 2H), 3.46 (t, *J*=6.5 Hz, 2H), 3.70–4.70 (m, 1H). *MS* (DC/I, *m/z*) 190 (MH^+).

tert-Butyl 2-[3-hydroxypropyl-(2,2,2-trifluoroacetyl)amino]acetate (2)

Compound 1 (17.65 g, 93.3 mmol, 1 eq.) was dissolved in DCM (100 ml), triethylamine (26 ml, 186.6 mmol, 2 eq.) was added, the mixture was cooled to 0°C in an ice-water bath, and Trifluoroacetic anhydride (21.5 g, 102.6 mmol, 1.1 eq.) was then added dropwise. The reaction mixture was stirred overnight at room temperature. The solution was washed with $NaHCO_3$ sat (3×50 ml), $KHSO_4$ 0.5 M (3×50 ml), and sat. NaCl (50 ml). The organic phase was dried over magnesium sulfate, filtrated, and concentrated to a pale yellow oil (24.7 g, 93%).

TLC (Rf=0.25, C_6H_{12}/EtOAc 1/1, ninhydrin or I_2/H_2SO_4, UV). *^1H NMR* (300 MHz, $(CD_3)_2SO$, δ ppm): (see Note 2) 1.44, 1.46 (2 s, 9H); 1.60–1.80 (m, 2H); 3.35–3.55 (m, 4H); 4.09, 4.25 (m, 2H); 4.57, 4.61 (t, *J*=5 Hz, 1H). *MS* (D/CI, *m/z*) 303 (M NH_4^+).

tert-Butyl 2-[3-bromopropyl-(2,2,2-trifluoroacetyl)amino]acetate (3)

To a stirred solution of 2 (10 g, 35 mmol, 1 eq.) and triphenylphosphine (12.4 g, 47.3 mmol, 1,35 eq.) in THF (150 ml), carbon tetrabromide (15.1 g, 45.6 mmol, 1.3 eq.) in acetonitrile (60 ml) was added dropwise. After 4 h the reaction medium was concentrated, taken up in ethyl acetate, and filtrated on a paper filter. The filtrate was concentrated to dryness, taken up in cyclohexane, and filtrated on sintered glass (no. 3). After a final concentration and purification on silica column (C_6H_{12}/EtOAc 8/2), a lightly yellow colored oil was obtained (10.4 g, 85%). *TLC* (Rf=0.6, C_6H_{12}/EtOAc 1/1, ninhydrin or I_2/H_2SO_4, UV); *^1H NMR* (300 MHz, $(CD_3)_2SO$, δ ppm) (see Note 2) 1.43, 1.45 (s, 9H); 2.05–2.20 (m, 2H); 3.50–3.65 (m, 4H), 4.11, 4.28 (m, 2H). *MS* (D/CI *m/z*) M NH_4^+.

2,2,2-Trifluoro-N-[3-[3-[(2,2,2-trifluoroacetyl)amino]propylamino]propyl]acetamide (4)

3,3′-Imino-bispropylamine (35 g, 266.7 mmol, 1 eq.) was dissolved in anhydrous THF (150 ml) under an argon atmosphere. The solution was cooled to 0°C in an ice bath and ethyl trifluoroacetate (65 ml, 546.8 mmol, 2.05 eq.) was added dropwise (see Note 3). After 3 h the reaction medium was warmed to room temperature and was left two more hours under argon. The insoluble was removed by filtration and the solution was concentrated. After drying overnight in a vacuum oven, a white powder was obtained (85.3 g, 99%). *TLC* (Rf=0.7, EtOH/NH_4OH 8/2, ninhydrin). *^1H NMR* (300 MHz, $CDCl_3$, δ ppm): 1.74 (m, *J*=6 Hz, 4H); 2.73 (t, *J*=6 Hz, 4H); 3.46 (m, 4H); 8.18 (m, 2H). *MS* (D/CI, *m/z*) 324 MH^+.

tert-Butyl 2-[3-[bis[3-[(2,2,2-trifluoroacetyl)amino]propyl]amino]propyl-(2,2,2-trifluoroacetyl)amino]acetate (5)

To a solution of compound 3 (26 g, 74.7 mmol, 1 eq.) and compound 4 (24.1 g, 74.7 mmol, 1 eq.) in acetonitrile (130 ml), potassium carbonate was added (30 g, 224 mmol, 3 eq.) and the mixture was heated to reflux during 6 h. The reaction medium was then filtrated and evaporated to dryness. Chromatography on silica (cyclohexane/EtOAc 8/2) yielded 5 as a pale yellow oil (16.6 g, 38%). *TLC* (Rf=0.35, EtOAc, I_2/H_2SO_4, UV). *^1H NMR* (300 MHz, $CDCl_3$, δ ppm): (see Note 2) 1.47, 1.48 (2 s, 9H); 1.65–1.85 (m, 6H); 2.40–2.60 (m, 6H); 3.40–3.55 (m, 6H); 3.97, 4.07 (m, 2H); 7.45–7.65 (m, 2H). *MS* (D/CI, *m/z*) 591 MH^+.

2-[3-[Bis[3-[(2,2,2-trifluoroacetyl)amino]propyl]amino]propyl-(2,2,2-trifluoroacetyl)amino]acetic acid (6)

Trifluoroacetic acid (50 ml) was added to a solution of compound 5 (15.8 g, 28.76 mmol) in DCM (50 ml). The mixture was stirred at room temperature during 2 h, and then concentrated. A faint yellow gum was obtained (18.7 g, 100%). *^1H NMR* (300 MHz, $CDCl_3$, δ ppm): (see Note 2) 1.80–2.10 (m, 6H); 3.12 (m, 6H); 3.29 (m, 4H); 3.50 (m, 2H); 4.13, 4.29 (m, 2H); 9.50–9.75 (m, 2H). *MS* (D/CI, *m/z*) 535 MH^+.

Ditetradecylamine Chlorohydrate (7)

Bromotetradecane (5 g, 18 mmol) and tetradecylamine (3.85 g, 18 mmol) were dissolved in absolute ethanol (30 ml). Sodium carbonate (4.8 g, 45 mmol, 2.5 eq.) was added and the mixture was refluxed overnight. The reaction medium was then evaporated, taken up in dichloromethane (100 ml), and washed successively with water (3×20 ml) and sat. NaCl (1×40 ml). The organic phase was dried over calcium chloride and concentrated. After salification (see Note 4) the product crystallized in isopropanol (3.3 g; 41%). *TLC* (Rf=0.55; DCM/MeOH 9/1; ninhydrin or I_2/H_2SO_4). *^1H NMR* (300 MHz, $CDCl_3$, δ ppm): 0.88 (t, *J*=7 Hz, 6H); 1.15–1.45 (m, 44H); 1.90 (m, 4H); 2.90 (m, 4H); 9.48 (m, 2H). *MS* (D/CI, *m/z*) 410 (MH^+).

tert-Butyl N-[2-[di(tetradecyl)amino]-2-oxo-ethyl]carbamate (8)

To a solution of ditetradecylamine chlorohydrate (3 g, 7.33 mmol) and Boc glycine (1.41 g, 8 mmol, 1.1 eq.) in dichloromethane (50 ml), diisopropylethylamine (6.2 ml, 36.65 mmol, 5 eq.) and BOP (3.56 g, 8 mmol, 1.1 eq.) were added. The solution was stirred at ambient temperature for 1 h and then concentrated. Ethyl acetate (150 ml) was added and the solution was washed successively with 0.5 M $KHSO_4$ (3×50 ml), sat. $NaHCO_3$ (3×50 ml), and sat. NaCl (50 ml); it was dried over Na_2SO_4, filtered, and concentrated to dryness (3.57 g, 86%). *TLC* (Rf=0.43; DCM/MeOH 99:1; ninhydrin or I_2/H_2SO_4); *^1H NMR* (400 MHz, $CDCl_3$, δ ppm): 0.87 (t, *J*=7 Hz, 6H); 1.16–1.35 (m, 44H); 1.46–1.64 (m, 4H); 3.13 (dd, *J*=8 Hz, 2H); 3.30 (dd, *J*=8 Hz, 2H); 3.94 (dd, *J*=4 Hz, 2H); 5.57 (m, 1H).

Amino-N,N-di(tetradecyl)acetamide; TFA Salt (9)

The above product (3.57 g; 6.3 mmol) was dissolved in an aqueous solution of trifluoroacetic acid (90%; 15 ml) and left at ambient temperature for 1 h. The solution was then evaporated under reduced pressure and the product crystallized in diethyl ether (3.6 g; 98%). *TLC* (Rf=0.1; DCM/MeOH 99:1; ninhydrin or I_2/H_2SO_4); *¹H NMR* (400 MHz, $CDCl_3$, δ ppm): 0.87 (t, *J*=7 Hz, 6H); 1.2–1.35 (m, 44H); 1.45–1.58 (m, 4H); 3.13 (dd, *J*=8 Hz, 2H); 3.31 (dd, *J*=8 Hz, 2H); 3.88 (m, 2H).

N-[3-[3-[[2-[[2-[Di(tetradecyl)amino]-2-oxo-ethyl]amino]-2-oxo-ethyl]-(2,2,2-trifluoroacetyl)amino]propyl-[3-[(2,2,2-trifluoroacetyl)amino]propyl]amino]propyl]-2,2,2-trifluoro-acetamide (10)

2-Amino-*N,N*-di(tetradecyl)acetamide 9 (3.6 g; 6.2 mmol) and 2-[3-[bis[3-[(2,2,2-trifluoroacetyl)amino]propyl]amino]propyl-(2,2,2-trifluoroacetyl)amino]acetic acid 6 (2.98 g; 5.58 mmol; 0.9 eq.) were dissolved in dichloromethane (20 ml). Diisopropylethylamine (5.3 ml, 31 mmol, 5 eq.) and BOP (3 g, 6.8 mmol, 1.1 eq.) were added. The solution was stirred at ambient temperature for 1 h, and then concentrated. The residue was redissolved in ethyl acetate (150 ml) and washed successively with 0.5 M $KHSO_4$ (3×20 ml), sat. $NaHCO_3$ (3×20 ml), water, and sat. NaCl solution. It was dried over Na_2SO_4, filtered, and concentrated to give 10 (4.67 g, 85%). *TLC* (Rf=0.1; DCM/MeOH 99:1; ninhydrin or I_2/H_2SO_4); *¹H NMR* (400 MHz, $CDCl_3$, δ ppm): 0.87 (t, *J*=7 Hz, 6H); 1.19–1.34 (m, 44H); 1.44–1.93 (m, 10H); 3.15 (m, 2H); 3.30 (m, 2H); 3.43 (m, 4H); 3.55 (m, 2H); 4.01–4.19 (m, 4H); 6.99 (m, 1H); 7.64 (m, 1H); 7.74 (m, 1H). *MS* (ES, *m/z*) 983.5 M.

3-[Bis(3-aminopropyl)amino]propyl-[2-[[2-[di(tetradecyl)amino]-2-oxo-ethyl]amino]-2-oxo-ethyl]ammonium; hydrochloride (11, DMAPAP)

The compound 10 (4 g, 4 mmol) was dissolved in THF (60 ml), 1N NaOH (60 ml) was added, and the mixture was stirred overnight at ambient temperature. The solution was evaporated to dryness, redissolved in a mixture of DCM/MeOH 1/1, and chromatographed on a short column of silica eluted with DCM/MeOH/NH_4OH (45:45:10). The fractions containing the non-protonated form of 11 were collected and evaporated. Subsequent treatment with 6N HCl in isopropanol, evaporation, and lyophilization gave 11 (2.8 g, 95%). *TLC* (Rf=0.2; DCM/MeOH/NH_4OH 45:45:10; ninhydrin or I_2/H_2SO_4); *¹H NMR* (400 MHz, DMSO d_6, δ ppm): 0.86 (t, *J*=7 Hz, 6H); 1.17–1.35 (m, 44H); 1.37–1.59 (m, 4H); 1.97–2.20 (m, 6H); 2.86–3.40 (m, 16H); 3.80 (m, 2H); 4.01 (m, 2H). *MS* (ES, *m/z*) 695.8 M, 696.8 M–H^+, 697.8 M–$2H^+$, 698.8 M–$3H^+$.

3.2. Synthesis of DDSTU Thiourea Lipid

tert-Butyl N-[2-hydroxy-1-(hydroxymethyl)ethyl]carbamate (12) (Scheme 3)

In a 250 ml round-bottom flask, 2-amino-1,3-propanediol (5 g, 54.9 mmol) was dissolved in ethanol (150 ml) and di-*tert*-butyldicarbonate (11.98 g, 54.9 mmol) was added at 0°C. The mixture was stirred at ambient temperature during 10 h. It was then evaporated under reduced pressure and the white residue taken up in a minimum of dichloromethane; heptane was then added until turbidity. The product crystallized at 0°C and was filtered on a sintered funnel (9 g; 86%). Mp 82°C. *TLC* (Rf = 0.48; DCM/MeOH 9:1; ninhydrin); *^1H NMR* (400 MHz, $CDCl_3$, δ ppm): 1.45 (s, 9H, CH_3), 3.24 (m, 2H, OH), 3.67–3.83 (m, 5H, CH, CH_2OH), 5.34 (d, 1H, *J* = 6.6 Hz, NH).

[2-(tert-Butoxycarbonylamino)-3-methylsulfonyloxy-propyl] methanesulfonate (13)

Compound 12 (4 g, 20.92 mmol) was dissolved in dichloromethane (47 ml), triethylamine (8.79 ml, 62.6 mmol) was added, and the solution was placed under a nitrogen atmosphere and cooled to 0°C. Methanesulfonyl chloride (3.9 ml, 50 mmol) was slowly added to maintain the temperature below 10°C. The mixture was stirred at ambient temperature during 3 h, whereupon no starting material could be detected (*TLC*: DCM/MeOH 8:2). The reaction medium was diluted with dichloromethane (100 ml); washed successively with a 10% citric acid solution (2 × 10 ml), water, and sat. NaCl; and then dried over sodium sulfate. After filtration and evaporation the light brown residue solidified on standing at 4°C (7 g, 97%). Mp 85°C. *TLC* (Rf = 0.84; DCM/MeOH 9:1; ninhydrin); *^1H NMR* (400 MHz, $CDCl_3$, δ ppm): 1.43 (s, 9H, CH_3), 3.06 (s, 6H, CH_3), 4.23–4.37 (m, 5H, CH, CH_2OH), 5.07 (d, 1H, *J* = 7.6 Hz, NH).

tert-Butyl N-[2-azido-1-(azidomethyl)ethyl]carbamate (14)

Compound 13 (8 g, 23.03 mmol) and sodium azide (14.97 g, 230 mmol) were placed in a 250 ml round bottom flask, DMF (50 ml) was added, and the mixture was stirred at 60°C for 2 h. A small aliquot was evaporated and controlled by TLC (DCM/MeOH 9:1). The reaction medium was evaporated under reduced pressure; the residue was dissolved in dichloromethane, washed with water (3 × 25 ml) and sat. NaCl (25 ml), and dried over Na_2SO_4. Evaporation of the solvent gave a yellow liquid (3.42 g; 61.6%). *TLC* (Rf = 0.95; DCM/MeOH 9:1; ninhydrin); *^1H NMR* (400 MHz, $CDCl_3$, δ ppm): 1.43 (s, 9H, CH_3), 3.40 (dd, *J* = 7.8 Hz, *J* = 16.6 Hz, CH_2), 3.51 (dd, *J* = 6.8 Hz, *J* = 16.6 Hz, CH_2), 3.85 (m, 1H, CH), 4.91 (d, 1H, *J* = 8.4 Hz, NH).

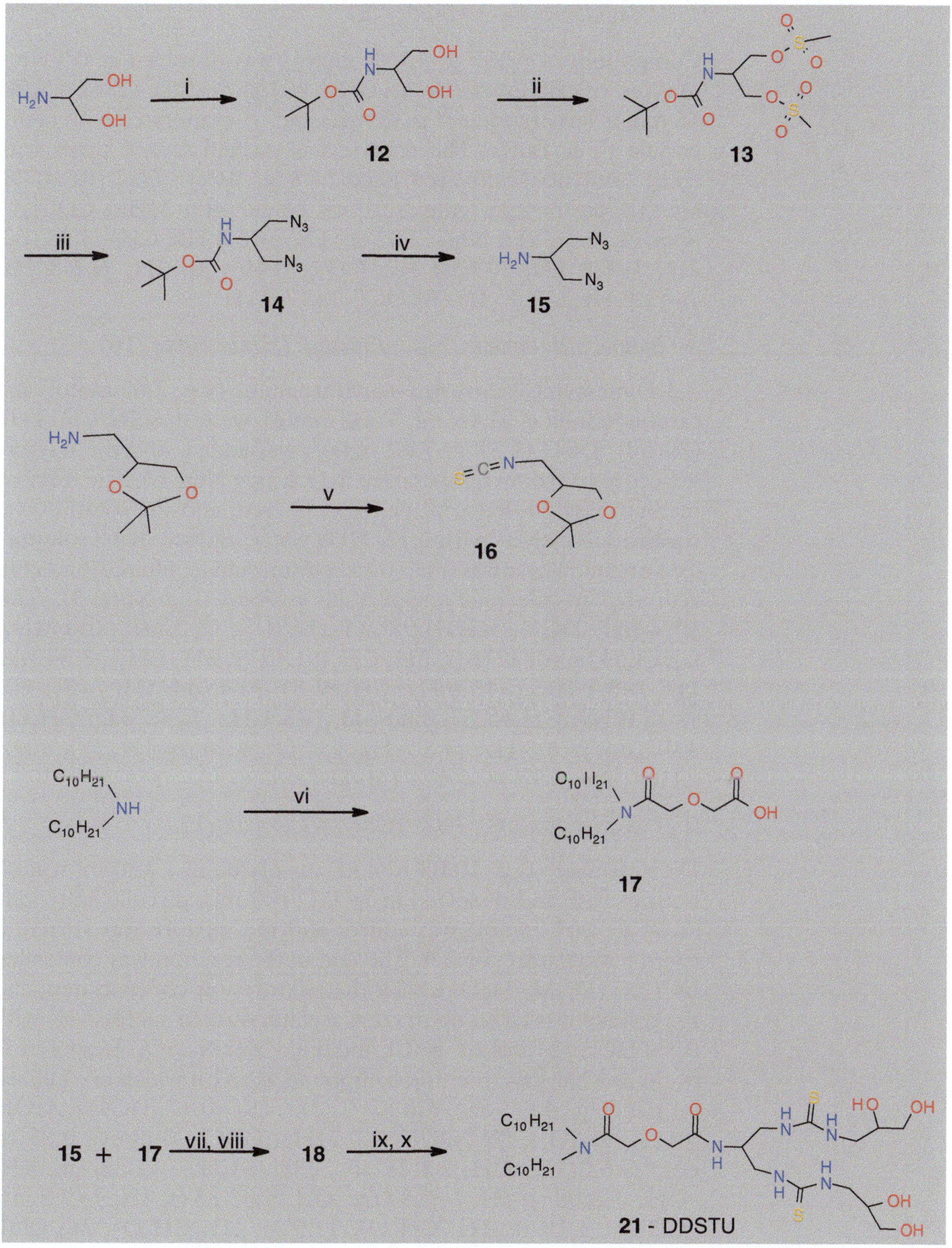

Scheme 3. *i*, Boc_2O, EtOH, 0°C; *ii*, CH_3SO_2Cl, Et_3N, DCM, 0°C; *iii*, NaN_3, DMF, 60°C; *iv*, TFA/H_2O (9:1), 0°C; *v*, CS_2, DCC, THF; *vi*, glycolic anhydride, DCM; *vii*, H_2, Pd/C, MeOH; *ix*, **16**+**19**, DCM; *x*, 1N HCl, CH_3CN.

1,3-Diazidopropan-2-amine (15)

Compound 14 (3.37 g, 13.97 mmol) was dissolved at 0°C in a mixture of trifluoroacetic acid (18 ml) and water (2 ml). After 15 min it was evaporated to dryness and co-evaporated with cyclohexane three times. The residue was washed several times with ether until no acidity remained (1.8 g; 91%). *TLC* (Rf = 0.78; DCM/MeOH 9:1; ninhydrin); *^1H NMR* (400 MHz, $CDCl_3$, δ ppm): 1.45 (s, 2H, NH), 3.02 (p, 1H, *J* = 5.8 Hz, CH), 3.32 (dd, 2H, *J* = 5.8 Hz, *J* = 12.0 Hz, CH_2), 3.40 (dd, 2H, *J* = 5.8 Hz, *J* = 12.0 Hz, CH_2). *MS* (EI DCI) 142 $(M+H)^+$.

4-(Isothiocyanatomethyl)-2,2-dimethyl-1,3-dioxolane (16)

2,2-Dimethyl-1,3-dioxolan-4-methanamine (1 g, 7.62 mmol) and carbon disulfide (0.46 ml, 7.62 mmol) were dissolved in THF (30 ml), DCC (1.57 g, 7.62 mmol) was added, and the mixture was stirred at room temperature for 1 h (see Note 5). The reaction medium was diluted with pentane, filtered, and the solution was washed successively with 0.1N HCl, water, and sat. NaCl solution. The organic phase was dried over sodium sulfate, filtered, and concentrated to the pure product (1 g, 75%) (see Note 6). *TLC* (Rf = 0.27; DCM/MeOH 98:2; I_2/H_2SO_4); *^1H NMR* (400 MHz, $CDCl_3$, δ ppm): 1.31 (s, 3H, CH_3), 1.42 (s, 3H, CH_3), 3.54 (dd, 1H, *J* = 5.1 Hz, *J* = 15.0 Hz, CH_2NCS), 3.64 (dd, 1H, *J* = 5.1 Hz, *J* = 15.0 Hz, CH_2NCS), 3.80 (dd, *J* = 5.1 Hz, *J* = 8.7 Hz, CH_2O), 4.07 (dd, *J* = 6.3 Hz, *J* = 8.7 Hz, CH_2O), 4.26 (q, 1H, J) 5.3 Hz, CHO). *MS* (ESI, *m/z*) 174 $(M+H)^+$.

2-[2-(Didecylamino)-2-oxo-ethoxy]acetic acid (17)

Didecylamine (3 g, 10.08 mmol) was placed in a 150 ml round-bottom flask and dissolved in DCM (100 ml), glycolic anhydride (1.64 g, 14.1 mmol) was added, and the mixture was stirred at room temperature for 2 h. The end of the reaction was controlled by TLC (DCM/MeOH 8:2); the mixture was concentrated; and the residue was taken up in ethyl acetate, washed successively with 0.1N HCl (×2) and sat. NaCl, and dried over Na_2SO_4. Evaporation of the solvent gave the title compound as an oil which crystallized on standing (4 g, 96%). Mp 55°C. *TLC* (Rf = 0.47; DCM/MeOH 9:1; I_2/H_2SO_4); *^1H NMR* (400 MHz, $CDCl_3$, δ ppm): 0.85 (t, 6H, *J* = 6.0 Hz, CH_3), 1.25 (m, 28H, $-CH_2-$), 1.53 (m, 4H, $-CH_2-$), 3.08 (t, 2H, *J* = 7.3 Hz, $-CH_2N$), 3.33 (t, 2H, *J* = 7.3 Hz, $-CH_2N$), 4.19 (s, 2H, CH_2O), 4.39 (s, 2H, CH_2O). *MS* (ESI, *m/z*) 414 $(M+H)^+$.

N-[2-Azido-1-(azidomethyl)ethyl]-2-[2-(didecylamino)-2-oxo-ethoxy]acetamide (18)

In a 100 ml round-bottom flask compound 17 (3.21 g, 7.76 mmol) and compound 15 (1.97 g, 7.76 mmol) were dissolved in dichloromethane (40 ml). Triethylamine (2 ml, 14.23 mmol) and BOP (3.43 g, 7.76 mmol) were added and the mixture was stirred for 1 h at room temperature (see Note 7). The solution was diluted with DCM (100 ml) and successively washed with 0.5 M $KHSO_4$ (3×20 ml), sat. $NaHCO_3$ (3×20 ml), water, and sat. NaCl solution. It was dried over Na_2SO_4, filtered, and concentrated, yielding 18 (4 g, 96%). *TLC* (Rf=0.24; DCM/MeOH 99:1; I_2/H_2SO_4); *^1H NMR* (400 MHz, $CDCl_3$, δ ppm): 0.88 (t, 6H, *J*=6.0 Hz, CH_3), 1.25 (m, 28H, $-CH_2-$), 1.51 (m, 4H, $-CH_2-$), 3.05 (t, 2H, *J*=7.3 Hz, $-CH_2N$), 3.31 (t, 2H, *J*=7.3 Hz, CH_2N), 3.47 (dd, 2H, *J*=6.0 Hz, CH_2N_3), 3.55 (dd, 2H, *J*=6.0 Hz, CH_2N_3), 4.10 (s, 2H, CH_2O), 4.22 (m, 1H, CH), 4.27 (s, 2H, CH_2O), 8.77 (d, 1H, *J*=8.4 Hz, NH). *MS* (ESI, *m/z*) 537 $(M+H)^+$, 559 $(M+Na)^+$.

*N-[2-Amino-1-(aminomethyl)ethyl]-2-[2-(didecylamino)-2-oxo-ethoxy]acetamide (*19*)*

Compound 18 (2.74 g, 5.1 mmol) was dissolved in methanol (50 ml) and cooled in an ice-water bath (see Note 8), and then 10% palladium on carbon (0.2 g) was added. The mixture was stirred at atmospheric pressure of hydrogen during 3 h (see Note 9). The solution was filtered on a sintered glass funnel coated with Celite (see Note 10) and concentrated under vacuum to give 19 (2.45 g, 99%). *TLC* (Rf=0.05; DCM/MeOH 8:2; ninhydrin or I_2/H_2SO_4); *^1H NMR* (400 MHz, $CDCl_3$, δ ppm): 0.84 (t, 6H, *J*=6.0 Hz, CH_3), 1.25 (m, 28H, $-CH_2-$), 1.48 (m, 4H, $-CH_2-$), 1.77 (m, 4H, NH_2), 2.54 (dd, 1H, *J*=6.0 Hz, *J*=12.0 Hz, CH_2NH_2), 2.71 (dd, 1H, *J*=5.8 Hz, *J*=12.6 Hz, CH_2NH_2), 3.05 (t, 2H, *J*=7.3 Hz, $-CH_2N$), 3.26 (m, 3H, $-CH_2N$, CH), 3.39 (s, 2H, CH_2O), 4.05 (s, 2H, CH_2O), 8.20 (m, 1H, NH).

*2-[2-(Didecylamino)-2-oxo-ethoxy]-N-[2-[(2,2-dimethyl-1,3-dioxolan-4-yl)methylcarbamothioylamino]-1-[[(2,2-dimethyl-1,3-dioxolan-4-yl)methylcarbamothioylamino]methyl]ethyl]acetamide (*20*)*

In a 100 ml round-bottom flask, compounds 19 (2.17 g, 4.48 mmol) and 16 (1.63 g, 9.4 mmol) were dissolved in DCM (45 ml). The solution was stirred overnight at room temperature. It was washed with water and sat. NaCl solution, dried over sodium sulfate, filtered, and concentrated. The crude product was chromatographed on silica gel with a gradient of cyclohexane/ethanol (0→40%, 60 min). The fractions containing compound 20 were evaporated to a clear syrup (2 g, 54%). *TLC* (Rf=0.6; DCM/MeOH 95:5; I_2/H_2SO_4); *^1H NMR* (400 MHz, $CDCl_3$, δ ppm): 0.85 (t, 6H, J) 6.6 Hz, CH_3), 1.23 (m, 28H, $-CH_2-$), 1.30 (s, 3H, CH_3), 1.39 (s, 3H, CH_3), 1.49 (m, 4H, $-CH_2-$), 3.06 (t, 2H, J) 7.2 Hz, $-CH_2NCO$), 3.27 (t, 2H, J) 7.2 Hz, $-CH_2-NCO$), 3.68 (m, 8H, CH_2NCS), 4.04 (m, 6H, CH_2O), 4.25 (m, 7H, CH_2O, CH, CHO). *MS* (ESI, *m/z*) 829 $(M-H)^-$, 831 $(M-H^+)$, 853 $(M-Na^+)$.

2-[2-(Didecylamino)-2-oxo-ethoxy]-N-[2-(2,3-dihydroxypropylcarbamothioylamino)-1-[(2,3-dihydroxypropylcarbamothioylamino)methyl]ethyl]acetamide (21; *DDSTU*)

Compound 20 (0.3 g, 0.36 mmol) was dissolved in acetonitrile (10 ml) and 1 N HCl was added (5 ml). The mixture was vigorously stirred for 1 h at room temperature, and then evaporated to dryness under high vacuum without heating. The residue was redissolved in DCM and purified on a silica column eluted first with DCM/MeOH (95:5), and then DCM/MeOH (8:2). Concentration of the eluate gave 21 (0.2 g, 57%). *TLC* (Rf = 0.17; DCM/MeOH 9:1; UV, I_2/H_2SO_4); *^1H NMR* (400 MHz, $CDCl_3$, δ ppm): 0.88 (t, 6H, J) 6.0 Hz, CH_3), 1.23 (m, 28H, $-CH_2-$), 1.54 (m, 4H, $-CH_2-$), 2.14 (s, 4H, OH), 3.09 (m, 2H, CH_2NCO), 3.30 (m, 2H, CH_2NCO), 3.63 (m, 8H, CH_2NCS), 3.92 (m, 1H, CH), 4.13 (m, 6H, CH_2O, CH_2OH), 4.31 (m, 7H, CH_2O, CHOH), 7.50 (m, 5H, NH). *HR-ESMS* calc for $C_{35}H_{70}N_6O_7NaS_2$: 773.4645. Found 773.4652. *MS* (ESI, *m/z*) 751 (M); 752 (MH^+).

3.3. Formulation

Lipoplexes are obtained by preformation of liposomes followed by a simple mixing with the nucleic acid of interest. All the methods are pretty straightforward. Ethanolic injection requires only two steps which are obviously crucial, the volume of dispersion should be well defined, and dropping the dissolved lipids in an aqueous medium should be regularly performed. Liposomes made out of DMAPAP and DOPE are either prepared as a film, the protocol of which has been previously described (8), or as an ethanolic injection which is described underneath. The cationic formulation is the sole formulation described, but non-cationic liposomes made of thiourea lipids are also prepared following this protocol.

Actually the main difficulty lies in the first step following the synthesis. How to suspend a newly synthesized lipid and how to formulate it? Does it need another lipid to be suspended?

Basically, amphiphilic lipids soluble in ethanol could be prepared by the protocol of ethanolic injection described underneath. However, they are not all soluble in ethanol.

Here are the main points to handle a lipid.

1. Lipids are very often hygroscopic molecules, take care of leaving them in dry environment, and do not leave them to hydrate for hours on your bench.
2. Look at the structure: If the lipid bears double bonds, specific conditions of storage should be used such as nitrogen conditioning to avoid oxidation.
3. Start with solubility studies:

Weight several flasks containing 1 mg of the lipid and dilute it in ethanol, acetone, and chloroform. If 1 mg is solubilized in less than 100 μl of EtOH, then the ethanolic injection is appropriate (see Note 11). Solubilization of the lipid in $CHCl_3$ indicates that the film method is more appropriate for liposome obtention. For a complete description see ref. 10.

3.4. Preparation of Liposomes by the Ethanolic Injection Method

1. Dissolve separately the lipids DMAPAP (10 μmol, 10 mg) and DOPE (10 μmol, 7.3 mg) in ethanol (limit of solubility for each lipid). Take care that the lipids are well dissolved separately before mixing them (see Note 12).
2. Mix them into an Eppendorf.
3. Note the total volume of ethanol required to perfectly solubilize both lipids.
4. Put milliQ filtered (0,22 μm) water in a round-bottom flask (see Note 13). The volume should be ten times the amount of organic solvent required to solubilize the lipids. Put the flask on a magnetic stirrer with a magnetic bar of an appropriate size. Check that the magnetic bar turns properly, fast, and continuously (see Note 14).
5. Drop the solubilized lipids on the stirring water (see Note 15). Leave the mixture for 5 h.
6. Remove the solvent with a rotary evaporator with a pressure control device. Be careful that the suspension does not foam (see Note 16).
7. Remove the flask from the evaporator when the suspension approximately reaches the volume you expect (an approximate 10–30 mM final concentration).
8. Pipet with a hand pipet or a syringe to determine the volume left in the flask.
9. Calculate the concentration of the lipid in your suspension according to the formula C=m/(M*V); m being the amount of lipid initially weighed, M the molecular mass of the lipid or the mean molecular mass if more than one lipid was used, V is the volume determined in 8.
10. Control the size by dynamic light scattering (see Note 13 and Subheading 3.6.2 for more details). For measurements on a nanoZS (Malvern Instruments), dilute 5 μl of the particles obtained in a 500 μl curve, and start the measure in the automatic mode.

3.5. Preparation of Cationic Lipoplexes

To obtain lipoplexes containing 1 μg of DNA in which DNA is fully associated, use a charge ratio of 4–8 (see Note 17). The charge ratio represents a molar ratio of cationic lipid to phosphate functions.

The protocol below is given for a charge ratio lipid/DNA = 6, which corresponds to a ratio total lipid to DNA = 12 as the cationic lipid only represents 50% of the total lipid content in the DMAPAP/DOPE mixture.

1. Dilute the DMAPAP/DOPE suspension to 1 mM total lipid in H_2O (see Note 18).
2. Dilute 1 μg plasmid DNA in 100 μl H_2O.
3. Dilute 12 μl of the 1 mM DMAPAP/DOPE suspension in 100 μl H_2O.
4. Add the plasmid DNA to the cationic liposome dropwise in a few seconds with constant vortexing (see Note 19).
5. Leave the sample for 1 h at room temperature to incubate before using it.

The preparation of thiourea lipoplexes is similar to the preparation of cationic lipoplexes. The difference lies in the amount of lipid required to condensate DNA. The interaction between thiourea lipid and phosphates occurs at a ratio of 1 as shown by fluorescence correlation spectroscopy (11); however, these lipids do protect DNA and transfect cells at a higher TU/PO ratio, which is why the lipoplexes are prepared at a ratio lipid/phosphate = 40. The complexes are usually prepared and left for a few hours before use. To evaluate DNA condensation, lipoplex loading on agarose gel has been widely proposed. Condensed DNA should not migrate into the gel due to the size of the complexes formed. You can also assess DNA accessibility using DNA intercalating agents and evaluate the fluorescence obtained as referred to free DNA as described underneath. Condensed DNA should not be accessible to the intercalating agents.

3.6. Lipoplex Characterization

3.6.1. DNA Complexation Checked by Fluorescence

1. Prepare the picogreen® solution as described by the provider (1/200 in Tris–EDTA buffer).
2. Load into a 96-well plate free DNA or complexed DNA (40 ng) in tripliquets.
3. Add 200 μl of the picogreen solution (Subheading 3.6.1) to each well filled with DNA and three more to obtain the picogreen background level.
4. Read the emission at 450 nm under an excitation at 350 nm on a multiplate reader able to measure fluorescence.
5. For the calculation, calculate the mean and the standard error on each tripliquets. Remove the picogreen background from the sample data. Calculate the percentage of fluorescence of each sample by dividing the sample data by the value of the free DNA taken as 100% fluorescence.

3.6.2. Lipoplex Size and Zeta Potential

The hydrodynamic diameter of the particles can be measured by quasi-elastic light scattering. The particles in suspension are submitted to the Brownian movement. When the particles are under a laser beam, they scatter the light in every direction. The variations of the light intensity as a function of time indicate the particle speed, which can be linked to their diameter by the Stokes–Einstein equation: $D = kT/6\pi R\eta$ where D is the particle scattering coefficient, T the temperature, k the Boltzmann constant, R the particle radius, and η the viscosity of the solvent.

The zeta potential is obtained through the measurement by the same technique of the electrophoretic mobility. It means that the curve used needs to be equipped with electrodes in order to provide an electric field which is proportional to the electrophoretic mobility. The zeta potential is obtained using the Smoluchowski law, $\zeta = \eta\mu_e/\varepsilon_r\varepsilon_o$, where ζ is the zeta potential, η the viscosity, μ_e the electrophoretic mobility, ε_r the dielectric constant of the dispersing medium, and ε_o the permittivity of free space.

1. Take a sample of your preparation and measure the size. The amount to be used depends on the system you are equipped with (see Note 20).

   Insure the reliability of the measure by assessing the autocorrelation function and the polydispersity index (see Note 21).
2. If the sample is sufficiently concentrated for size measurement, it should be possible to measure the zeta potential on a similar sample. However, a conductivity medium should be used such as 20 mM NaCl to provide ion displacement during the electrophoresis (see Note 22).

   An example of how the results should be presented is given in Table 1.

3.7. Lipoplex Stability in Culture Medium

Two criteria can be evaluated:

- The particle stability towards serum in terms of particle size or protein association.
- DNA release or protection towards enzymatic degradation.

3.7.1. Particle Stability

1. Take a sample of lipoplexes (10 μl, ten times more concentrated than previously described as you would use for in vivo injection 0.1 g/l DNA).
2. Dilute it in 200 μl culture medium supplemented or not with 10% serum.
3. Increase the temperature of the DLS system to 37°C.

Table 1
Lipoplex characterization using dynamic light scattering

	Lipid/PO-ratio	Size by intensity (nm)	Polydispersity index (PDI)	Zeta potential (mV)	Conductivity (mS/cm)
DMAPAP/DOPE	N/P 4	69.1 ± 0.4	0,18 ± 0.02	22 ± 1	0.22
DDSTU	TU/P 40	74.5 ± 2.0	0.13 ± 0.02	8 ± 4	0.23

4. Take a measure of the particle size every 2 min at 37°C.
5. Trace the evolution of the particles in terms of size, polydispersity index, and counts as a function of time.

3.7.2. DNA Protection from Enzymatic Degradation

1. Prepare the lipoplexes with a DNA concentration of 0.1 g/l as described above in Subheading 3.5.
2. To 50 μl of the lipoplex solution, add 50 μl of culture medium supplemented with 10–50% of murine fresh serum (see Note 23).
3. Incubate the samples at 37°C.
4. Every hour or so, take 10 μl sample and freeze it at −20°C.
5. After 24 h, take all the samples out of the freezer, add 2% sodium dodecyl sulfate (SDS) (5 μl), EDTA (2 μl, 0.5 M), and bromophenol blue (3 μl) to each sample (10 μl).
6. Load the mixture onto 1% agarose gel containing 0.05% SDS and put under 80 V/cm voltage.
7. After 24 h of rinsing the gel in water, plunge the gel in a solution of ethidium bromide and visualize it under UV light to reveal DNA. For an example see Fig. 1 in ref. 6.

4. Notes

1. It should be noted that the synthetic routes have been designed to permit the syntheses of numerous structural analogs as well. The syntheses of compounds 1–7 were already described in a previous issue of Methods in Molecular Biology (12).
2. A mixture of rotamers was observed by the doubling of some NMR signals.
3. Partial hydrolysis of the ethyl trifluoroacetate generates trifluoroacetic acid: It is mandatory that the pH of the ester is neutral; otherwise the amine would protonate and fail to react. The ester should be treated with dry sodium (or potassium) carbonate prior to using (CAUTION: CO_2 gas release).
4. The crude residue was dissolved in a warm mixture of isopropanol (600 ml) and a 5 M HCl solution in isopropanol (300 ml), which induced crystallization of the product as a white flaky powder. It was thoroughly washed with isopropanol and dichloromethane and dried.
5. After 1 h the pH of the solution was neutral, which attested the total consumption of the amine and the end of the reaction.
6. The product is evaporable and therefore temperature and duration of evaporation should be controlled.
7. pH should be >8; otherwise more Et_3N has to be added.

8. The solution should be cooled in an ice-water bath before adding the catalyst; otherwise the methanol vapors may catch fire on contact with palladium.
9. The end of reaction could be simply monitored by sampling the solution and observe the disappearance of the azide band in IR.
10. Care should be taken not to trash the dry catalyst to avoid possible fire. Instead it should be wetted and disposed of properly.
11. If not, try to heat it or sonicate it. Additional energy might help, but pay attention to an eventual reversible process. If sonication of the solution leads to a suspension of the lipid (cloudy solution), then try another solvent, first in acetone 50 μl for instance, and then dilute it in EtOH.
12. Solubility of the lipids should be checked with intensive care since the presence of non-soluble entities will reduce particle homogeneity after evaporation and might cause aggregation. A lipid used at the limit of its solubility might precipitate directly when dropped on H_2O or during the evaporation and volume reduction. In this case, a co-lipid bearing a larger hydrophilic head might be required to insure a correct interaction with the aqueous buffer and maintain the colloidal stability of the suspension.
13. All buffers and water used should be filtered on 0.22 μm filters since any dust might interfere with light scattering experiments.
14. Nonhomogeneous stirring would lead to a polydisperse population of liposomes.
15. Dropping can be performed via a peristaltic pump for a better homogeneity of the dispersion.
16. Formation of a suspension of micelles in the formulation could lead to a foam during the evaporation process; reduce the pressure cautiously.
17. The ratio between the lipid amines and DNA phosphates is fully dependent on the amine substitution and the conditions used (10).
18. The protocol is described with H_2O but can be changed for NaCl 150 mM or cellular medium. Obviously, the buffer will influence the aggregation state of the lipoplexes. Basically, all ions which will interact with the charges will reduce their availability for the interaction and enlarge the range of aggregation (13).
19. In order to maintain an excess of cationic charges and hence avoid precipitation by going through a charge ratio (±) equal to 1, DNA should be added on the cationic lipid and not the opposite order.
20. The concentration to be used depends on the sensitivity of the system and the angle used to detect the sample. The case

of multiple diffusion is rare as usually the samples are not too concentrated. If you are equipped with a Zeta Sizer NanoSeries Malvern (Malvern Instruments, Venissieux, France), the concentration of the samples can be approximately 0.1 mg/ml in H_2O.

21. Pay attention to the data obtained. Very often, a value will be given without reproducibility. The value is reliable if the polydispersity index is below 0.2, and if the data obtained in terms of intensity, volume, and number are identical and reproducible. Basically if the peak moves between three measurements and does not give similar results between volume, number, and intensity, you more than probably have a polydisperse sample which gives different results due to the low volume of measurement.
22. Reminder: Only samples exhibiting similar conductivity and pH can be compared in terms of zeta potential value.
23. Serum should be fresh to insure a strong enzymatic activity. Serum preleved from mice is more active than lyophilized serum from Sigma and more active than repetitively frozen sera.

References

1. Bangham A, Standish M, Watkins J (1965) Diffusion of univalent ions across the lamellae of swollen phospholipids. J Mol Biol 13: 238–252
2. Nicolazzi C, Garinot M, Mignet N, Scherman D, Bessodes M (2003) Cationic lipids for transfection. Curr Med Chem 10:1263–1277
3. Byk G, Wetzer B, Frederic M, Dubertret C, Pitard B, Jaslin G, Scherman D (2000) Reduction-sensitive lipopolyamines as a novel nonviral gene delivery system for modulated release of DNA with improved transgene expression. J Med Chem 43:4377–4387
4. Tranchant I, Thompson B, Nicolazzi C, Mignet N, Scherman D (2004) Physicochemical optimization of plasmid delivery by cationic lipids. J Gene Med. (Suppl. 1):S24–S35
5. Tranchant I, Mignet N, Crozat E, Chain J, Girard C, Scherman D, Herscovici J (2004) DNA complexing lipopolythiourea. Bioconjug Chem 15:1342–1348
6. Leblond J, Mignet N, Largeau C, Seguin J, Scherman D, Herscovici J (2008) Lipopolythiourea transfecting agents: lysine thiourea derivatives. Bioconjug Chem 19:306–314
7. Breton M, Leblond J, Seguin J, Midoux P, Scherman D, Herscovici J, Pichon C, Mignet N (2010) Comparative gene transfer between cationic and thiourea lipoplexes. J Gene Med 12:45–54
8. Thompson B, Mignet N, Hofland H, Lamons D, Seguin J, Nicolazzi C, de la Figuera N, Kuen R, Meng Y, Scherman D, Bessodes M (2005) Neutral post-grafted colloidal particles for gene delivery. Bioconjug Chem 16: 608–614
9. Leblond J, Mignet N, Largeau C, Spanedda MV, Seguin J, Scherman D, Herscovici J (2007) Lipopolythioureas: a new non-cationic system for gene transfer. Bioconjug Chem 18:484–493
10. Mignet N, Scherman D (2010) Anionic pH sensitive lipoplexes. Methods Mol Biol 605:435–444
11. Kral T, Leblond J, Hof M, Scherman D, Herscovici J, Mignet N (2010) Lipopolythiourea/DNA interaction: a biophysical study. Biophys Chem 148:68–73
12. Bessodes M, Scherman D (2010) Acid-labile liposome/pDNA complexes. Methods Mol Biol 605:405–423
13. Turek J, Dubertret C, Jaslin G, Antonakis K, Scherman D, Pitard B (2000) Formulations which increase the size of lipoplexes prevent serum-associated inhibition of transfection. J Gene Med 2:32–40

Chapter 7

Histidine-Rich Cationic Amphipathic Peptides for Plasmid DNA and siRNA Delivery

Antoine Kichler, A. James Mason, Arnaud Marquette, and Burkhard Bechinger

Abstract

Amphipathic, pH-responsive, membrane-active peptides such as LAH4 and derivatives thereof have the ability to effectively deliver genes and small interfering RNA (siRNA) into mammalian cells. Their ability to bind and protect nucleic acids and then disrupt membranes when activated at low pH enables them to harness the endocytic machinery to deliver cargo efficiently and with low associated toxicity. This chapter describes protocols for the chemical synthesis of transfection peptides of the LAH4 family, complex formation with nucleic acids, and their use for the in vitro delivery of either plasmid DNA or siRNA into mammalian cell lines.

Key words: Gene therapy, Nonviral delivery system, Cationic amphipathic peptides, Plasmid DNA, siRNA, RNA interference, Endosomal release, Histidine, Cell-penetrating peptide

1. Introduction

A large number of mono- and poly-cationic compounds have been developed in recent years with the aim of overcoming the bottlenecks for the development of nonviral nucleic acid-based therapeutics and/or laboratory toolkits. Ideally, these compounds should be multifunctional; in particular they should stabilize the nucleic acid and protect it from nucleases, promote cellular attachment and uptake, and ensure that the nucleic acid reaches the desired intracellular compartment (Fig. 1). In the case of plasmid DNA, nuclear delivery is required but for small interfering RNAs (siRNAs), which are used for inducing RNA interference (RNAi), delivery to the cytosol is sufficient. In both cases, complexes are commonly taken up through nonspecific endocytosis and it is the

Manfred Ogris and David Oupicky (eds.), *Nanotechnology for Nucleic Acid Delivery: Methods and Protocols*, Methods in Molecular Biology, vol. 948, DOI 10.1007/978-1-62703-140-0_7,

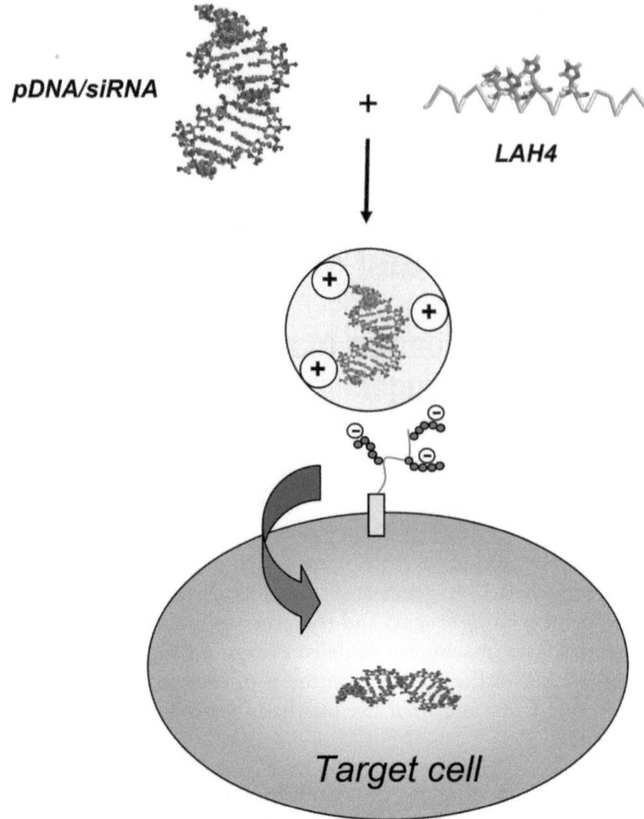

Fig. 1. Schematic representation of LAH4-mediated transfection of eukaryotic cells. Addition of an excess of LAH4 to plasmid DNA or siRNA generates complexes which exhibit a small excess of positive charges on their surface. Particles interact with the cell membrane, preferentially with proteoglycans, via nonspecific ionic interactions. The LAH4/nucleic acid complexes enter the cell through endocytosis. During acidification of the endosome, histidine residues become protonated and LAH4 interacts with the endosomal membranes in an in-plane alignment. Membrane destabilization occurs, followed by the release of DNA/siRNA into the cytosol.

ability to escape from endosomes that often determines the success of a cationic formulation.

One way that has been found to increase the efficiency of endosomal escape of the nucleic acids consists in using the "proton sponge" mechanism. This process involves a transporter-induced swelling of the endosomal compartment during acidification which leads to disruption of the endosomal membrane (1). Various compounds have been described that react to the change in pH encountered in the endosomes (2). Many of these compounds incorporate the imidazole group found in histidine (3) which has a p*K*a around 6.1. At increasingly acidic pH such groups are protonated leading to enhanced cationicity of the polyplexes and, potentially, some release of the vector compound and hence such groups may constitute effective triggers for nucleic acid delivering activities.

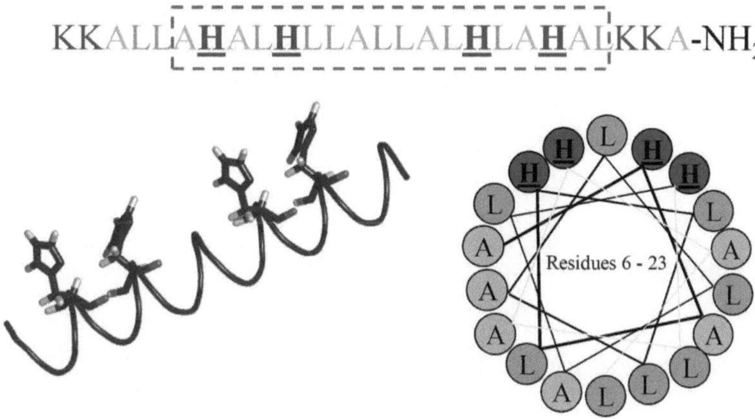

Fig. 2. Helical wheel diagram for LAH4-L1 (residues 6–23). The primary sequence (*top*) and three-dimensional model (*left*) of LAH4-L1 indicate the positions of the histidine residues and how these are located in the amphipathic helix conformation that the peptide adopts when bound either to DNA or target membranes. The helical wheel projection (*right*) allows one to visualize the distribution of hydrophobic and polar residues with respect to the helical axis. The angle of the positively charged helix face (histidine residues are in *bold* and *underlined*), at pH 5, for the LAH4-L1 peptide is 80°.

Our efforts have focussed on the development of pH-responsive, cationic amphipathic peptides (4–9). These peptides have both nucleic acid binding and membrane disruptive capabilities with their amphipathic α-helix motif being inspired by antimicrobial peptides whose main mode of action involves membrane disruption. The concept uses histidine pH triggers to release membrane-destabilizing peptides from the peptide–nucleic acid complex allowing them to bind to endosomal membranes. The original sequence, LAH4 (KKALLALALHHLAHLALHLALA LKKA) (4), and later variants, including LAH4-L1 (KKALLAH ALHLLALLALHLAHALKKA) (5, 7), have been designed to adopt an α-helix conformation when bound to nucleic acids and, particularly, lipid bilayers (Fig. 2). The LAH peptides exhibit a hydrophobic surface, comprising a mixture of alanine and leucine residues, and a hydrophilic surface in which the four (or more) histidine residues are located. The segregation of the two types of surfaces and the positioning of the histidine residues (Fig. 2) are crucial to the effective disordering of the membranes and the efficacy of delivery mediated by the peptide (4, 5).

The membrane disordering and disruption are the final elements in a strategy that has proven to be remarkably effective, with both DNA and siRNA delivery that matches and often outperforms commercially available cationic nonviral vectors (4, 9). Furthermore, we have observed only low cytotoxicity associated with the pH-responsive peptide delivery system. We believe that this system has a number of advantages that will make it attractive to researchers who rely on effective delivery of nucleic acid cargoes with few side effects and, accordingly, are pleased to provide detailed protocols for the production, characterization, and use of this emerging delivery system.

2. Materials

2.1. Peptide Synthesis

The peptide LAH4 (MW 2,777) with the sequence KKALLALALHHLAHLALHL ALALKKA-NH_2 (4) or derivatives such as LAH4-L1 (5) can be purchased from a number of peptide synthesis services (e.g., NeoMPS, Strasbourg, France; GenScript, Piscataway, NJ; GeneCust, Luxembourg; and many others). We would recommend requesting a custom synthesis and purchasing the peptide as desalted grade as the most cost-effective solution. Alternatively, it can be synthesized and prepared in any laboratory that has access to a solid-phase peptide synthesizer (e.g., ABI433 automatic peptide synthesizers, Weiterstadt, Germany) or the expertise and setup to go through the synthetic cycles manually.

Materials needed for peptide synthesis:

1. Automatic peptide synthesizer (e.g., ABI433, Weiterstadt, Germany).
2. Access to preparative reverse-phase HPLC instrumentation and MALDI mass spectrometric analysis. HPLC is performed for example on a 300 Å Prontosil column (Bischoff, Leonberg, Germany) of 20 mm diameter and 250 mm length (or similar). We have purified LAH4 peptides also on columns of dimension 7.8 × 100 mm and the conditions change accordingly. Furthermore, during the acetonitrile shortage during the economic crisis in 2008 it has proven advantageous that methanol can be used instead of acetonitrile.
3. Solvents and Fmoc-protected amino acids (Fmoc-Ala-OH, Fmoc-Leu-OH, Fmoc-Lys(Boc)-OH, and Fmoc-His(Trt)-OH, e.g., from Novabiochem Merck Darmstadt, Germany; Bachem, Heidelberg, Germany; or Applied Biosystems, Weiterstadt, Germany) according to the instructions given for your automatic peptide synthesizer, including HBTU (2-(1H-benzo-triazole-1-yl)-1,1,3,3-tetra-methyluroniumhexafluorophosphate, Novabiochem, Merck, Darmstadt, Germany), Dipea (*N*-ethyldiisopropylamine, Sigma, St. Louis, MO, USA), TentaGel R RAM, or TentaGel S RAM resin (load 0.18–0.25 mmol/g; Rapp Polymer, Tübingen, Germany).
4. 5% H_2O, 1% triethylsilane (97%, Sigma) in trifluoroacetic acid (TFA) (99.9% Carl Roth, Karlsruhe, Germany) for the cleavage reaction.

5. Diethyl ether (stabilized with BHT for analysis, sds Solvents Documentation Syntheses, Peypin, France).
6. HPLC-grade acetonitrile, water, and TFA. When using methanol as an alternative to acetonitrile make sure that the impurities do not absorb at 214 nm, thereby interfering with the detection of chromatogram.

If the peptide is prepared by automated peptide synthesis it is recommended to follow the detailed protocols that are provided by the manufacturers of such instrumentation for the setup of the synthetic cycles, including the solvents, resins, and Fmoc-protected amino acid residues. Furthermore access to preparative and analytical reverse-phase (C4, C8, or C18) HPLC and mass spectrometric characterization of the final product are necessary.

As solid-phase peptide synthesis is an elaborate science, which fills many textbooks and libraries, discussing the details and pitfalls of the technique is beyond the scope of this paper. Nevertheless, in Note 1 a short outline of the fundamental idea is provided. So far the LAH4 peptides and their derivatives could be obtained in good yields (several 100 mg of crude product) using standard synthesis cycles of automated peptide synthesizers. In our hands the crude product contains about 90% of the LAH4 sequence and can be used after desalting for preliminary transfection assays without apparent loss in efficiency when compared to the purified product. However, it should be noted that even a 99% yield of each individual cycle results in truncation products, where one residue is missing and an overall yield of the full sequence reduces to about 80% (0.99^{25}). Therefore, the shorter sequences can be separated from the desired product using preparative HPLC and identified by MALDI mass spectrometric analysis.

2.2. Gel Mobility Shift Assay

1. Agarose (ultrapure agarose from Invitrogen, Cergy Pontoise, France).
2. TBE buffer: 89 mM Tris–HCl, 89 mM boric acid, 2 mM EDTA, pH 8.0.
3. Loading buffer 6×: 0.25% Bromophenol blue, 0.25% xylene cyanol, and 30% glycerol in water.
4. SYBRSafe DNA gel stain (Invitrogen, Cergy Pontoise, France).
5. Electrophoresis unit.

2.3. Plasmid DNA and siRNA

1. pCMV-Luc is an expression plasmid encoding the firefly luciferase gene under the control of the human cytomegalovirus (CMV) immediate-early promoter. Plasmid DNA is resuspended in sterile water at 1 mg/mL and is stored at −20°C.
2. The siRNA-Luc (from Sigma-Aldrich, Saint-Quentin Fallavier, France) sequence that we use for knock down of luciferase has

been previously published (10): sense oligonucleotide: 5′-CGUACGCGGAAUACUUCGATT-3′; antisense oligonucleotide: 5′-UCGAAGUAUUCCGCGUACGTT-3′. Aliquots of a 10 μM stock solution are stored at −20°C.

3. The sequence of the siRNA-eGFP (Sigma-Aldrich, Saint-Quentin Fallavier, France) that we use as control siRNA has been published previously by Caplen et al. (11): sense oligonucleotide: 5′-GCAAGCUGACCCUGAAGUUCAU-3′; antisense oligonucleotide: 5′-GAACUUCAGGGUCAGCUUGC CG-3′. Aliquots of a 10 μM stock solution are stored at −20°C.
4. Limit the freeze–thaw cycles of each tube. To this end, make small aliquots of DNA and siRNA.

2.4. Dynamic Light Scattering

1. Dynamic light scattering (DLS) measurement system. We have chosen the Zetasizer Nano-S instrument (Malvern Instruments, UK) since it has the ability to measure molecular systems, from some hundreds of Daltons to micrometer size aggregates.
2. Low-volume cuvettes. High-quality quartz cuvettes are the best for the purpose of the described measurements since they minimize the light scatted at the air- and liquid–window interfaces.
3. Suspension of 60 nm latex nanospheres (Duke Scientific) for calibration of the DLS instrument.

2.5. Cell Culture

1. Human embryonic retinoblasts (cell line 911) and 911 cells that have been stably transfected with a CMV-luciferase plasmid (911-Luc cells). 911-Luc cells are cultured in the presence of 0.1 mg/mL of geneticin (GIBCO, Cergy Pontoise, France).
2. Dulbecco's modified Eagle's medium (DMEM; from GIBCO, Cergy Pontoise, France) supplemented with 100 units/mL penicillin, 100 μg/mL streptomycin (we use as stock solution penicillin–streptomycin from GIBCO containing 5,000 units of penicillin and 5,000 μg of streptomycin/mL), and 10% of fetal calf serum (GIBCO, Cergy Pontoise, France).
3. Trypsin (for example TrypLE Express from GIBCO, Cergy Pontoise, France, which is a solution containing a stable trypsin-like enzyme).

2.6. Cell Transfection

1. 24-Well culture plates (from Costar).
2. DMEM supplemented with 100 units/mL penicillin and 100 μg/mL streptomycin.
3. Sterile (0.2 μM filtered) 150 mM NaCl solution.

2.7. Luciferase Assay

1. Lysis buffer: 25 mM Tris–phosphate, 8 mM $MgCl_2$, 1 mM dithiothreitol, 1 mM EDTA, 15% glycerol, 1% Triton X-100, pH 7.8.

2. Assay buffer: 25 mM Tris–phosphate, 8 mM $MgCl_2$, 1 mM dithiothreitol, 1 mM EDTA, 15% glycerol, and 2 mM of ATP—this latter component of the buffer has to be added just before performing the luciferase assay (store aliquots of 1 mL of ATP at 40 mM in water at −20°C).
3. Make aliquots of 50 mL of the lysis and assay buffer and store them at −20°C.
4. D-Luciferin sodium salt (Invitrogen, Cergy Pontoise, France): 167 μM in water. Make aliquots of 10 mL, protect the tubes from light, and store them at −20°C.
5. Luminometer (for example Victor^2, Perkin Elmer, Courtaboeuf, France).

2.8. Protein Assay

1. Detergent-compatible protein determination kit (for example the Bradford protein assay from Bio-Rad, Marnes-la-Coquette, France).
2. Transparent 96-well plate.
3. A 96-well plate spectrophotometer able to read at 595 nm.

2.9. MTT Cell Viability Assay

1. 3-(4,5-Dimethylthiazol-2-yl)-2,5-diphenyl-tetrazolium bromide (MTT; from Sigma) at 5 mg/mL in PBS (GIBCO).
2. A 96-well plate reader capable of measuring optical absorbance at $\lambda = 570$ nm.

2.10. Flow Cytometry

1. Fluorescent Fitc-siRNA-Luc: A fluorescein group is coupled to the 3′ end of the antisense strand of the duplex (such a modification can be made by suppliers as for example Sigma).
2. Phosphate-buffered saline (PBS).
3. Flow cytometer (for example FACSCalibur from Becton Dickinson, Le Pont-De-Claix, France).

3. Methods

3.1. Peptide Synthesis

1. We have prepared LAH4 and related peptides by solid-phase peptide synthesis on either Millipore 9050 or ABI433 automatic peptide synthesizers using Fmoc (9-fluorenylmethyloxycarbonyl) chemistry which has also allowed us to introduce amino acid residues labeled with stable isotopes for NMR structural studies, amino acid replacements, or the addition of optical dyes (12, 13). As the detailed protocols and the solvents used depend on the synthesizer they shall not be specified here in detail but reference is made to the corresponding user manuals.

In short, LAH4 is typically prepared following the standard protocols of the automatic synthesizer using a fourfold excess of Fmoc-protected amino acids during chain elongation with activation by HBTU and Dipea using for example a TentaGel R RAM or a TentaGel S RAM resin (0.18–0.25 mmol/g). The typical scale of the synthesis is 0.25 mmol. After cleavage of the peptide product by exposing the resin to TFA and in the presence of scavengers such as triethylsilane the crude product (several 100 mg) is precipitated in ice-cold diethylether and LAH4 purified by preparative HPLC using an acetonitrile/water gradient typically ranging from 5 to 60% acetonitrile in the presence of 0.1% TFA and a detection of the absorbance at $\lambda = 214$ nm. The peptide typically elutes at 40% acetonitrile using for example a C4 reversed-phase column of 2 cm diameter and 25 cm length at a flow between 7 and 10 mL/min (see Note 2). At the end of the run the acetonitrile concentration is raised to 95% to clean the column.

2. The identity and high purity of the product should be verified by MALDI mass spectrometry and analytical HPLC.
3. After lyophilization the TFA counterions are exchanged by at least three cycles of dissolving the peptide in 5% acetate (v/v) and lyophilization (see Note 3).
4. Peptides are resuspended in sterile water at 1 mg/mL (see Note 4).

3.2. Gel Mobility Shift Assay

The agarose gel mobility shift assay is a method to determine the capacity of a given compound to complex nucleic acids. More precisely, it allows one to determine the minimal amount of compound required to retard the migration of plasmid DNA or siRNA during agarose gel electrophoresis. Indeed, the complexation leads to the formation of particles which are unable to migrate through the agarose mesh (6, 9).

1. Prepare a 1% agarose gel (≥1.3% for siRNA) by dissolving agarose in 100 mL of TBE buffer and boiling the suspension at 100°C using a microwave oven. After cooling down to about 50°C, add 8 μL of SYBRSafe to the suspension.
2. Fill the electrophoresis unit, and position a comb into the gel. Using a Pasteur pipette remove bubbles from the gel and then wait until agarose has solidified. Place the gel in the electrophoresis unit and remove the comb. Take care to add enough TBE buffer to cover the gel.
3. Samples containing 500 ng pDNA or 750 ng of siRNA are prepared using the same buffer conditions than those used for transfection. Therefore, dilute the pDNA/siRNA in 10 μL of 150 mM NaCl.

4. Take a tube containing a 1 mg/mL solution of LAH4 (or derivative) from the freezer. After thawing, vortex vigorously the tube—an additional short sonication using a bath sonicator may be performed—before withdrawing the desired amount of peptide from the tube.
5. Dilute increasing amounts of peptide in 10 μL of NaCl 150 mM (typically, use peptide/nucleic acids weight/weight ratios ranging from 0.5/1 to 10/1).
6. Mix the nucleic acids and the peptide, and incubate for about 15 min at room temperature.
7. Add 3 μL of loading buffer to each sample, mix, and then load the samples into the wells of the agarose gel. Also include a sample containing nucleic acids without peptide.
8. Electrophoresis conditions: 100 V for 20–30 min.
9. Visualize the agarose gel stained with SYBRSafe using an UV illuminator. An example of gel is shown in Fig. 3: the results indicate that addition of a cationic peptide such as LAH4 inhibits in a dose-dependent manner the migration of DNA (see Note 5). In this experiment, complete retardation was obtained at an LAH4/DNA w/w ratio of 2.5. This ratio of 2.5 corresponds to 0.59 LAH4 peptide per base pair or expressed as a positive/negative charge ratio to +1.5 (see Notes 6 and 7).

3.3. Plasmid DNA

Plasmid DNA is purified using a commercially available kit (for example NucleoBond from Macherey-Nagel, Hoerdt, France). The concentration of the DNA solution is determined by spectrophotometry at $\lambda = 260$ nm taking into account that one absorbance unit corresponds to 50 μg/mL of double-stranded DNA. The optical density is also taken at $\lambda = 280$ nm (see Note 8).

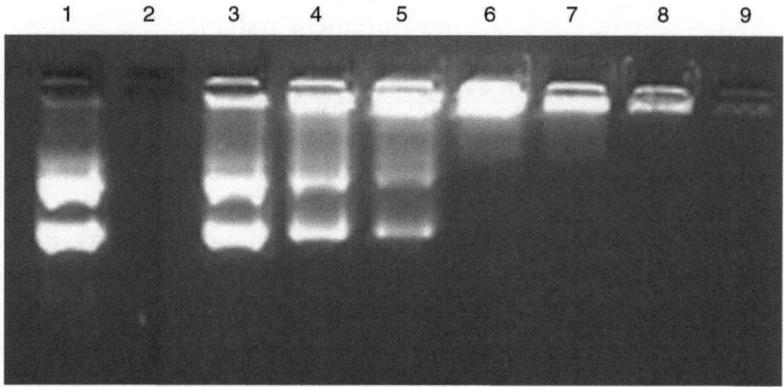

Fig. 3. Gel retardation experiment with LAH4/DNA complexes. The following conditions were used: *Lanes 1–9*: (1) naked DNA, (2) loading buffer only, (3–9) LAH4/DNA complexes with w/w ratios equal to 0.5, 1, 1.5, 2.5, 5, 7.5, 10, respectively.

3.4. Dynamic Light Scattering

The protocol described below outlines (1) the preparation of the peptide (here LAH4 as an example) and the DNA stock solutions, (2) the measurement of the size of the transfection complexes, and (3) the critical analysis of the recorded data.

3.4.1. Preparation of LAH4 and DNA Stock Solutions

The procedure for preparation of LAH4 and DNA stock solutions constitutes a major factor for the success of size measurement of LAH4–DNA complexes and should be done step by step as described below.

1. Weight the appropriate amount of peptide and prepare a 1 mg/mL stock solution of LAH4 in the desired buffer (e.g., 10 mM acetate, pH 5) (see Notes 4 and 9). Mix by gently tapping and shaking the tube. Sonication of the LAH4 solution for periods of about 10–30 s will help to remove any visible aggregated powder.
2. Perform a first size measurement on the LAH4 stock solution by DLS (as explained in the following sections) in order to confirm the solubility of the peptide at low pH. If this is not the case, ultracentrifugation of the solution at 100,000 × g for 1 h has to be performed in order to remove the remaining aggregates and impurities bigger than about 100 nm (see Note 9).
3. Prepare a 1 mg/mL stock solution of DNA following the same protocol described for LAH4 with the exception of the sonication step.

3.4.2. Size Measurement of LAH4–DNA Complexes

1. Prepare a minimum of 40 μL of complexes by mixing the right amount of the peptide and the DNA stock solutions in an Eppendorf tube. Note that the concentrations of the stock solution were chosen high enough so that the complexes could be detected by our experimental system.
2. Transfer the LAH4–DNA mixture in the low-volume quartz cuvette. Place the cuvette in the sample holder of the DLS measurement system.
3. Set the working temperature and the related physical parameters needed by the program to perform the mathematical procedure of inversion of the data. Refractive index of the buffer and of the complexed molecules at $\lambda = 633$ nm have to be known. At 25°C, the index of refraction of pure water ($n = 1.33$) and of pure proteins ($n = 1.45$) can be considered as good approximations for the solvent and the complex, respectively. Prior to the measurements the geometry of the cuvette and the viscosity of the buffer have also to be entered in the program.
4. Record a first measurement after equilibration of the sample in the cuvette. Perform at least four more acquisitions to ensure

the reproducibility of the measurements and the stability of the sample over time.

5. After measurements, the sample can be recovered for use in other experiments.

3.4.3. Analyzing the Quality of the Data

A careful analysis of the results has to be performed in order to avoid artifacts and to prevent mis- and/or overinterpretation of the data. For this, it is useful to recall some basics of the functioning of the DLS instrument and of the mathematical algorithm applied to the data.

The software controlling the Zetasizer Nano S performs an analysis of the time autocorrelation function of the scattered light to extract the size distribution of the dissolved particles. The basic idea behind the experiment is that the scattering of a light beam depends on the size of the particles in solution and consequently their velocity. If the particles are small, they diffuse very fast and the "memory" is only short lived. In other words the correlation in the diffraction properties between two intervals separated by a given period of time is low. If a particle is large, it diffuses slowly and the correlation remains even if the separating time period is extended. According to the general theory, the intensity of autocorrelation function is proportional to the square of the electric field–field time autocorrelation function, denoted as $g^{(1)}(t)$. The latter takes complex forms depending on the size, shape, and composition of the scattering particles.

The averaged intensity hydrodynamic radius $<R_H>$ and the poly dispersity index (PDI) values (an estimate of the distribution width around $<R_H>$) are calculated from a mathematical analysis implemented into the Malvern software. For this, the correlation functions are compared to a mono-exponential decay $g^{(1)}(t)=\exp(-Gt)$, supposing a monodisperse population of spherical particles. In this case, the particle's hydrodynamic radius is proportional to the invert of the time relaxation of the decay. The decay rate is equal to $G= Dq^2$, where D is the diffusion coefficient of the particles and q the magnitude of the scattering wave vector that depends on λ_0, θ, and the refractive index of the solvent. The Stokes–Einstein relation, $D = k_B T / 6\pi\eta\langle R_H \rangle$, where k_B is Boltzmann's constant and h is the dynamic viscosity, then relates the diffusion coefficient to $<R_H>$. PDI values are calculated from the deviation of the data from the mono-exponential decay. With this background in mind, analyze carefully the experimental data in the following manner:

1. Compare the average intensity over time of the signal arising from the transfection complexes (known as SLS, for static light scattered) with that of a "blank measurement" recorded with the buffer only, in the same quartz cuvette. If the intensity of the latter is much higher, the molecules in the former are probably strongly aggregated involving very large size particles (up

to several μm). In this condition special attention has to be given to the analysis of the data and to possible artifacts generated by the inversion algorithm.

2. Inspect the autocorrelation function measured on your samples. If the value at $t \approx 0$ μs is close to zero, the density of the complexes is too low to be measured. The peptide and DNA concentrations of the stock solutions have to be increased in order for the oligomers to be detected. Alternatively sedimentation of large size aggregates might have led to the depletion of the molecules in solution that cannot longer be observed by the system.
3. The PDI indicates the deviation of the sample when compared to an ideal monodispersed sample. For PDI values from 0 to 0.2, the sample solution can be considered as monodisperse. If PDI > 0.7, the sample is highly polydisperse and the values of the hydrodynamic radius given by the instrument are probably meaningless. Short periods of sonication of the sample may help to homogenize the solution (14).

3.5. Cell Culture

1. The day prior to transfection, use trypsin to detach actively dividing cells from the culture flask (Corning).
2. After addition of 15 mL of DMEM supplemented with 10% of serum, the cells are counted.
3. Finalize the protocol of the experiment in order to determine how many wells are needed. Notably, transfection experiments are usually done using duplicates or triplicates.
4. The required volume of cell suspension to obtain a confluence of 50–80% the following day is then calculated (for 911 and 911-Luc cells, we use 200,000 cells/well) (see Note 10). Dilute the cell suspension with a volume of culture medium which allows the addition of 1 mL of culture medium into each well of the 24-well plate.
5. The cells are then incubated at the appropriate cell culture conditions (a humidified tissue culture incubator at 37°C and 5% CO_2).

3.6. Cell Transfection

Before starting the DNA/siRNA transfection experiment, the cells should be examined using a microscope. This observation allows for verification of the confluence of the cells and also of the (good) shape of the cells. The peptide/DNA or siRNA complexes are generated just before transfection of the cells.

For a transfection performed in duplicates, the following protocols are used.

3.6.1. DNA Transfection

1. The tube containing the plasmid DNA is thawed and gently vortexed. Take 4 μg of plasmid DNA (for one duplicate) and dilute them in 100 μL of sterile (0.20 μM filtered) 150 mM NaCl.

2. The tube containing a 1 mg/mL solution of LAH4 is thawed. The tube is then vigorously vortexed—a short sonication using a bath sonicator may be additionally performed—before withdrawing the desired amount of peptide (see Notes 11 and 12). Complete to 100 μL with sterile 150 mM NaCl.
3. Mix the DNA with the peptide, centrifuge very shortly to pull down all the drops, and let the tube at room temperature for about 15 min.
4. Add to the mixture serum-free medium (if applicable) to obtain a final volume of 1 mL; 0.5 mL of the transfection mixture are then put in each well of the duplicate.
5. Remove the culture medium from the cells by aspiration. If the cells that are used adhere well, they may be rinsed with PBS in order to remove all serum containing proteins from the wells (see Note 13).
6. Add 0.5 mL of LAH4/DNA transfection medium into each well of the duplicate and incubate at 37°C for 2–4 h.
7. Remove carefully the transfection medium and replace it with 1 mL of complete culture medium (DMEM supplemented with 100 units/mL penicillin, 100 μg/mL streptomycin, and 10% of fetal calf serum).
8. Incubate at 37°C for 28–48 h, depending on the cell type.
9. Analyze the cells for luciferase expression (see Notes 14 and 15) and protein content or proceed for the MTT cell viability assay (see below).

3.6.2. siRNA Transfection

1. Thaw tubes containing the siRNA-Luc. Take 0.46 μg (3.5 μL of the 10 μM siRNA solution) of siRNA-Luc (to obtain a final concentration in the wells of 50 nM). Complete to 30 μL with NaCl 150 mM.
2. The tube containing a 1 mg/mL solution of LAH4 is thawed. The tube is then vigorously vortexed—a short sonication step using a bath sonicator may be additionally applied—before withdrawing the desired amount of peptide (see Notes 11 and 12). Complete to 30 μL with NaCl 150 mM (see Note 14).
3. Of note, transfection complexes of related composition should be prepared with a control siRNA such as a siRNA-GFP. This control is important in order to ascertain that the silencing is gene specific.
4. Mix the siRNA with the peptide, centrifuge very shortly to pull down all the drops, and let the tube at room temperature for about 15 min.
5. Add to the mixture serum-free medium (if possible) to obtain a final volume of 0.7 mL; 0.350 mL of the transfection mixture are then added to each well of the duplicate.

6. Remove the culture medium from the cells by aspiration. If the cells that are used adhere well, the cultures may be rinsed with PBS in order to remove serum containing proteins from the wells.
7. Add 0.350 mL of LAH4/siRNA transfection medium into each well of the duplicate and incubate at 37°C for 2–4 h.
8. Remove carefully the transfection medium and replace it with 1 mL of complete culture medium (DMEM supplemented with 100 units/mL penicillin, 100 μg/mL streptomycin, and 10% of fetal calf serum).
9. Incubate at 37°C for 28–48 h, depending on the cell type.
10. Analyze the cells for luciferase expression and protein content or proceed for the MTT cell viability assay (see below) (see Notes 16 and 17).

3.7. Luciferase Assay

To determine luciferase activity, the following protocol is used:

1. Remove carefully the culture medium from the 24-well plates.
2. Add 250 μL of lysis buffer to each well.
3. After 10 min, the cell lysate is recovered and transferred into 1.5 mL Eppendorf tubes.
4. Centrifuge the tubes for 5 min at 10,000 × g to pellet debris.
5. Take 50 μL of the supernatant of each tube and transfer them into the wells of a white 96-well plate.
6. Measure the bioluminescence using a luminometer which automatically injects 100 μL of buffer assay and 100 μL of luciferin solution.
7. Read luminescence over 10 s.
8. Remove the luciferase background (obtained with the lysate from non-transfected cells) from each value.
9. Calculate the light units for 10 s/250 μL of sample.
10. After having measured the protein content (see below), express the efficiency as light units/10 s/mg (or μg) of protein.

3.8. Protein Content

The protein determination procedure described here is valid for the Bradford protein assay from Bio-Rad (see Note 18).

1. Transfer 2 μL of cell lysate to a transparent 96-well plate. Of note, add a control with 2 μL of lysis buffer in order to obtain the background value.
2. Add 200 μL of Bradford reagent which was diluted five times in water.
3. Mix gently.

4. Read absorbance at λ = 595 nm and determine the protein content in each sample by using a bovine serum albumin standard curve.

3.9. MTT Cell Viability Assay

This assay is used to measure the cell viability (see Note 18). It therefore allows for an evaluation of the cytotoxicity of the transfection complexes.

1. The transfection should be performed using the same experimental conditions as those used for the luciferase assay (DNA or siRNA).
2. Instead of removing the medium and lysing the cells after 28–48 h of transfection, add to the medium MTT reagent at a final concentration of 0.5 mg/mL per well.
3. Cells are incubated at 37°C and 5% CO_2 for 3 h.
4. Remove the medium from the cells.
5. Add 500 μL DMSO to each well to dissolve the formazan crystals.
6. Transfer 100 μL from each sample into a transparent 96-well plate and measure absorption at λ = 570 nm.
7. Untreated cells serve as control (=100% cell viability).

3.10. Flow Cytometry for the Evaluation of the Delivery Efficiency of the siRNA

By using a fluorescently labeled siRNA and flow cytometry it is possible to quantify the amount of nucleic acid that is associated with or endocytosed by the cells.

1. Complexes are made fluorescent by using a fluorescently labeled siRNA-Luc (an Fitc group is conjugated to the 3′ end of the antisense strand; Sigma).
2. After preparation of the complexes as described for cell transfection, they are added to 911-Luc cells (final siRNA concentration = 50 nM).
3. After 3 h of incubation at 37°C, cells are washed with cold PBS and harvested in 1 mM EDTA in PBS and 10,000 cells are then analyzed by flow cytometry.

4. Notes

1. Solid-phase peptide synthesis is based on a cycle of repetitive (and therefore potentially automated) steps where one amino acid after the other is connected to the growing peptide chain. The synthesis starts with the most C-terminal residue of the sequence which is attached to beads of a solid resin and progresses towards the N-terminus (i.e., reverse to the conventional

reading of polypeptide sequences). By attaching the intermediate and final products to the solid support it is possible to freely exchange the solvents simply by washing steps where the resin is retained by a ceramic filter funnel (or column) without the need of recrystallization or other types of intermediate purification.

A synthetic cycle starts with one amino acid being attached to the resin by its carboxyl group and the amine being protected (e.g., Fmoc). In a first step the protection group is cleaved from the chain which liberates the amino group for the following reaction. In parallel the amino acid N-terminal ($n-1$) to the one already on the resin (n) is activated chemically; thus its carboxyl group reacts with the amine of the residue already coupled to the solid support. The amine group of this amino acid is protected (e.g., by an Fmoc group); thus it cannot polymerize with its like. By mixing an excess of this activated amino acid with the chain on the resin and incubation of this reaction mixture by several minutes the polypeptide chain grows by one residue, and a new cycle can start by removal of the protection group.

Ideally the reaction yield of each step is ≥99%, but critical steps may occur where the cleavage of the protection group or the addition of the next amino acid residues is hindered (e.g., sterically). In such cases double couplings, extended couplings, a change in chemistry and/or solvents, or the use of pseudoproline residues may be helpful (e.g., (15) and references cited therein).

2. The optimal flow to obtain best resolution on this type of column would be about three times increased, but this is not always possible as the resulting back pressure of the HPLC system and column exceeds the capacity of the instrument.

3. More recent studies suggest that a more reliable method consists in at least three cycles of dissolving 1 mg/mL peptide in 2 mM HCl (respecting a few-fold excess over the TFA ions) and lyophilization (16). If the peptide is purchased from a commercial source make sure you have information about the counterions. Note that the molecular weight to be considered during the preparation of a solution is much different if you have TFA ($2{,}777+9\times113$), acetate ($2{,}777+9\times59$), or Cl^- ($2{,}777+9\times36$) counterions to nine cationic sites of the polypeptide sequence.

4. We recommend the storage of the peptide powder over longer time periods at ≤−20°C. Therefore, before usage the container and the powder have to be first equilibrated thoroughly to reach room temperature, which can take many minutes, even hours. In order to avoid the condensation of water into the

peptide powder and its recipient, it is important that during this process the vial remains tightly sealed. Thereafter, weight the appropriate amount of peptide and resuspend LAH4 at a final concentration of, e.g., stock solution of 1 mg/mL. Before transferring the vial with the peptide back into the cold assure that it is tightly sealed with, e.g., Parafilm. Avoid repeated freezing and thawing of the stock solution. Prepare rather small aliquots (500 μL in 1.5 mL Eppendorf tubes).

5. If the condensation of the nucleic acids is very efficient, then the nucleic acids are no more accessible to SYBRSafe. This, in turn, results in the absence of a nucleic acid stain in the agarose gel. This is for example the case in Fig. 3 with the highest LAH4/DNA ratio.

6. The peptide LAH4 has five positive charges at neutral pH: indeed, the C-terminus is amidated and therefore it does not present a negative charge. Also of note, at acidic pH, the imidazole groups of the histidine residues become protonated and this adds four positive charges to the peptide (net charge = +9 at pH 5).

7. In our hands, 2.5 to 5 times more peptide of the LAH4 family is required to retard the migration of siRNAs when compared with plasmid DNA. This indicates that complex formation and interactions of the cationic peptides with plasmid DNA are different when compared to siRNA (9).

8. The quality of plasmid DNA may influence the transfection process and only siRNA and DNA of high quality should be used. It is recommended that the optical density ratio OD260/OD280 is ≥1.8.

9. For biophysical experiments we prefer to use an LAH4 stock solution at low pH, where the peptide is monomeric (14), and therefore, mixes with DNA in the most homogenous manner. In this manner, changing back and forth between acidic and neutral pH gives reproducible DLS measurements for the transfection complexes.

10. It is usually recommended to transfect cells at a confluence of 50–80%. However, the optimal cell density for efficient DNA or siRNA transfection may vary between cell types. In any case, it is important to keep the same seeding protocol for a given cell type between the experiments. It is also important, especially when using primary cells, to use cells having approximately the same number of passages.

11. The optimal peptide/nucleic acid ratio for transfection has to be determined for each new cell line.

12. The optimal peptide/nucleic acid ratio may be different for a given cell line for plasmid DNA and siRNA. For example, in

our hands using 911 cells, the optimal weight/weight peptide/nucleic acid ratio for siRNA delivery is higher than the one for plasmid DNA transfection.

13. The DNA and siRNA transfection process with LAH4 and derivatives is reduced in the presence of serum (7). Therefore, when possible, avoid addition of serum—or at low %—during the 2–3 h of incubation.
14. Optimization of each step of the transfection process should be undertaken if the DNA/siRNA transfection efficiency is low. Besides testing various peptide–nucleic acid ratios, other parameters may be changed: concentration of nucleic acids per well; duration of incubation of the complexes with the cells; buffer conditions employed during complexation (for example, water seems to give better results with the LAH4 peptides for the delivery of siRNA (9)); confluence of the cells; percentage of serum in the medium during the 2–4 h of incubation … Another way to improve the transfection consists in not replacing the transfection medium after the 2–4 h incubation step. In this case, add culture medium containing serum into the wells.
15. It has been shown for PEI/DNA complexes that just after addition of the complexes to the cells, a short centrifugation step of the culture plates at 1,000 rpm (=138 g) for 5 min increases the transfection efficiency (17). This step may also be performed in order to increase the transfection of DNA/siRNA by LAH4 and derivatives.
16. When using cells expressing a reporter gene such as luciferase for measuring the efficiency of siRNA knockdown, one has to take into account the fact that the efficiency may vary from one cellular clone to another. Indeed, the number of integrated copies probably influences the level of expression and thus also the efficiency of silencing.

    Cells expressing a reporter gene are very useful for a first evaluation of the siRNA transfection efficiency of compounds. However, for further experiments, the best manner to evaluate the siRNA delivery efficiency of a system consists in targeting endogenous genes such as GAPDH.
17. For siRNA experiments, it is important to choose the amount of peptide which gives the highest transfection efficiency associated with the lowest cytotoxicity. Indeed, cell toxicity may lead to nonspecific knockdown of genes.
18. For determination of the amount of protein, use a detergent-compatible detection kit. Notably, the total cellular protein content is also an indicator of cell viability.

Acknowledgments

We thank Christopher Aisenbrey for his assistance for the section concerning the chemical synthesis of the peptides. We are most grateful for the financial support by the Agence Nationale de la Recherche (TRANSPEP, project ANR-07-PCVI-0018), Vaincre la Mucoviscidose, the University of Strasbourg (PPF RMN), the Region Alsace, the French Ministry of Research, the CNRS, and the Medical Research Council (New Investigator Research Grant to AJM).

References

1. Zuber G et al (2001) Towards synthetic viruses. Adv Drug Deliv Rev 52:245–253
2. Gao W, Chan JM, Farokhzad OC (2011) pH-responsive nanoparticles for drug delivery. Mol Pharm 7:1913–1920
3. Midoux P et al (2009) Chemical vectors for gene delivery: a current review on polymers, peptides and lipids containing histidine or imidazole as nucleic acids carriers. Br J Pharmacol 157:166–178
4. Kichler A et al (2003) Histidine rich amphipathic peptide antibiotics promote efficient delivery of DNA into mammalian cells. Proc Natl Acad Sci USA 96:1564–1568
5. Mason AJ et al (2006) The antibiotic and DNA transfecting peptide LAH4 selectively associates with, and disorders, anionic lipids in mixed membranes. FASEB J 20:320–322
6. Prongidi-Fix L et al (2007) Self-promoted cellular uptake of peptide/DNA transfection complexes. Biochemistry 46:11253–11262
7. Mason AJ et al (2007) Optimising histidine rich peptides for efficient DNA delivery in the presence of serum. J Control Release 118:95–104
8. Lan Y et al (2010) Incorporation of 2,3-diaminopropionic acid in linear cationic amphipathic peptides produces pH sensitive vectors. ChemBioChem 11:1266–1272
9. Langlet-Bertin B et al (2010) Design and evaluation of histidine-rich amphipathic peptides for siRNA delivery. Pharm Res 27:1426–1436
10. Elbashir SM et al (2001) Duplexes of 21-nucleotide RNAs mediate RNA interference in cultured mammalian cells. Nature 411:494–498
11. Caplen NJ et al (2001) Specific inhibition of gene expression by small double-stranded RNAs in invertebrate and vertebrate systems. Proc Natl Acad Sci USA 98:9742–9747
12. Aisenbrey C, Bechinger B, Grobner G (2008) Macromolecular crowding at membrane interfaces: adsorption and alignment of membrane peptides. J Mol Biol 375:376–385
13. Georgescu J, Bechinger B (2010) NMR structures of the histidine-rich peptide LAH4 in micellar environments: membrane insertion, pH-dependent mode of antimicrobial action and DNA transfection. Biophys J 99: 2507–2515
14. Marquette A, Mason AJ, Bechinger B (2008) Aggregation and membrane permeabilizing properties of designed histidine-containing cationic linear peptide antibiotics. J Pept Sci 14:488–495
15. Harzer U, Bechinger B (2000) The alignment of lysine-anchored membrane peptides under conditions of hydrophobic mismatch: A CD, 15 N and 31 P solid-state NMR spectroscopy investigation. Biochemistry 39:13106–13114
16. Andrushchenko VV, Vogel HJ, Prenner EJ (2007) Optimization of the hydrochloric acid concentration used for trifluoroacetate removal from synthetic peptides. J Pept Sci 13:37–43
17. Boussif O, Zanta MA, Behr JP (1996) Optimized galenics improve in vitro gene transfer with cationic molecules up to 1000-fold. Gene Ther 3:1074–1080

Chapter 8

Synthesis of Polyethylenimine-Based Nanocarriers for Systemic Tumor Targeting of Nucleic Acids

Wolfgang Rödl, David Schaffert, Ernst Wagner, and Manfred Ogris

Abstract

Nucleic acid-based therapies offer the option to treat tumors in a highly selective way, while toxicity towards healthy tissue can be avoided when proper delivery vehicles are used. We have recently developed carrier systems based on linear polyethylenimine, which after chemical coupling of proteinous or peptidic ligands can form nanosized polyplexes with plasmid DNA or RNA and deliver their payload into target cells by receptor-mediated endocytosis. This chapter describes the synthesis of linear PEI (LPEI) from a precursor polymer and the current coupling techniques and purification procedure for peptide conjugates with linear polyethylenimine. A protocol is also given for the formation and characterization of polyplexes formed with LPEI conjugate and plasmid DNA.

Key words: Polyethylenimine, Polyethylene glycol, Molecular conjugates, EGF receptor, Targeting, Gene delivery

1. Introduction

The standard treatment for solid cancers is usually surgery, followed by radiotherapy and treatment with chemotherapeutic drugs. Dose-limiting toxicity and resistance mechanisms often preclude a successful treatment of relapsing disease. Nucleic acid-based therapeutics offer the possibility to develop highly specific, tailor-made therapies for the treatment of malignant diseases taking into account the genetic aberrations occurring in tumor cells compared to healthy body tissue. For gene therapy approaches, the gene of interest is cloned into an appropriate expression cassette and can be incorporated into a viral vector, for example the widely used adenovirus; as an alternative, plasmids are cloned and produced in *Escherichia coli* for nonviral delivery approaches. Several physical

Manfred Ogris and David Oupicky (eds.), *Nanotechnology for Nucleic Acid Delivery: Methods and Protocols*, Methods in Molecular Biology, vol. 948, DOI 10.1007/978-1-62703-140-0_8, © Springer Science+Business Media, LLC 2013

delivery methods for plasmid are also applicable in vivo, like electroporation, particle bombardment, or ultrasound-enhanced delivery with microbubbles (see also Chapter 15 in this book, Vlaskou et al.). For systemic delivery, particle-mediated systems are commonly used, based on lipids, polycations, or combinations thereof. For polycation-based transfection systems, polyethylenimines (PEIs) represent a kind of "golden standard" (1). PEIs are polymers with one of the highest charge densities: 1 mg of PEI contains approximately 23 μmol potentially protonable amines, of which approximately 50% are protonated at pH 7 (2). Due to their high positive charge density, nanosized particles, the so-called polyplexes, can be formed by electrostatic interaction after mixing PEI with nucleic acids containing a negatively charged phosphate backbone. PEI polyplexes, usually carrying a positive surface charge, bind to negatively charged cell surfaces mainly by interaction with proteoglycans and are thereafter internalized by adsorptive endocytosis (3). After internalization into endosomes and acidification by ATP-driven proton pumps, additional amines become protonated leading to the so-called proton sponge effect (4): protons absorbed by PEI trigger the influx of chloride ions, which in turn leads to attraction of water molecules and the osmotic imbalance causes vesicle disruption and subsequent release of its payload into the cytoplasm. Besides branched PEI, linear PEI (LPEI) has been used for transfection studies in vitro and in vivo (5, 6). Compared to branched PEI, LPEI exhibits a clearly improved transfection performance both in vitro and in vivo (7). The synthesis of LPEI is usually carried out by hydrolysis of the precursor polymer polyethyloxazoline under highly acidic conditions (8, 9). In order to obtain a product with fully biofunctional LPEI, care has to be taken that complete hydrolysis of the precursor is achieved, as residual *N*-acyl groups negatively affect the transfection efficiency (10). Albeit being nonbiodegradable, LPEI-based vectors can be designed in a way that renders them well biocompatible. When polyplexes between LPEI and plasmid DNA are formed, they usually exhibit a positive surface charge and excess of free, not polyplex-bound, LPEI (11). After intravenous injection, LPEI polyplexes rapidly interact with blood components and aggregate within the first vascular bed encountered, namely, the lung (5, 6). This makes LPEI an excellent transfection reagent to achieve high transgene expression levels in the lung, where after crossing the endothelial barrier mostly pulmonary cells at the basolateral site are transfected (12). At least two important parameters for this efficient transfection of lung tissue have been identified, namely, the aggregation with blood platelets (13) and the presence of free, non-polyplex-bound LPEI (11). Although similar aggregation occurs with polyplexes based on branched PEI (BPEI), the latter polyplexes are far less efficient in lung transfection compared to LPEI (7). Major reasons for this effect are the differences in the aggregation behavior and the dissociation behavior between PEI and plasmid DNA. For optimal transfection via the systemic route, LPEI polyplexes have to be rather small. After intravenous injection, LPEI polyplexes rapidly aggregate in the bloodstream, which causes their entrapment in the lung. On the cellular level, a reduced binding strength towards plasmid DNA of LPEI compared to BPEI has been observed: LPEI polyplexes are dissociated after endocytosis within intracellular vesicles (14) and release intact plasmid (15), which then is accessible for the translation machinery. In vivo, LPEI polyplexes initially aggregating in the lung redistribute to a considerable extent to the liver within the first minutes after injection (16), whereas BPEI polyplexes do not. To reduce the interaction with blood components and aggregation in blood, LPEI can be, similarly as described for stealth liposomes, chemically modified with the hydrophilic polymer polyethyleneglycol (PEG) (17). Coupling of PEG to PEI is on the one hand beneficial when it comes to systemic application of polyplexes in vivo, where the PEG component significantly reduces protein binding and allows blood circulation and passive accumulation in well-vascularized tumors (16, 18–20). After cellular internalization, excessive PEGylation can be disadvantageous, as it negatively affects the endosomal release of polyplexes (17). Such limitations can be overcome by designing pH-responsive PEI conjugates, e.g., by coupling PEG via chemical bonds which are cleaved after acidification of the endosome (21, 22). Alternatively, rather short PEG molecules can be used, which prevent aggregation in blood and still allow transfection of tumor cells in vivo (23). Similar as for PEG, protein ligands like transferrin or EGF can hamper endosomal release of targeted polyplexes when coupled to PEI polyplexes, but also interfere with proper particle condensation (24–26). Hence, we developed a platform for the development of targeted LPEI polyplexes, where short, peptidic ligands are utilized (27). Here, the peptidic ligand is coupled to LPEI via a rather short 2 kDa PEG spacer. To ensure the formation of LPEI–PEG–peptide conjugates without cross-linking of LPEI molecules, a heterobifunctional PEG linker 3-(2-pyridyldithio)propionamide-PEG-NHS ester (short: NHS-PEG-OPSS) is used. In the first coupling step, the NHS group reacts with one of the secondary amines in the LPEI chain forming a stable amide bond. In order to improve the reactivity of the NHS ester with secondary amines and to reduce ester hydrolysis, this reaction step is best carried out under water-free conditions in absolute ethanol or other suitable solvents. After purification by cation exchange chromatography, which removes unreacted PEG, the distal OPSS group on the PEG linker is available for coupling to free thiols by forming a reducible disulfide bond. The thiopyridone group released during this reaction strongly absorbs at 343 nm, allowing UV-control of the reaction. After a second cation exchange chromatography step, unreacted peptide and thiopyridone are removed. The resulting LPEI–PEG–peptide conjugate forms nanosized polyplexes with plasmid DNA, but also with RNA. With such polyplexes, targeted delivery and high transfection efficiency can be obtained both in vitro and in vivo in tumors after intratumoral (27) or intravenous polyplex administration (23). For a further preclinical development it is of note that LPEI-based polyplexes have already been applied in clinical trials (28), and that LPEI is available in GMP grade from the company Polyplus-transfection in France (www.polyplus-transfection.com).

2. Materials

With the exception of the materials mentioned below, all reagents can be obtained from standard lab suppliers. Please choose the quality grade "per synthesis" or "per analysis" if available. For all reactions in aqueous solution use desalted and highly purified water.

The precursor polymer for LPEI, poly(2-ethyl-2-oxazolin), 50 kDa, is obtained from Aldrich (Aldrich Cat. No. 37,284-6). Alternatively, LPEI with a molecular weight of 22 kDa can be purchased as ExGen 500 from Fermentas (Burlington, Canada) or as JetPEI from Polyplus-transfection (Illkirch, France). The heterobifunctional 2 kDa PEG linker NHS-PEG-OPSS has been synthesized by Rapp Polymere GmbH (Tübingen, Germany), a 3 kDa version is available from IRIS Biotech (Marktredwitz, Germany). Peptides used in this study were synthesized by standard fmoc solid-phase synthesis and obtained with greater than 95% purity from Biosyntan (Berlin, Germany). Ion exchange resin MacroPrep HighS was purchased from Biorad (Munich, Germany).

For LPEI synthesis, the following equipment is needed:

Round-bottom flask (NS 19, 250 mL).
Reflux condenser (NS 19, 200–300 mm length).
Conical joint clip (NS 19).
Heating mantel fitting the round-bottom flasks.
Magnetic stirrer and stir bar.
Büchner funnel.
Erlenmeyer flask (250–500 mL volume) with side arm suitable for vacuum.
Vacuum pump.
Suitable pressure tubing.
A rubber plug with hole fitting the Erlenmeyer flask and the Büchner funnel, respectively.
A lyophilization system.

For conjugate synthesis the following lab equipment is needed:

Lab-shaker with controllable temperature.
Vortex mixer.
Sterile polypropylene tubes (2 and 15 mL).

Conjugates are purified on an HPLC/FPLC system running under aqueous conditions. A gradient mixer with atleast two channels, a multiline UV/VIS detector, and a fraction collector are needed. The pH is checked by using a micro-pH electrode; for photometric measurements a standard UV/VIS photometer is sufficient.

3. Methods

The following assays are used for quantification of LPEI conjugate components.

3.1. Quantification of LPEI (Copper Assay)

This assay has been first described by Ungaro et al. (29). After adding a solution of Cu^{2+} to PEI solutions a dark blue cuprammonium complex is formed, which absorbs strongly at 285 nm enabling quantification of the LPEI content in an aqueous solution.

1. Prepare the Cu^{2+} solution by dissolving 23 mg of $CuSO_4$ in 0,1 M Na-acetate buffer (pH 5.4) and stir until the solution is clear.
2. Dilute the LPEI standard with water to a final volume of 100 μL with concentrations ranging from 10 to 100 μg/mL; use 100 μL water as blank.
3. Dilute your LPEI sample to be analyzed also to a final volume of 100 μL with water. Use duplicates and different amounts of sample.
4. Add 100 μL Cu^{2+} solution to the blank and the samples, mix, and incubate at room temperature for 5 min.
5. Set the absorption wavelength to 285 nm on a standard photometer and set the absorption of the blank to zero.
6. Measure the absorption of the standard LPEI samples and the LPEI samples to be analyzed and calculate the LPEI concentration with the help of the standard curve (see Note 1).

3.2. Determination of OPSS Content

After addition of excess dithiothreitol (DTT) (1 M stock), the dithiopyridone group is cleaved and induces an increased UV absorption at 343 nm (ε_{343nm} = 8,080/M/cm).

1. Dilute the OPSS containing sample in 150 μL water, add 15 μL 1 M DTT solution, mix well, and incubate for 10 min.

2. Transfer to a microcuvette suitable for UV measurement and measure the absorption at 343 nm; use 15 μL 1 M DTT diluted with 150 μL water as blank.

   Calculate the OPSS concentration with $\varepsilon_{343nm} = 8{,}080/M/cm$ taking into account the dilution factor.

3.3. Quantification of LPEI/PEG Ratio by ¹H NMR

In addition to the copper and DTT assay, ^1H NMR analysis allows nondestructive determination of the molar ratio between LPEI and PEG in a conjugate. By comparison of the distinct proton signals of ethyleneglycol and ethyleneimine units the yield of the PEGylation reaction can be calculated.

1. Dialyze 5 mg LPEI–PEG–OPSS dissolved in 1 mL water using a dialysis tubing (MWCO 14 kDa) against water (2 L) overnight at 4°C (see Note 2).
2. Transfer the dialyzed product to a 15 mL polypropylene tube, freeze-dry for 1 day, and dissolve it in 1 mL D_2O.
3. Adjust the pH of the solution to pH 7.0 using 1 M stock solutions of DCl or NaOD (see Note 3).
4. Analysis is then carried out on a ≥200 MHz NMR with 3-(tri-methyl silyl) propionic-2,2,3,3-d acid (TSP) as internal reference. Alternatively the residual solvent peak can be used to correct the spectrum.

The spectrum is characterized by two major signals, a rather broad PEI signal at 2.9–3.1 ppm and a sharp PEG signal at 3.7 ppm (see Note 4). The signals of the OPSS moiety can be observed above 7 ppm, but with lower PEGylation rates these signals are normally too weak to allow quantification.

To quantify the degree of PEGylation use the 22 kDa PEI-signal integral (2.9–3.1 ppm) to normalize the other signals. Set the PEI integral to 2,558 ($N = 22{,}000(M_{w(PEI)})/43(M_{w(ethyleneimine\ monomer)}) \times 5$) (protons per ethyleneimine unit) and calculate the number of protons in the PEG-chain using $N = (PEG(M_{w(PEG)})/44) \times 4$ (for a 3 kDa PEG this yields 273). Divide the integral value of the PEG-signal (3.7 ppm) by this value to obtain the average number of PEG chains per PEI molecule. An example for NMR analysis and calculation of the PEG/LPEI ratio is shown in Fig. 1.

3.4. Quantification of Free Thiols (Ellman's Assay)

This assay is based on the thiol-specific reactivity of Ellman's reagent (5,5′-dithiobis-(2-nitrobenzoic acid), short DTNB) forming mixed disulfides. During the reaction, 2-nitro-5-thiobenzoate (TNB) is released, which can be quantified by measuring the absorption at 412 nm (30).

1. Dissolve 2 mg DTNB in 0.1 M HEPES pH 7.4 (see Note 5).
2. Dilute 5 μL of the solution from step 1, fill up to 500 μL with 0.1 M HEPES pH 7.4, and use as a blank on a photometer set to an absorption wavelength of 412 nm.

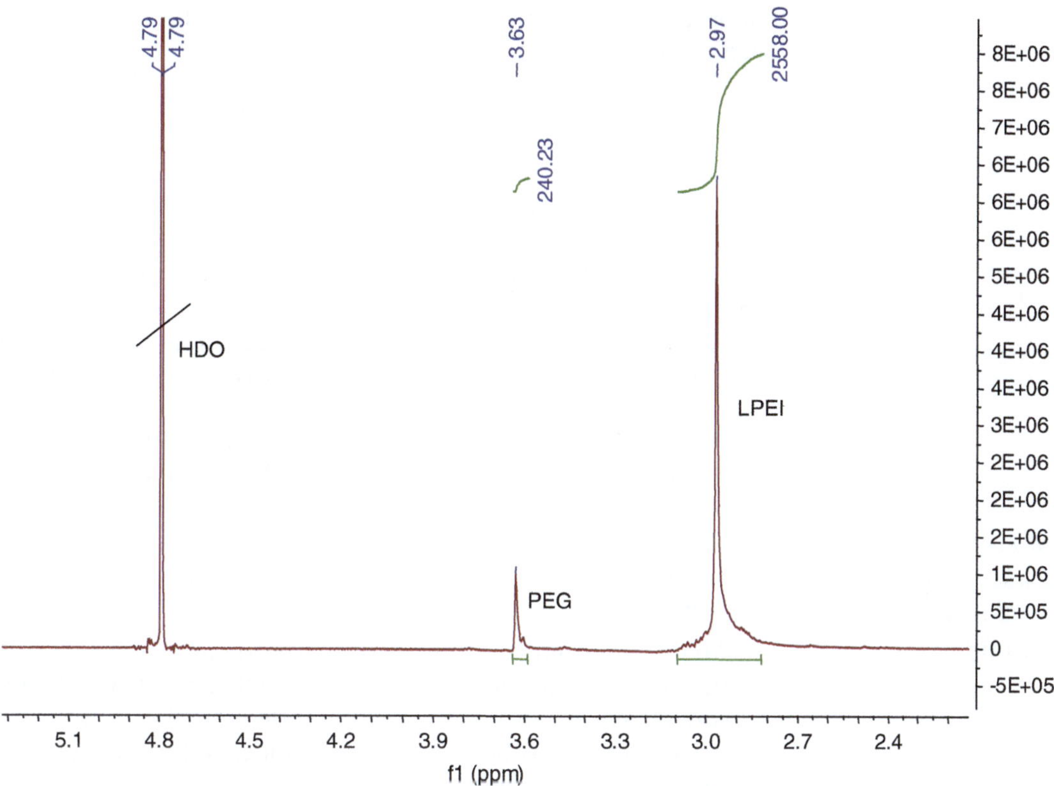

Fig. 1. ^{1}H NMR profile of LPEI–PEG–OPSS. NHS-PEG-OPSS (MW 2 kDa) has been coupled to LPEI as described above and the LPEI integral (*green line* at 2.97 ppm) set to 2,558. The number of protons in PEG (2 kDa) is (2,000/44) × 4 = 182. The molar ratio PEG/LPEI is calculated by dividing the integral value of PEG (*green line* at 3.63 ppm) by the number or protons in PEG: 240/182 = 1.3 $PEG_{2kDa}/LPEI_{22kDa}$ M/M.

3. Dilute your sample (using different amounts) in the solution from step 1, incubate for 20 min at ambient temperature, and measure the absorption at 412 nm.
4. Calculate the thiol content using the molar extinction coefficient ($\varepsilon_{412nm} = 14{,}100/\text{Mol} \times \text{cm}$).

3.5. Peptide Quantification by A_{280} Measurement

Prior to coupling, the absorption coefficient of the peptide has to be estimated, as the peptide content in conjugates is calculated according to the absorption of aromatic amino acids within the peptide. For this purpose, a useful online tool from the Expasy Bioinformatics tool portal (expasy.org) can be utilized (see http://web.expasy.org/protparam/). The molar extinction coefficient epsilon (ε) is calculated with the following formula:

$$\varepsilon_{(\text{peptide})} = \text{number(Tyr)} \times \varepsilon_{(\text{Tyr})} + \text{number(Trp)} \times \varepsilon_{(\text{Trp})} + \text{number(Cys)} \times \varepsilon_{(\text{Cys})}.$$

For proteins in water measured at 280 nm use the following extinction coefficients:

$$\varepsilon_{(\mathrm{Tyr})} = 1{,}490 / \mathrm{Mol} \times \mathrm{cm}.$$

$$\varepsilon_{(\mathrm{Trp})} = 5{,}500 / \mathrm{Mol} \times \mathrm{cm}.$$

$$\varepsilon_{(\mathrm{Cys})} = 125 / \mathrm{Mol} \times \mathrm{cm}.$$

For the GE11 peptide (CYHWYGYTPQNVI) the absorption coefficient $\varepsilon_{280\mathrm{nm}} = 9{,}970/\mathrm{Mol} \times \mathrm{cm}$ is calculated.

Nevertheless, it is of note that the theoretical calculation of $\varepsilon_{280\mathrm{nm}}$ has to be amended by absorption measurements at 280 nm with the actual peptide used.

In the following subsections, the synthesis of LPEI from the precursor molecule and the conjugate synthesis is described.

3.6. Synthesis of LPEI 22 kDa Free Base from Poly(2-Ethyl-2-Oxazolin)

1. Dissolve 5 g poly(2-ethyl-2-oxazolin) 40 kDa (125 μmol) in 50 mL HCl (30%, v/v) in a 100 mL round-bottom flask (see Note 6).
2. Attach the round-bottom flask to the reflux condenser with the conical joint clip and attach the reflux condenser to the cooling system. Apply the heat mantle with the temperature set to 104°C and boil the reaction under reflux and constant stirring for 48 h (see Note 7).
3. Lay out a round filter paper into the Büchner funnel and attach it via the rubber plug to the Erlenmeyer flask. Then attach the rubber tubing and apply vacuum using the vacuum pump.
4. Transfer the content of the round-bottom flask including the fine, white precipitate to the Büchner funnel and soak through the liquid. The isolated precipitate is LPEI as HCl salt.
5. Carry out three repeated washes of the precipitate with 100 mL 30% HCl per washing cycle as in step 4 and remove the filtrate from the Erlenmeyer flask between the washing steps if necessary (see Note 8).
6. Let the precipitate air-dry overnight, then dissolve it in 200 mL water, and lyophilize the product (see Note 9).
7. Transfer 5 g LPEI-HCl to a round-bottom flask and resuspend it in 50 mL 1 M NaOH.
8. Attach the round-bottom flask to the reflux condenser with the conical joint clip and the reflux condenser to the cooling system. Apply the heat mantle with the temperature set to 104°C and boil the reaction under reflux and stirring. Carefully add 100–200 mL 1 M NaOH (in small portions) to the still hot solution until the solution is clear (see Note 6); then switch off the heating.

9. Let the mixture cool down to room temperature, until a gel-like precipitate forms. Then filter the precipitate as described in step 4 and wash 3× with 100 mL 1 M NaOH and thereafter 5× with 100 mL water to remove residual NaOH.
10. Transfer the precipitate to a 50 mL polypropylene tube, cover with 5 mL water, and freeze-dry for 3 days.
11. The resulting LPEI can be stored at room temperature in a desiccator protected from light.

3.7. Synthesis and Purification of LPEI–PEG–OPSS

1. Dissolve 74.8 mg (3.4 μmol) LPEI (free base) in 1.5 mL absolute ethanol using a 2 mL polypropylene reaction vial with lid and incubate on a standard lab mixer for 15 min at 800 rpm and 35°C (see Note 10).
2. Dissolve NHS-PEG-OPSS linker (17 μmol) in 100 μL DMSO (water free), add to the reaction mixture from step 1, and incubate again for 3 h at 35°C with 800 rpm.
3. Transfer the reaction mixture to a 15 mL polypropylene tube and add approximately 2 mL of a 20 mM HEPES pH 7.4 solution and NaCl (3 M stock) so that in the final volume of 5 mL the NaCl reaches 0.5 M (the starting NaCl concentration during ion exchange purification, see below).
4. Adjust the pH with concentrated hydrochloric acid (HCl) and check the pH with a micro-pH probe until pH 7 is reached (see Note 11). Thereafter fill up the reaction mix to 5 mL with 20 mM HEPES (see Note 12).
5. Equip the HPLC system with a column (HR10/10, i.e., 10 cm in length, 10 mm diameter) filled with cation exchange resin (MacroPrep High S). Set the UV detector setting measuring the absorption at 240, 280, and 343 nm. Equilibrate the system with 83.3% solution A (20 mM HEPES pH 7.4) and 16.7% solution B (3 M NaCl, 20 mM HEPES pH 7.4) for at least 1 h at a flow of 0.5 mL/min; this corresponds to a concentration of 500 mM NaCl and 20 mM HEPES pH 7.4 (see Note 13).
6. Program a gradient (flow 0.5 mL/min) with 16.7% A and 83.3% B from 0 to 25 min, and linear change to 100% B over 40 min (25–65 min), followed by 100% B for 20 min.
7. Load the product from step 4 onto the column and run the gradient. Fractions eluting during the first 25 min (at 500 mM NaCl, 20 mm HEPES pH 7.4) contain unreacted NHS-PEG-OPSS and by-products from the reaction between the NHS ester in PEG and amines in LPEI. A chromatogram is shown in Fig. 2. Fractions eluting between 2 and 2.8 M NaCl contain LPEI modified with PEG (LPEI–PEG–OPSS); these fractions are pooled and subsequently dialyzed against 5 L HBS (20 mM HEPES pH 7.4, 150 mM NaCl) at 4°C under constant stirring overnight (see Note 14).

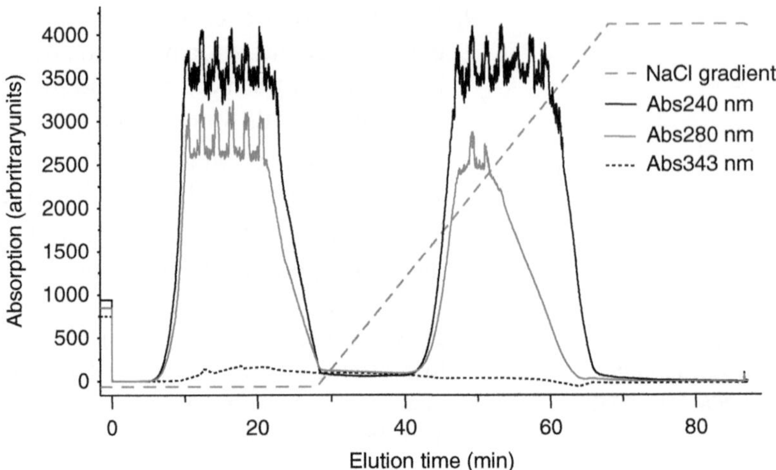

Fig. 2. Cation exchange chromatography profile of LPEI–PEG–OPSS. 5 mL reaction mixture from step 4, Subheading 3.7, were loaded onto an HR10/10 cation exchange column and purified as described in the text. Fractions eluting from 40 to 60 min were pooled and further processed as described.

3.8. Coupling of Peptide to LPEI–PEG–OPSS

The peptide CYHWYGYTPQNVI (GE11, see ref. 27) used in this study was synthesized by standard Fmoc solid-phase peptide synthesis and purified on a C18 reversed-phase HPLC column. The product was eluted with an acetonitrile gradient (A: 0.05% (v/v) TFA in water, B: 0.05% TFA (v/v) in 80% acetonitrile in water), 0.6 mL/min flow, linear gradient 2.5% B/minute, detection at 220 nm, the peptide eluted at 13.1 min, and thereafter lyophilized (see Note 15).

1. Dissolve the peptide in 30% acetonitrile/water/0.1% TFA (2.2 μmol peptide in 100 μL).
   To ensure that the thiol residue has not been oxidized or dimerized, quantify the free thiol content by Ellman's assay. For this purpose, dilute 1 μL peptide sample 1:100 with water and use 15 μL dilution for the assay; proceed as described in Subheading 3.4.
2. Incubate LPEI–PEG–OPSS (in HBS) and peptide at a molar ratio of OPSS/SH of 1/2. The final concentration of LPEI in the reaction mix should be between 4 and 5 mg/mL.
3. For online reaction monitoring, check the absorption of the LPEI–PEG–OPSS solution at 343 nm prior to mixing.
4. Add 2.2 μmol peptide (in 100 μL, see above) to LPEI–PEG–OPSS (containing 1.1 μmol OPSS) in HBS, immediately after addition carefully sparge argon gas though the sample (see Note 16).

5. Thereafter (ca 1 min) take a 150 μL sample from total 5.1 mL reaction mix, transfer to a cuvette, and measure A_{343}. The A_{343} is measured every 30 min, until there is no more increase in absorption observed. This is a measure for the completeness of the reaction.
6. Calculate the thiopyridone released as described in Subheading 3.2.
7. Add NaCl (3 M stock) to obtain a final concentration of 500 mM NaCl.
8. Load the reaction mixture onto a 10/10 MacroPrep column (MacroPrep High S; HR10/10; BioRad, München, Germany).
9. Apply a salt gradient from 0.5–3 M NaCl in 20 mM HEPES pH 7.4 and 10% (v/v) acetonitrile. The product elutes between 2.0 and 2.8 M NaCl. Flow: 0.5 mL/min; Solution A: 20 mM HEPES pH 7.4 and 10% (v/v) acetonitrile; Solution B: 20 mM HEPES pH 7.4, 3 M NaCl, and 10% (v/v) acetonitrile. A chromatogram from this purification step is depicted in Fig. 3.
10. Pool the fractions eluting between 2 and 2.8 M NaCl and dialyze them against 5 L HBS (20 mM HEPES pH 7.4, 150 mM NaCl) at 4°C under constant stirring overnight.
11. Analyze the LPEI content in the resulting conjugate by copper assay.

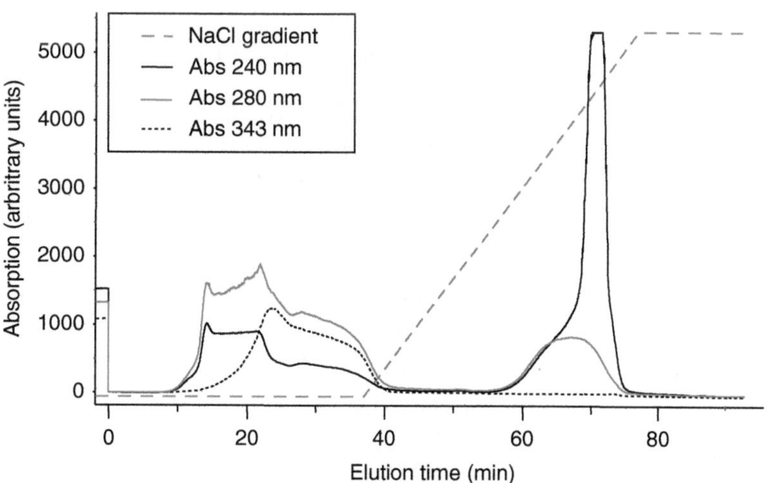

Fig. 3. Cation exchange chromatography profile of LPEI-PEG-GE11. 10 mL reaction mixture from step 8, Subheading 3.8, were loaded onto an HR10/10 cation exchange column and purified as described in the text. Fractions eluting from 57 to 77 min were pooled and further processed as described.

12. Calculate the peptide concentration as described in Subheading 3.5. For the measurement, Take a 150 μL aliquot of your sample (undiluted, can be reused, so use sterile, clean cuvettes) and measure the absorption at 280 nm.
13. For further storage freeze conveniently sized aliquots (snap freezing) and store at −80°C.

3.9. Polyplex Formation

The LPEI–PEG–peptide conjugates have been developed for local or systemic delivery of nucleic acids in vivo. For this purpose, polyplexes are generated in a low-salt buffer (HBG: HEPES-buffered glucose, 20 mM HEPES pH 7.4, 5% glucose, w/v). This allows the formation of rather small, colloidal stable polyplexes.

The HBG buffer is sterile filtered (0.4 μm pore size) and stored in aliquots either at 4°C or frozen at −20°C. Plasmid DNA is usually produced in suitable *E. coli* strains and the plasmid isolated after alkaline lysis of bacterial cells using commercially available purification kits. Please make sure to use kits which result in a low endotoxin contamination in the purified plasmid. Alternatively, plasmids can be produced by companies specialized in plasmid production (see Note 17).

LPEI/DNA polyplexes are defined by their molar ratio of phosphate in plasmid and nitrogen in PEI (N/P ratio). The N/P ratio is transformed into a w/w ratio by the following formula:

$$\mu\text{g PEI} = \mu\text{g DNA} \times 43 \times (\text{N} / \text{P ratio}) / 330.$$

43 is the molecular weight of one repeating [CH_2–CH_2–NH] unit in PEI (N) and 330 is the average molecular weight of one nucleotide (P). Polyplexes with N/P ratios ranging from 5 to 10 are recommended.

Here, an example is given to obtain 500 μL polyplexes containing 100 μg plasmid DNA at an N/P ratio of 6.

1. Prepare 250 μL plasmid diluted in HBG (final plasmid concentration 400 μg/mL).
2. Prepare 250 μL LPEI–PEG–peptide conjugate solution in HBG (final LPEI concentration $400 \times 43 \times 6/330 = 313$ μg/mL).
3. Transfer the LPEI–PEG–peptide dilution to the plasmid dilution and immediately pipette the solution ten times up and down (see Note 18).
4. For quality control, measure polyplex size by standard laser light scattering. The average particle size should be below 300 nm.
5. Store the polyplexes at ambient temperature for no longer than 20 min prior to use.

4. Notes

1. The absorption value of your sample should be between 0.1 and 0.8 to ensure linear correlation between absorption and LPEI concentration.
2. Removal of HEPES is necessary, which otherwise interferes with the NMR analysis.
3. Adjusting the pH to 7.0 with deuterated hydrochloric acid (DCl) or deuterated sodium hydroxide (NaDO) is important at this step to obtain comparable results, as acidic pH can cause a shift of the LPEI peak towards 3.5 ppm.
4. One has to be aware of the possibility of additional peaks due to incomplete deprotection of the LPEI-precursor; especially signals of propionamidyl residues (1.2–1.3 ppm; 2.1–2.3 ppm; 3.5–3.8 ppm) can lead to an overestimation of PEGylation (31).
5. Always prepare the Ellman's reagent fresh; do not exceed storage for longer than 1 day.
6. Take safety measures (protective goggles, gloves, lab coat) when handling highly concentrated acids or bases.
7. Carry out the reaction in a fume hood and take proper safety measures when carrying out the reaction overnight. The reaction mixture appears first clear; after approximately 5 h a white precipitate is formed (LPEI-HCl).
8. Washing steps have to be repeated until the filtrate is clear and odorless. The residual proprionic acid can develop a strong odor.
9. LPEI-HCl is well soluble in water at pH > 2, but remains insoluble in >25% HCl. At this step, LPEI-HCl can be neutralized with concentrated NaOH to pH 7 and thereafter used as transfection reagent. To calculate the LPEI, content, please note that LPEI-Cl has a molecular weight of 79.6 Da per repeating unit ($[CH_2\text{–}CH_2\text{–}NH] * Cl^-$).
10. LPEI dissolves fast, but constant mixing is necessary to achieve a homogeneous solution, which is highly viscous.
11. When adding in 10 μL aliquots HCl (total approximately 100 μL needed), do this under constant vortexing to avoid local low pH, which could result in LPEI precipitation. Cave: the solution heats up if HCl is added too fast.
12. This is necessary to reduce the EtOH concentration prior to loading of the mix onto the ion exchange columns, where otherwise a too high EtOH concentration results in pressure increase and compression of the column material.

13. Solutions A and B are sparged with argon for at least 15 min prior to use. Constant sparging with argon during purification is preferable.
14. At this step, the resulting LPEI–PEG–OPSS conjugate can be snap frozen in liq. N_2 and stored at −80°C until further use (do not exceed storage times of a month).
15. Peptides containing free thiols, like the terminal Cys residue, should be stored under argon at −80°C for no longer than a month, depending on the peptide. After extended storage, dimerization or oxidation of thiol groups occurs.
16. Make sure that the final reaction mixture contains at least 10% (v/v) acetonitrile; otherwise the peptide can precipitate.
17. Elevated levels of high-molecular-weight bacterial genomic DNA can be present in the plasmid preparation using commercialized isolation kits (32), especially when using for example low copy plasmid (33). In such cases, preparation methods using additional purification steps are recommended (32). For example, the company PlasmidFactory (www.plasmidfactory.com) offers plasmid in "ccc" grade, i.e., supercoiled plasmid structure and absence of bacterial genomic DNA impurities. Besides biological effects, the presence of high-molecular-weight DNA impurities can lead to excessive particle aggregation during the mixing process. For the generation of well-defined larger batches of polyplexes, methods using controllable mixing devices are recommended (see ref. 34 and Chapter 10 in this book, Kasper et al.).
18. Mixing should be done immediately to avoid formation of aggregates; the appearance of aggregates correlates positive with an increase in the final concentration of polyplexes and the salt concentration in the dilution buffer; there is a negative correlation with the increase in the N/P ratio (see also ref. 35).

Acknowledgments

This work was supported by the Center for Nanoscience (CeNS) and the German Research Foundation (SFB824) to M.O., and the Nanosystems Initiative Munich (NIM) to E.W.

References

1. Boussif O, Lezoualc'h F, Zanta MA, Mergny MD, Scherman D, Demeneix B, Behr JP (1995) A versatile vector for gene and oligonucleotide transfer into cells in culture and in vivo: polyethylenimine. Proc Natl Acad Sci U S A 92:7297–7301
2. Tang MX, Szoka FC (1997) The influence of polymer structure on the interactions of cationic polymers with DNA and morphology of the resulting complexes. Gene Ther 4:823–832
3. Kopatz I, Remy JS, Behr JP (2004) A model for non-viral gene delivery: through syndecan adhesion molecules and powered by actin. J Gene Med 6:769–776
4. Sonawane ND, Szoka FC Jr, Verkman AS (2003) Chloride accumulation and swelling in endosomes enhances DNA transfer by polyamine-DNA polyplexes. J Biol Chem 278:44826–44831
5. Ferrari S, Moro E, Pettenazzo A, Behr JP, Zacchello F, Scarpa M (1997) ExGen 500 is an efficient vector for gene delivery to lung epithelial cells in vitro and in vivo. Gene Ther 4:1100–1106
6. Goula D, Benoist C, Mantero S, Merlo G, Levi G, Demeneix BA (1998) Polyethylenimine-based intravenous delivery of transgenes to mouse lung. Gene Ther 5:1291–1295
7. Wightman L, Kircheis R, Rossler V, Carotta S, Ruzicka R, Kursa M, Wagner E (2001) Different behavior of branched and linear polyethylenimine for gene delivery in vitro and in vivo. J Gene Med 3:362–372
8. Brissault B, Kichler A, Guis C, Leborgne C, Danos O, Cheradame H (2003) Synthesis of linear polyethylenimine derivatives for DNA transfection. Bioconjug Chem 14:581–587
9. Ogris M, Wagner E (2008) Linear polyethylenimine: synthesis and transfection procedures for in vitro and in vivo. In: Friedmann T, Rossi J (eds) Gene transfer: delivery and expression of DNA and RNA, a laboratory manual. Cold Spring Harbor Laboratory Press, Cold Spring Harbor, NY, pp 521–526
10. Thomas M, Lu JJ, Ge Q, Zhang C, Chen J, Klibanov AM (2005) Full deacylation of polyethylenimine dramatically boosts its gene delivery efficiency and specificity to mouse lung. Proc Natl Acad Sci U S A 102:5679–5684
11. Boeckle S, von Gersdorff K, van der Piepen S, Culmsee C, Wagner E, Ogris M (2004) Purification of polyethylenimine polyplexes highlights the role of free polycations in gene transfer. J Gene Med 6:1102–1111
12. Goula D, Becker N, Lemkine GF, Normandie P, Rodrigues J, Mantero S, Levi G, Demeneix BA (2000) Rapid crossing of the pulmonary endothelial barrier by polyethylenimine/DNA complexes 965. Gene Ther 7:499–504
13. Chollet P, Favrot MC, Hurbin A, Coll JL (2002) Side-effects of a systemic injection of linear polyethylenimine-DNA complexes. J Gene Med 4:84–91
14. Itaka K, Harada A, Yamasaki Y, Nakamura K, Kawaguchi H, Kataoka K (2004) In situ single cell observation by fluorescence resonance energy transfer reveals fast intra-cytoplasmic delivery and easy release of plasmid DNA complexed with linear polyethylenimine. J Gene Med 6:76–84
15. de Bruin KG, Fella C, Ogris M, Wagner E, Ruthardt N, Brauchle C (2008) Dynamics of photoinduced endosomal release of polyplexes. J Control Release 130:175–182
16. Zintchenko A, Susha AS, Concia M, Feldmann J, Wagner E, Rogach AL, Ogris M (2009) Drug nanocarriers labeled with near-infrared-emitting quantum dots (quantoplexes): imaging fast dynamics of distribution in living animals. Mol Ther 17:1849–1856
17. Kursa M, Walker GF, Roessler V, Ogris M, Roedl W, Kircheis R, Wagner E (2003) Novel shielded transferrin-polyethylene glycol-polyethylenimine/DNA complexes for systemic tumor-targeted gene transfer. Bioconjug Chem 14:222–231
18. Ogris M, Brunner S, Schuller S, Kircheis R, Wagner E (1999) PEGylated DNA/transferrin-PEI complexes: reduced interaction with blood components, extended circulation in blood and potential for systemic gene delivery. Gene Ther 6:595–605
19. Schwerdt A, Zintchenko A, Concia M, Roesen N, Fisher KD, Lindner LH, Issels RD, Wagner E, Ogris M (2008) Hyperthermia induced targeting of thermosensitive gene carriers to tumors. Hum Gene Ther 19:1283–1292
20. Smrekar B, Wightman L, Wolschek MF, Lichtenberger C, Ruzicka R, Ogris M, Rodl W, Kursa M, Wagner E, Kircheis R (2003) Tissue-dependent factors affect gene delivery to tumors in vivo. Gene Ther 10:1079–1088
21. Fella C, Walker GF, Ogris M, Wagner E (2008) Amine-reactive pyridylhydrazone-based PEG reagents for pH-reversible PEI polyplex shielding. Eur J Pharm Sci 34:309–320
22. Walker GF, Fella C, Pelisek J, Fahrmeir J, Boeckle S, Ogris M, Wagner E (2005) Toward synthetic viruses: endosomal pH-triggered

deshielding of targeted polyplexes greatly enhances gene transfer in vitro and in vivo. Mol Ther 11:418–425

23. Klutz K, Schaffert D, Willhauck MJ, Grunwald GK, Haase R, Wunderlich N, Zach C, Gildehaus FJ, Senekowitsch-Schmidtke R, Goke B, Wagner E, Ogris M, Spitzweg C (2011) Epidermal growth factor receptor-targeted (131)I-therapy of liver cancer following systemic delivery of the sodium iodide symporter gene. Mol Ther 19:676–685
24. Kircheis R, Wightman L, Schreiber A, Robitza B, Rossler V, Kursa M, Wagner E (2001) Polyethylenimine/DNA complexes shielded by transferrin target gene expression to tumors after systemic application. Gene Ther 8:28–40
25. Ogris M, Steinlein P, Carotta S, Brunner S, Wagner E (2001) DNA/polyethylenimine transfection particles: Influence of ligands, polymer size, and PEGylation on internalization and gene expression. AAPS Pharm Sci 3:E21
26. Ogris M, Walker G, Blessing T, Kircheis R, Wolschek M, Wagner E (2003) Tumor-targeted gene therapy: strategies for the preparation of ligand-polyethylene glycol-polyethylenimine/ DNA complexes. J Control Release 91:173–181
27. Schafer A, Pahnke A, Schaffert D, Van Weerden WM, de Ridder CM, Rodl W, Vetter A, Spitzweg C, Kraaij R, Wagner E, Ogris M (2011) Disconnecting the yin and yang relation of epidermal growth factor receptor (EGFR) mediated delivery: a fully synthetic, EGFR-targeted gene transfer system avoiding receptor activation. Hum Gene Ther 22:1463–1473
28. Sidi AA, Ohana P, Benjamin S, Shalev M, Ransom JH, Lamm D, Hochberg A, Leibovitch I (2008) Phase I/II marker lesion study of intravesical BC-819 DNA plasmid in H19 over expressing superficial bladder cancer refractory to bacillus calmette-guerin. J Urol 180: 2379–2383
29. Ungaro F, De Rosa G, Miro A, Quaglia F (2003) Spectrophotometric determination of polyethylenimine in the presence of an oligonucleotide for the characterization of controlled release formulations. J Pharm Biomed Anal 31:143–149
30. Eyer P, Worek F, Kiderlen D, Sinko G, Stuglin A, Simeon-Rudolf V, Reiner E (2003) Molar absorption coefficients for the reduced Ellman reagent: reassessment. Anal Biochem 312: 224–227
31. Jeong JH, Song SH, Lim DW, Lee H, Park TG (2001) DNA transfection using linear poly(ethylenimine) prepared by controlled acid hydrolysis of poly(2-ethyl-2-oxazoline). J Control Release 73:391–399
32. Schleef M, Schmidt T (2004) Animal-free production of ccc-supercoiled plasmids for research and clinical applications. J Gene Med 6(Suppl 1):S45–S53
33. Magnusson T, Haase R, Schleef M, Wagner E, Ogris M (2011) Sustained, high transgene expression in liver with plasmid vectors using optimized promoter-enhancer combinations. J Gene Med 13:382–391
34. Kasper JC, Schaffert D, Ogris M, Wagner E, Friess W (2011) The establishment of an up-scaled micro-mixer method allows the standardized and reproducible preparation of well-defined plasmid/LPEI polyplexes. Eur J Pharm Biopharm 77:182–185
35. Ogris M, Steinlein P, Kursa M, Mechtler K, Kircheis R, Wagner E (1998) The size of DNA/transferrin-PEI complexes is an important factor for gene expression in cultured cells. Gene Ther 5:1425–1433

Chapter 9

Synthesis of Bioreducible Polycations with Controlled Topologies

Ye-Zi You, Jun-Jie Yan, Zhi-Qiang Yu, and David Oupicky

Abstract

Bioreducible polycations, which possess disulfide linkages in the backbone, have appeared as promising gene delivery carriers due to their high stability in extracellular physiological condition and bioreduction-triggered release of genetic materials, as well as reduced cytotoxicity because intracellular cytosol is a reducing environment containing high level of reducing molecules such as glutathione. Here, we describe the syntheses of bioreducible polycations, and the methods for control over their topology are also presented.

Key words: Bioreducible polycations, Poly(2-(dimethylamino)ethyl methacrylate), Polyethylenimine, Poly(amido amine)s, Controlled topologies, Michael-addition polymerization

1. Introduction

Synthetic polycations have received increasing attention in gene and drug delivery (1–20). Despite the high stability and low immunogenicity, polycationic vectors are plagued by some problems, such as low gene transfection efficiency and high cytotoxicity (6–10). In order to achieve efficient DNA delivery, different kinds of disulfide-containing polycations have been designed for improving the efficacy of the gene delivery and reducing cytotoxicity. The polyplexes, prepared by introducing disulfide bonds into the structure of the polycations, show the capability to release the therapeutic nucleic acids selectively in the subcellular reducing space. The redox-sensitive polyplexes have already proven to be suitable for delivering a variety of nucleic acids, including plasmid DNA,

Manfred Ogris and David Oupicky (eds.), *Nanotechnology for Nucleic Acid Delivery: Methods and Protocols*, Methods in Molecular Biology, vol. 948, DOI 10.1007/978-1-62703-140-0_9, © Springer Science+Business Media, LLC 2013

mRNA, antisense oligonucleotides, and siRNA (7–27). On the other hand, the basicity and degree of protonation of polycationic vectors depend on the amount of primary, secondary, and tertiary amines and their topology, which greatly influence the cytotoxicity, the escape of polyplexes from lysosome, and the transfection efficiency. It has been reported that the introduction of irregular branched structure into polycation might improve the transfection and the topology of polycation has a great impact on the gene transfection efficiency (28–30). Therefore, the disulfide bonds and topology of polycations play very important role in gene transfection. Here, we describe selected methods to prepare reducible polycations and to control their topologies.

The most studied bioreducible polycations include bioreducible polyethylenimine (PEI), bioreducible poly(2-(dimethylamino) ethyl methacrylate) (PDMAEMA), bioreducible poly(amido amine)s and poly(amino ester)s. Bioreducible PEIs are prepared via functionalizing with reducible cross-linking agents (14, 15), consecutive thiolation and oxidation (16, 17), or Michael-addition reaction (18), while their topologies (linear or network) can be controlled by the oligomer topology (linear or branched). Bioreducible PDMAEMA can be prepared via RAFT polymerization, while its topology can be controlled by a proper selection of a RAFT agent and cross-linking agent used in the polymerization. RAFT polymerization of DMAEMA using difunctional RAFT agent to obtain α,ω-dithioester-functionalized PDMAEMA, followed by aminolysis and oxidation of the synthesized oligomer produced linear disulfide-containing PDMAEMA (9). Likewise, RAFT polymerization of DMAEMA mediated by 1,2-bis(2-(3-methylbuta-1,3-dien-2-yloxy)ethyl) disulfane in the presence of bioreducible disulfide-based dimethacrylate (DSDMA) produced hyperbranched PDMAEMA with each branching point in the polymer linked by a disulfide bond (31). Bioreducible poly(amido amine)s and poly(amino ester)s are mainly prepared by Michael-addition polymerization of disulfide-based diacrylamides and diacrylates and amine monomers (12, 13, 19–27). Their topologies can be finely tuned by the polymerization conditions due to the fact that the primary and secondary amines have different reactivities under different polymerization conditions (32).

2. Materials

Aqueous solutions are prepared using ultrapure water (prepared by purifying deionized water to attain a resistivity of 18.4 MΩ cm at 25°C) and all reagents are analytical grade.

3. Methods

3.1. Bioreducible Polyethylenimine

PEI is the most widely used polycation in gene delivery due to its excellent transfection efficiency. However, high cytotoxicity stimulated the demand for synthesis of biodegradable PEI. Bioreducible PEIs are prepared via modifications with reducible cross-linking agents (14, 15), consecutive thiolation and oxidation (16, 17), or Michael-addition reaction (18). Bioreducible PEI polycations can be also obtained by covalently linking low-molecular-weight PEI to bioreducible polymers via disulfide bonds. The topologies (linear, network) of bioreducible PEI are mainly controlled by the structure of its oligomer. The detailed procedure for preparing linear reducible PEI is shown in Scheme 1 (33).

1. Syntheses of 5, 6, 7, and 8 (step a): 4-(Dimethylamino)-pyridine (10 mmol), triethylamine (92 mmol), and *p*-toluene-sulfonyl chloride (44 mmol) are added into the mixture containing 40 mL of dichloromethane and 10 mmol of 1 (2–4). After the mixture is stirred at 0°C for 24 h, dichloromethane is removed by evaporation and the mixture is redissolved in 100 mL of chloroform. The solution is washed three times with 0.1 M HCl. The organic layer is dried with anhydrous $MgSO_4$. The final products are obtained as white or pale yellow powder.
2. Syntheses of 2, 3, and 4 (step b): Ethanolamine (10 equiv., 10 mmol) is slowly added into the mixture containing 20–30 mL of DMF and 1 mmol of 5 (6, 7). After the reaction mixture is stirred at 50°C for 1–2 days, solvent is removed by evaporation; the mixture is redissolved in DMSO and evaporated again for the removal of the remaining ethanolamine.
3. Syntheses of 9, 10, 11, and 12 (step c): Potassium thioacetate (2.5 equiv.) is added into the mixture containing 50 mL of DMF and 10 mmol of 5 (6–8) and stirred at room temperature overnight. DMF is evaporated and the residue is poured into 100 mL of water. The mixture is extracted three or four times with chloroform, and the organic layer is collected and dried with $MgSO_4$. The final products are obtained via evaporating the organic solvent.
4. Tosyl and acetyl deprotection (step d): The mixture containing 9 (1 mmol) and phenol (0.41 mmol) in 33% HBr/AcOH (15.4 mL) is refluxed for 24 h. Then, the reaction mixture is filtered, and the resulting pale brown solid is washed with ethanol. After refluxing for 1 h with ethanol, the mixture is filtered and the resulting solid is washed twice with diethyl ether to obtain the pale brown powder (13). The syntheses of 14, 15, and 16 are carried in the same way.

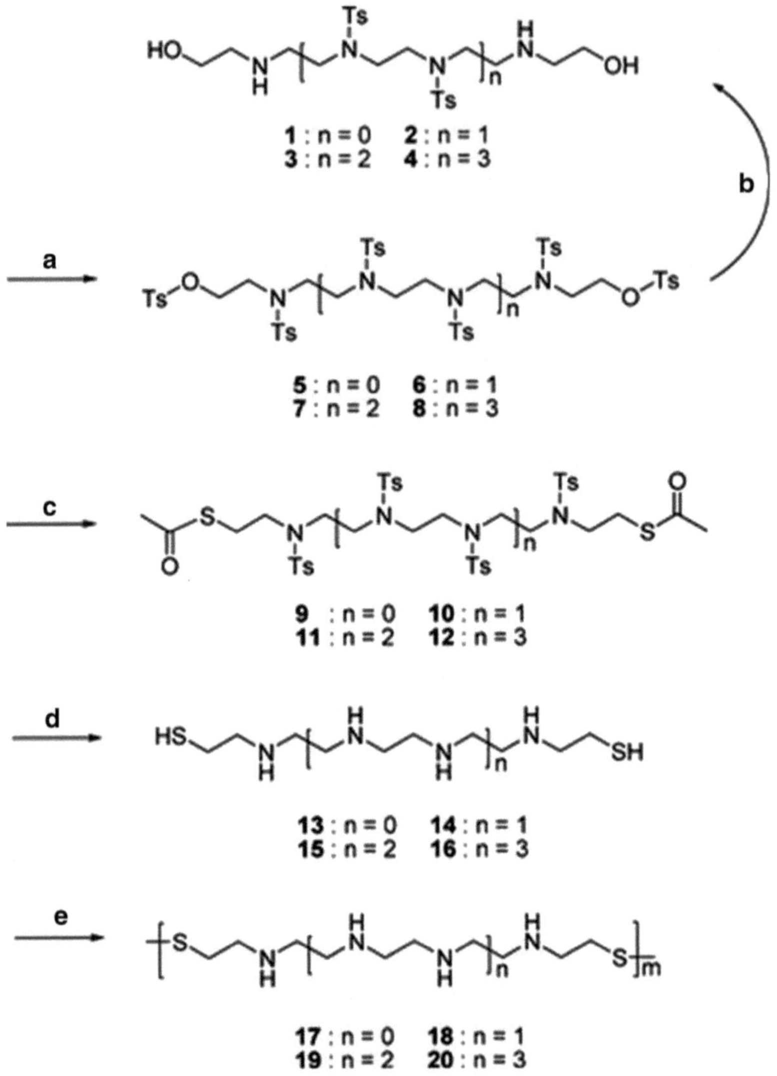

Scheme 1. The outline of preparing linear reducible PEI (reproduced from ref. 33 with permission of ACS).

5. Polymerization (step e): A solution of 13 (or 14, or 15, or 16) (100 mM) in water is stirred at room temperature for 2 days under oxygen pressure (1 atm), then water is removed, the residue is washed with diethyl ether, and dried to give the final polymer products.

The outline for preparing cross-linked reducible PEI is shown in Scheme 2 (34); the detailed procedure is as follows:

1. Synthesis of thiolated PEI: PEI (1.0 g, 800 Da) is dissolved in 5 mL of deionized water, and then hydrochloric acid (0.1 M) is added dropwise to the PEI solution until pH is 7.2. The yellow solid is obtained via removal of water, then dissolved in

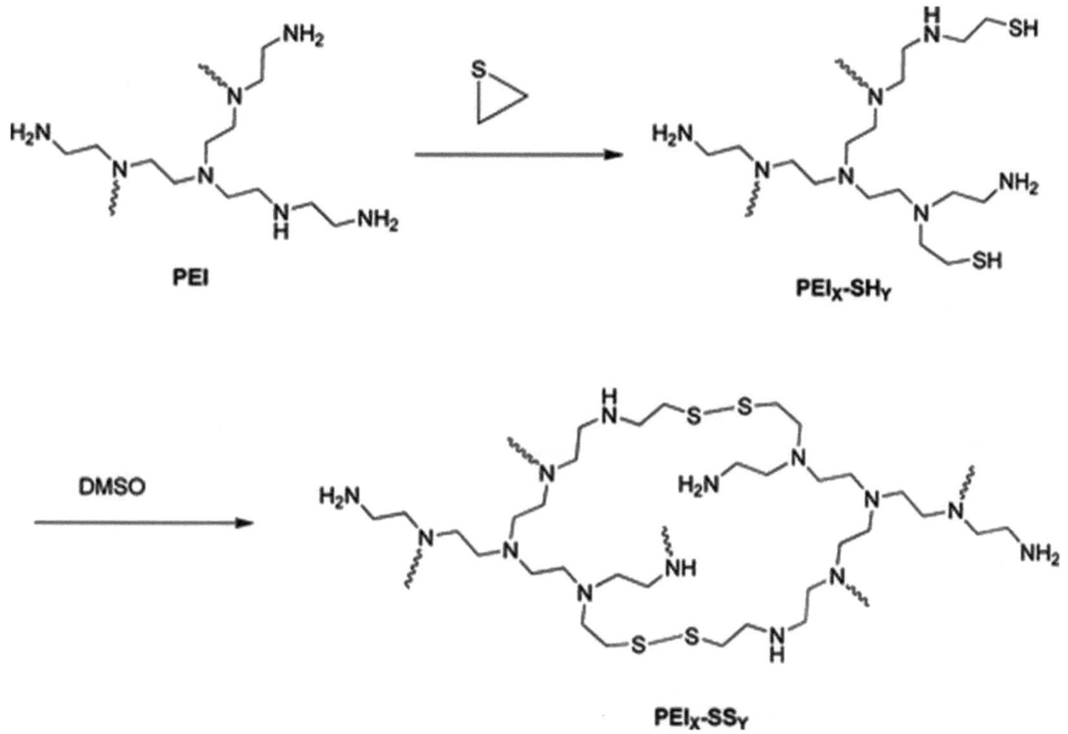

Scheme 2. The outline of preparing cross-linked linear reducible PEI (reproduced from ref. 34 with permission of ACS).

3 mL of methanol, and transferred to an ampule. After the container is purged with argon for 5 min, a calculated amount of thiirane is added. The ampule is sealed and kept in a 50°C oil bath for 24 h. Then the mixture is transferred to a flask, evaporated to dryness, and stored under argon.

2. Synthesis of disulfide cross-linked PEI: Thiolated PEI (0.5 g) is dissolved in 3 mL of methanol and 1.5 mL of dimethyl sulfoxide (DMSO) and is stirred for 48 h. The reaction solution is precipitated in diethyl ether three times and the product is dried under high vacuum to give a light yellow viscous liquid or solid.

3.2. Synthesis of Linear and Branched PDMAEMA

Methacrylate- and acrylate-based polycations (such as PDMAEMA) have been widely investigated due to convenient synthesis by free-radical polymerization and the possibility to optimize their properties by copolymerization of a number of ionic and nonionic comonomers. However, high cytotoxicity impeded PDMAEMA prospects in gene delivery and prompted the development of bioreducible PDMAEMA. Bioreducible PDMAEMA can be prepared as follows (9): 1,4-bis(2-(thiobenzoylthio)prop-2-yl)benzene (BTBP)-mediated RAFT polymerization of DMAEMA produces α,ω-dithioester-functionalized PDMAEMA, which is then converted

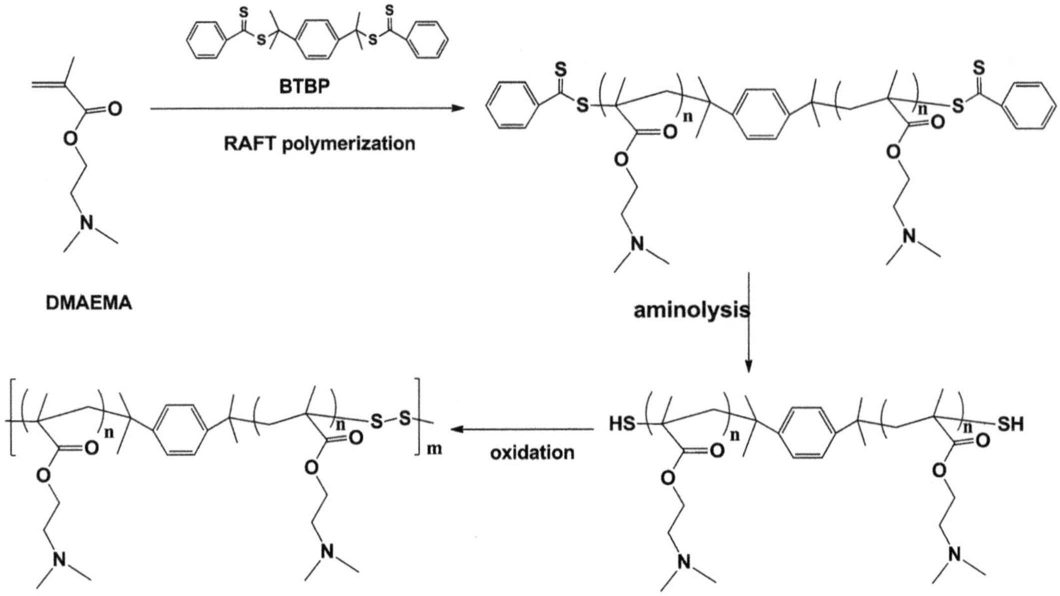

Scheme 3. Synthesis of linear reducible PDMAEMA (reproduced from ref. 9 with permission of Elsevier).

into α,ω-dithiol-ended PDMAEMA oligomer by aminolysis, and bioreducible PDMAEMA is synthesized by oxidation of the terminal thiol (as shown in Scheme 3). The typical procedure is as follows:

1. Synthesis of the α,ω-dithioester-functionalized PDMAEMA: BTBP-mediated RAFT polymerization of DMAEMA is performed in THF at 60°C. The polymerization solution containing DMAEMA (1.0 g), BTBP (80.0 mg), and AIBN (AIBN:BTBP = 1:20) is added into a glass ampoule, thoroughly deoxygenated, sealed under vacuum, and placed in a thermostated water bath at 60°C for 48 h. The polymer, α,ω-dithioester-terminated PDMAEMA, is obtained by precipitation into hexane and isolated by filtration.
2. Synthesis of the α,ω-dithiol-ended PDMAEMA oligomer: The α,ω-dithioester-functionalized PDMAEMA (0.5 g) is dissolved in THF (4 mL) containing a few drops of aqueous sodium bisulfite ($Na_2S_2O_4$), and hexylamine (0.2 mL) is added after the reaction mixture is purged of oxygen by bubbling with N_2 for 30 min. The reaction mixture is stirred for 3 h under N_2 and then added dropwise to a tenfold hexane and the polymer is collected by filtration.
3. Synthesis of linear reducible PDMAEMA: Hexylamine (0.2 mL) and DMSO (0.2 mL) are added to solution of the α,ω-dithiol-ended PDMAEMA oligomer in methanol, and the reaction mixture is stirred at room temperature under oxygen atmosphere. The solvent is then removed and bioreducible

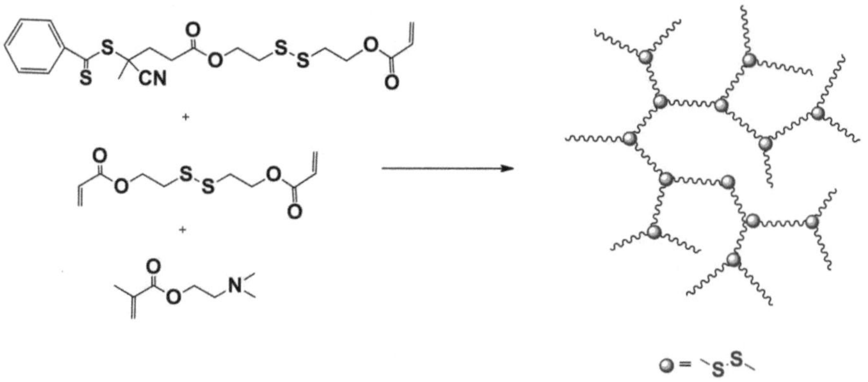

Scheme 4. Synthesis of hyperbranched reducible PDMAEMA (reproduced from ref. 31 with permission of ACS).

PDMAEMA is dissolved in THF and then added dropwise to hexane; the polymer is collected by filtration (Scheme 4).

4. Synthesis of hyperbranched bioreducible PDMAEMA: 1,2-Bis(2-(3-methylbuta-1,3-dien-2-yloxy)ethyl) disulfane-mediated RAFT polymerization of DMAEMA in the presence of biodegradable DSDMA produces PDMAEMA with hyperbranched structure; each branch point in the polymer is linked by a disulfide bond, and the degree of branching (DB) can be tuned by the amount of DSDMA (31). The detailed procedure is as follows: DMAEMA (316 mg, 2.0 mmol), 1,2-bis(2-(3-methylbuta-1,3-dien-2-yloxy) ethyl) disulfane (24 mg, 0.05 mmol), DSDMA (29 mg, 0.1 mmol), and AIBN (1.64 mg, 0.01 mmol) are dissolved in 3.0 mL of dimethylacetamide. Aliquots are transferred to three different vials, which are then sealed with rubber septa. Each vial is deoxygenated by purging with nitrogen for 30 min prior to placement in water bath at 70°C. The vials are taken out at 20, 42, and 65 h. Immediate cooling is applied via an ice-water bath and radical quenching is applied via air exposure. The polymers are collected after precipitation twice from dichloromethane to hexane and then dried under vacuum.

3.3. Syntheses of Bioreducible Poly(Amido Amine)s and Poly(Amino Ester)s Via Michael-Addition Polymerization

Via Michael-addition polymerization between amines and cystamine bisacrylamide (CBA), a series of linear and hyperbranched bioreducible poly(amido amine)s have been developed. Polymerization conditions can control the topology of the produced polymers.

3.3.1. Control of the Topology of Reducible Poly(Amino Ester)s by Temperature (32)

Generally, when diamine, triamine, or multiamine monomers react with CBA, linear polymers are obtained. But, in some cases, Michael-addition polymerization of disulfide-based diacrylate and equimolar *N*-methylethylenediamine forms ABB′-type intermediates first; 2° amines (formed) are inactive below 40°C, leading to the formation of linear poly(amino ester) via AB-type intermediates (B do not participate in the reaction). However, elevated temperature activates 2° amines (formed), and they participate in the addition reaction, which results in the formation of hyperbranched bioreducible poly(amino ester)s via ABB -type intermediates (both B and B participate in the reaction). The degree of branching of the hyperbranched polymers obtained increases with the increase of temperature. Therefore, bioreducible polymer topology from linear to hyperbranched can be tuned simply by varying the polymerization temperature. The detailed polymerization procedure is as follows: disulfide-based diacrylate (0.524 g, 2.0 mmol) and *N*-methylethylenediamine (0.148 g, 2.0 mmol) are added into 2 mL of chloroform. After the polymerization is performed in the dark at 25°C (or 50°C) for a certain time, the reaction mixture is poured into 50 mL of diethyl ether under vigorous stirring. The polymer is collected and purified by reprecipitation from a chloroform solution into diethyl ether followed by drying under vacuum for 1 day at room temperature (Schemes 5 and 6).

3.3.2. Control of the Topology of Reducible Poly(Amido Amine)s by Monomer Feed Ratio

The amino units in a trifunctional amine have different reactivity in Michael-addition polymerization of trivalent amine with bisacrylamide/bisacrylate; hence the topology of the produced polymers can be tuned simply via varying reaction conditions such as the molar ratio of trivalent amine to bisacrylamide/bisacrylate. Generally, in Michael-addition polymerization of 1-(2-aminoethyl) piperazine (AEPZ) with equal molar *N*,*N'*-CBA, AB-type intermediate

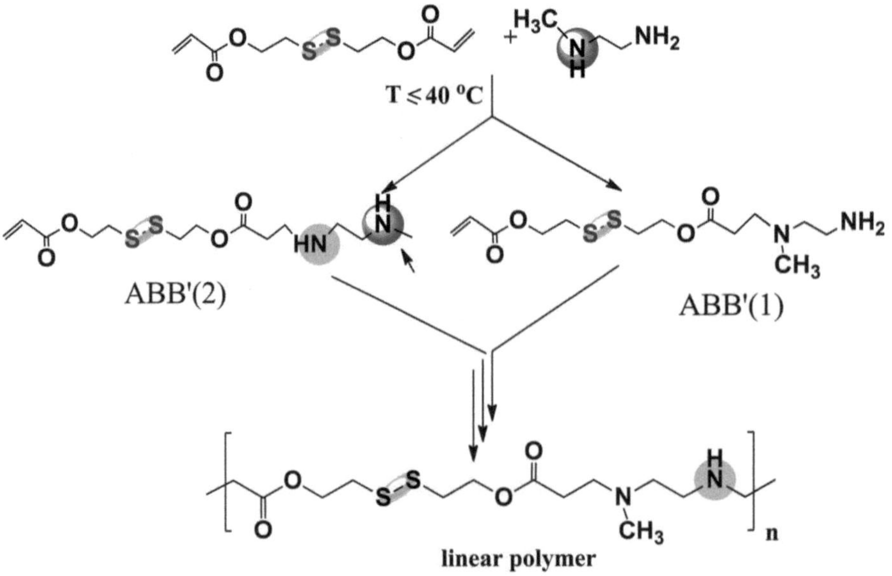

Scheme 5. Synthesis of linear reducible poly(amino ester) (reproduced from ref. 32 with permission of ACS).

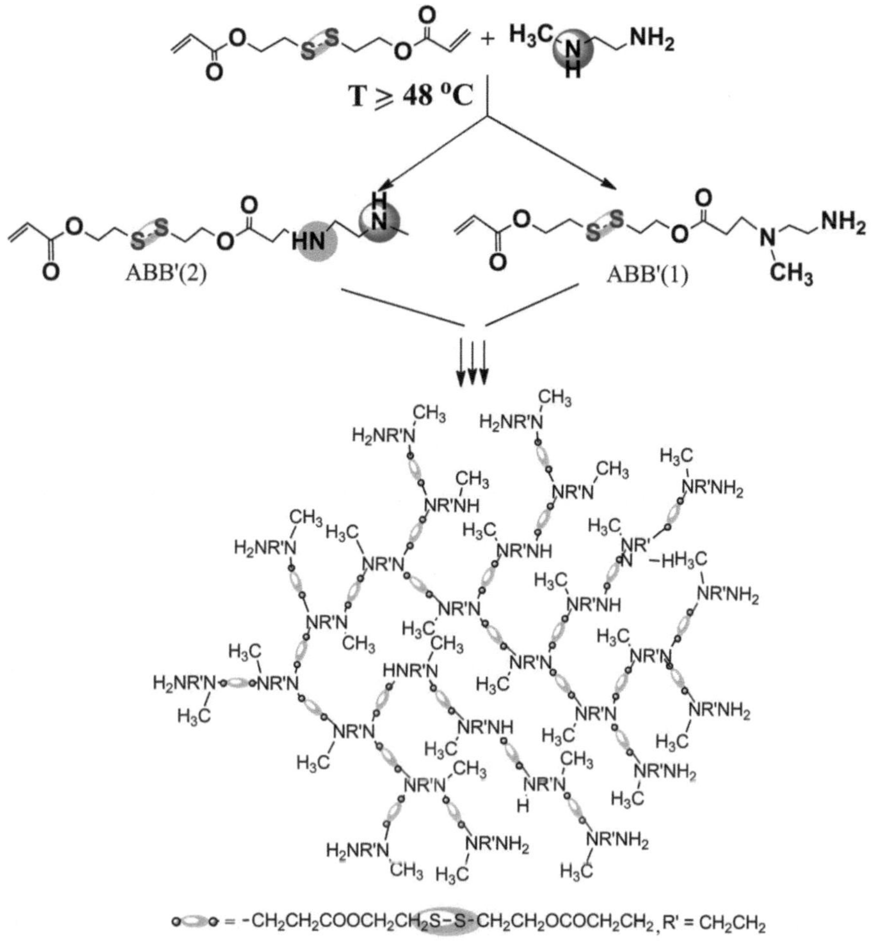

Scheme 6. Synthesis of hyperbranched reducible poly(amino ester) (reproduced from ref. 32 with permission of ACS).

forms at the starting stage and 2° amine (produced during the polymerization) does not take part in reaction due to high steric hindrance, which results in the formation of linear poly(amido amine) as shown in Scheme 7. The detailed procedure is as follows: CBA (260 mg, 1.0 mmole) and AEPZ (129.2 mg, 1.0 mmole) are added into a vial and dissolved in methanol/water mixture (3.0 mL, 8/2, v/v), and then the polymerization is performed in the dark at 50°C. The reaction is allowed to proceed for 5 days yielding a viscous solution. The resulting bioreducible polymer is obtained via precipitating in cool acetone and drying under vacuum for 4 h at room temperature.

However, in Michael-addition polymerization of AEPZ with double molar CBA, A_2B-type intermediate is produced at the starting stage, and 2° amine (produced during the polymerization) participates

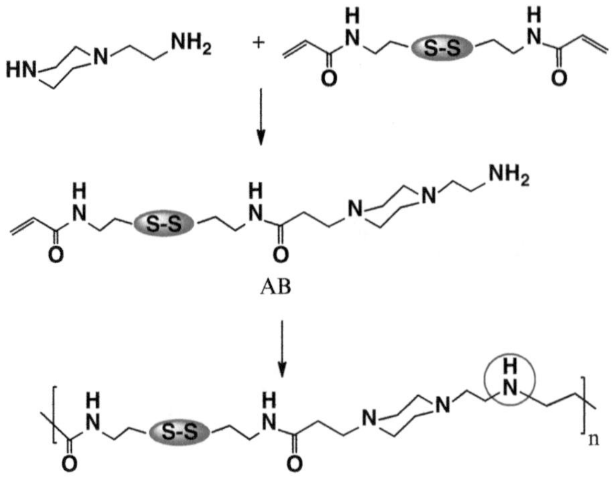

Scheme 7. Synthesis of linear reducible poly(amido amine) (reproduced from ref. 35 with permission of ACS).

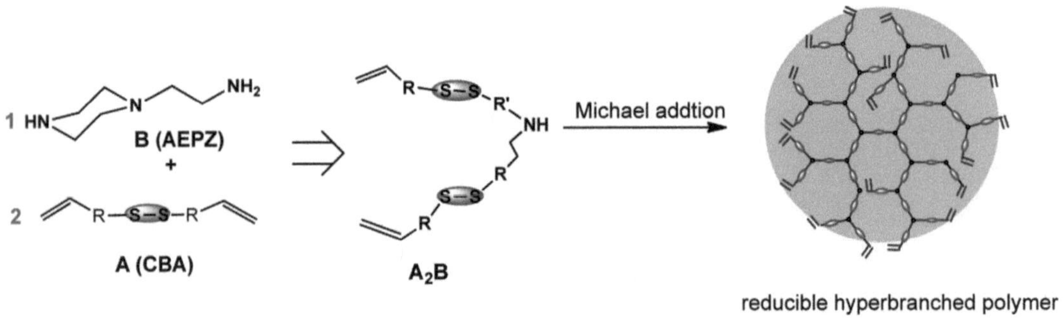

Scheme 8. Synthesis of hyperbranched reducible poly(amido amine) (reproduced from ref. 35 with permission of ACS).

in the polymerization in later stages, which leads to the formation of hyperbranched reducible poly(amido amine) (Scheme 8).

In a typical experiment, CBA (520 mg, 2.0 mmole) and AEPZ (129 mg, 1.0 mmole) are added into a vial and dissolved in methanol/water mixture (5.0 mL, 8/2, v/v). After the polymerization has been performed in the dark at 50°C for 120 h, the polymerization is stopped via decreasing the temperature to room temperature. The resulting reducible polymer is obtained via precipitating in cool acetone and drying under vacuum for 3 h at room temperature.

4. Notes

1. In preparing linear reducible PDMAEMA, the molar ratio of BTBP/AIBN should be over 20; otherwise the end units of some PDMAEMAs are not thioester (9).

2. In preparing hyperbranched reducible PDMAEMA, the molar percentage of bioreducible DSDMA should be below 50%; otherwise, cross-linked PDMAEMA forms (31).
3. In Michael-addition polymerization, the monomer concentration should not be too high (<300 mg/mL); otherwise cross-linking could take place (21, 23, 25).
4. In the case of monomer feed ratio control of the topology of bioreducible poly(amido amine)s, the use of polar solvents may, under equivalent feed ratio, also produce hyperbranched polymers (30).
5. In control over the topology of bioreducible poly(amido amine) via monomer feed ratio part, in some cases, especially under high temperature, equivalent feed ratio also produces hyperbranched polymers.
6. Bioreducible poly(amido amine)s must be kept under 20°C. Higher temperature will lead to cross-linking (21, 23, 25).

References

1. Li S, Huang L (2000) Nonviral gene therapy: promises and challenges. Gene Ther 7:31–34
2. Han S, Mahato RI, Sung YK, Kim SW (2000) Development of biomaterials for gene therapy. Mol Ther 2:302–317
3. Merdan T, Kopecek J, Kissel T (2002) Prospects for cationic polymers in gene and oligonucleotide therapy against cancer. Adv Drug Deliv Rev 54:715–758
4. Pack DW, Hoffman AS, Pun S, Stayton PS (2005) Design and development of polymers for gene delivery. Nat Rev Drug Discov 4:581–593
5. Akinc A, Anderson DG, Lynn DM, Langer R (2003) Synthesis of poly(beta-amino ester)s optimized for highly effective gene delivery. Bioconjug Chem 14:979–988
6. Kim YH, Park JH, Lee M, Park TG, Kim SW (2005) Polyethylenimine with acid-labile linkages as a biodegradable gene carrier. J Control Release 103:209–219
7. Christensen LV, Chang CW, Yockman JW, Conners R, Jackson H, Zhong ZY, Feijen J, Bu DA, Kim SW (2007) Reducible poly(amido ethylenediamine) for hypoxia-inducible VEGF delivery. J Control Release 118:254–261
8. Jeong JH, Christensen LV, Yockman JW, Zhong ZY, Engbersen JFJ, Kim WJ, Feijen J, Kim SW (2007) Reducible poly(amido ethylenimine) directed to enhance RNA interference. Biomaterials 28:1912–1917
9. You YZ, Manickam DS, Zhou QH, Oupicky D (2007) Reducible poly(2-dimethylaminoethyl methaerylate): synthesis, cytotoxicity, and gene delivery activity. J Control Release 122:217–225
10. Wong SY, Pelet JM, Putnam D (2007) Polymer systems for gene delivery-past, present, and future. Prog Polym Sci 32:799–837
11. Lin C, Zhong ZY, Lok MC, Jiang XJ, Hennink WE, Feijen J, Engbersen JFJ (2007) Random and block copolymers of bioreducible poly(amido amine)s with high- and low-basicity amino groups: study of DNA condensation and buffer capacity on gene transfection. J Control Release 123:67–75
12. Lin C, Zhong ZY, Lok MC, Jiang XL, Hennink WE, Feijen J, Engbersen JFJ (2006) Linear poly(amido amine)s with secondary and tertiary amino groups and variable amounts of disulfide linkages: synthesis and in vitro gene transfer properties. J Control Release 116:130–137
13. Lin C, Zhong ZY, Lok MC, Jiang XL, Hennink WE, Feijen J, Engbersen JFJ (2007) Novel bioreducible poly(amido amine)s for highly efficient gene delivery. Bioconjug Chem 18:138–145
14. Wang YX, Chen P, Shen JC (2006) The development and characterization of a glutathione-sensitive cross-linked polyethylenimine gene vector. Biomaterials 27:5292–5298

15. Gosselin MA, Guo WJ, Lee RJ (2001) Efficient gene transfer using reversibly cross-linked low molecular weight polyethylenimine. Bioconjug Chem 12:989–994
16. Peng Q, Zhong ZL, Zhuo RX (2008) Disulfide cross-linked polyethylenimines (PEI) prepared via thiolation of low molecular weight PEI as highly efficient gene vectors. Bioconjug Chem 19:499–506
17. Son S, Singha K, Kim WJ (2010) Bioreducible BPEI-SS-PEG-cNGR polymer as a tumor targeted nonviral gene carrier. Biomaterials 31:6344–6354
18. Sun YX, Zeng X, Meng QF, Zhang XZ, Cheng SX, Zhuo RX (2008) The influence of RGD addition on the gene transfer characteristics of disulfide-containing polyethyleneimine/DNA complexes. Biomaterials 29:4356–4365
19. Ou M, Xu RZ, Kim SH, Bull DA, Kim SW (2009) A family of bioreducible poly(disulfide amine)s for gene delivery. Biomaterials 30:5804–5814
20. Kim TI, Lee M, Kim SW (2010) A guanidinylated bioreducible polymer with high nuclear localization ability for gene delivery systems. Biomaterials 31:1798–1804
21. Chen J, Wu C, Oupicky D (2009) Bioreducible hyperbranched poly(amido amine)s for gene delivery. Biomacromolecules 10:2921–2927
22. Lin C, Blaauboer CJ, Timoneda MM, Lok MC, van Steenbergen M, Hennink WE, Zhong ZY, Feijen J, Engbersen JFJ (2008) Bioreducible poly(amido amine)s with oligoamine side chains: synthesis, characterization, and structural effects on gene delivery. J Control Release 126:166–174
23. Wan L, You Y, Zou Y, Oupicky D, Mao GZ (2009) DNA release dynamics from bioreducible poly(amido amine) polyplexes. J Phys Chem B 113:13735–13741
24. Lin C, Engbersen JFJ (2008) Effect of chemical functionalities in poly(amido amine)s for non-viral gene transfection. J Control Release 132:267–272
25. Blacklock J, You YZ, Zhou QH, Mao GZ, Oupicky D (2009) Gene delivery in vitro and in vivo from bioreducible multilayered polyelectrolyte films of plasmid DNA. Biomaterials 30:939–950
26. Piest M, Lin C, Mateos-Timoneda MA, Lok MC, Hennink WE, Feijen J, Engbersen JFJ (2008) Novel poly(amido amine)s with bioreducible disulfide linkages in their diamino-units: structure effects and in vitro gene transfer properties. J Control Release 130:38–45
27. Meng FH, Hennink WE, Zhong ZY (2009) Reduction-sensitive polymers and bioconjugates for biomedical applications. Biomaterials 30:2180–2198
28. Wang RB, Zhou LZ, Zhou YF, Li GL, Zhu XY, Gu HC, Jiang XL, Li HQ, Wu JL, Guo XQ, Zhu BS, Yan DY (2011) Synthesis and gene delivery of poly(amido amine)s with different branched architecture. Biomacromolecules 11:489–495
29. Stiriba SE, Frey H, Haag R (2002) Dendritic polymers in biomedical applications: from potential to clinical use in diagnostics and therapy. Angew Chem Int Ed 41:1329–1334
30. You YZ, Yu ZQ, Cui MM, Hong CY (2010) Preparation of photoluminescent nanorings with controllable bioreducibility and stimuli-responsiveness. Angew Chem Int Ed 49:1099–1102
31. Tao L, Liu JQ, Tan BH, Davis TP (2009) RAFT synthesis and DNA binding of biodegradable, hyperbranched poly(2-dimethylamino)ethyl methacrylate. Macromolecules 42:4960–4962
32. Hong CY, You YZ, Wu DC, Liu Y, Pan CY (2007) Thermal control over the topology of cleavable polymers: from linear to hyperbranched structures. J Am Chem Soc 129:5354
33. Lee Y, Mo H, Koo H, Park JY, Cho MY, Jin GW, Park JS (2007) Visualization of the degradation of a disulfide polymer, linear poly(ethylenimine sulfide), for gene delivery. Bioconjug Chem 18:13–18
34. Peng Q, Hu C, Cheng J, Zhong ZL, Zhuo RX (2009) Influence of disulfide density and molecular weight on disulfide cross-linked polyethylenimine as gene vectors. Bioconjug Chem 20:340–346
35. You YZ, Hong CY, Pan CY (2009) Facile One-Pot approach for preparing dually responsive core-shell nanostructure. Macromolecules 42:573–575

Chapter 10

Lyophilization of Synthetic Gene Carriers

Julia Christina Kasper, Sarah Küchler, and Wolfgang Friess

Abstract

Lyophilization, also known as freeze-drying, is a widely used method for stabilization, improvement of long-term storage stability, and simplification of the handling of drugs and/or carrier systems. Lyophilization is time- and energy-consuming; hence, optimized processes are required to avoid time loss and higher costs without compromising product stability. Since the last decade nonviral, synthetic carriers for gene delivery are of increasing interest. However, these systems suffer from poor physical stability in aqueous solution or suspension. Hence, to ensure long-term storage stability lyophilization of the gene carrier systems is favored. Though, lyophilized products retrieving original carrier size and transfection efficiency after reconstitution are mandatory. This chapter gives an overview of the basic steps and troubleshooting for successful lyophilization of synthetic gene carriers. Furthermore the required excipients and their mechanism of action are summarized.

Key words: Lyophilization, Freeze-drying, Synthetic gene carriers, Stabilization, Formulation, Long-term stability

1. Introduction

Synthetic gene delivery systems open up new treatment options for various diseases such as cancer or hereditary disorders. A great variety of gene carrier systems like lipoplexes, DEMA-polymers, polyplexes, etc. are described in the literature (1–3); for review see ref. 4. However, these nonviral vectors suffer from physical instability in an aqueous environment and are highly susceptible to aggregation and particle size growth over time which impairs their transfection efficiency. By freeze-drying, long-term stability of the formulations can be obtained and reproducible reconstitution is possible by simply adding water.

Lyophilization, also known as freeze-drying, is very complex and the process and formulation need to be adjusted individually

Manfred Ogris and David Oupicky (eds.), *Nanotechnology for Nucleic Acid Delivery: Methods and Protocols*, Methods in Molecular Biology, vol. 948, DOI 10.1007/978-1-62703-140-0_10, © Springer Science+Business Media, LLC 2013

Table 1
Commonly used excipients in lyophilization processes

Stabilizers	Function	Examples (most commonly used)
Cryoprotectants	Protection during freezing	Sugars/polyols (sucrose, trehalose, etc.; conc. ≥0.3 M), polymers (polyethylene glycol, dextran, etc.), amino acids (glycin, proline, etc.)
Lyoprotectants	Protection during drying and/or storage	Sugars/polyols (sucrose, glucose), polymers (polyvinylpyrrolidone (PVP), etc.), amino acids (proline, alanine, etc.)
Surfactants	Prevention of physical and/or chemical instabilities such as agglomeration, bindings, etc.	Polysorbate 20, polysorbate 80, poloxamer
Bulking agents	Providing pharmaceutical elegant cake structure, prevention of "cake collapse," required for low doses/mass per vial	Mainly mannitol, hydroxylethyl starch, trehalose, sorbitol, lactose, etc.
Antioxidants	Prevention of oxidation damage	Ascorbic acid, cysteine hydrochloride, sulfites
Tonicity modifiers	Providing isotonicity for increased local and systemic tolerability (approximately 280 mOsm/L)	Sugars: Sucrose, mannitol, glycine, sodium chloride (caution: not recommended due to potential interference with charge interactions stabilizing the carrier)

for each delivery system. Freeze-drying generates stress especially during freezing and dehydration which can influence the particle stability. Stabilizers are capable to prevent aggregation and particle size growth during lyophilization, reduce freezing (cryoprotectant) and drying (lyoprotectant) stress, increase storage stability, and maintain transfection rates comparable to freshly prepared preparations (5, 6). An overview of typical stabilizers is provided in Table 1. It is noteworthy that some excipients exhibit different stabilizing functions simultaneously such as sugars which are cryo- and lyoprotectants at the same time. However, sometimes a combination of excipients may be necessary for sufficient stabilization (7, 8). Important parameters to ensure optimal results after lyophilization are the choice of stabilizers and the excipient/delivery system ratio. To identify the ideal stabilizer, freeze–thaw studies should be performed prior to lyophilization. The mechanisms of protection are not completely understood yet; a number of hypotheses are being discussed such as the vitrification of sugars at the glass transition temperature T_g' providing a glassy matrix which retards molecular motion and decreases particle interaction (9). T_g' of an amorphous substance is the critical temperature at which the material changes its behavior from a hard and relatively brittle state into an elastic/ rubberlike state. It is one of the most important parameters for the optimization of a lyophilization process.

The typical lyophilization process can be divided into three stages: freezing (solidification), primary drying (sublimation), and secondary drying (desorption). During freezing, the solvent is separated from the solutes to form ice and the solutes become highly concentrated. This phase takes up to several hours. Primary drying (or ice sublimation) is induced by reducing the chamber pressure below the vapor pressure of ice and increasing the shelf temperature to provide the energy that is needed for sublimation. Thereby, ice is transferred from the product to the condenser by sublimation and crystallization on the condenser. This is the most time-consuming stage. During secondary drying, residual water is desorbed at increased shelf temperature and low pressure aiming for the optimal residual moisture content which is typically <1% (10, 11).

2. Materials

This section lists the materials required for successful lyophilization (see also Table 1). All solutions should be prepared using purified water and analytical grade reagents. As the composition of an effective formulation might vary between different synthetic gene carriers the listed excipients are only suggestions; variations might be necessary. In general, depending on the intended application, e.g., in clinical trials, approved pharmaceutical excipients must be used.

1. Freeze-drier: The freeze-drier should comprise a set of shelves that can be cooled to −50°C or below and heated above ambient temperature for primary and secondary drying. Defined temperature ramps should be possible; a condenser usually cooled to −60°C or below with sufficient capacity for the collection of the sublimed ice and a high-performance vacuum pump for pressure reduction (see Note 1). Temperature probes (T-type thermocouples or resistance thermodetectors) for endpoint detection and process monitoring are recommended.
2. Lyophilization trays that are consistent in size with the dimensions of the shelves (see Note 2).
3. 2R lyophilization glass vials with matching lyophilization stoppers (see Note 3).
4. Formulation buffer: For example, 10 mM histidine buffer pH 6.2 (see Note 4).
5. Synthetic gene carrier stock solution: Prepare a stock solution of the synthetic gene carrier (see Notes 5 and 6) in the formulation buffer.

6. Stabilizer stock solution: Prepare a stabilizer stock solution (e.g., 20% [m/v] sucrose) in the formulation buffer. The preselection of excipients for cryoprotection needs to be evaluated previously in a freeze–thaw study (see Subheading 3.1).

3. Methods

In a first step, freeze–thaw studies should be performed to evaluate the sensitivity of the synthetic gene carrier against freeze-thawing-induced stresses and to preselect suitable cryoprotectants. The following protocol describes the lyophilization of synthetic gene carriers using a simple standard formulation and a conservative freeze-drying cycle. Synthetic gene carriers are highly diverse; thus, some gene delivery systems can be lyophilized easily, and other more stress-sensitive carrier systems require more scientific effort and process/formulation optimization. For further formulation and process variations the reader is referred to the note section. Figure 1 shows a schematic depiction of the work flow when synthetic gene carriers are lyophilized for the first time.

3.1. Freeze–Thaw Studies

1. Prior to lyophilization, suitable cryoprotectants need to be evaluated by freeze–thaw studies. As mentioned above mono-, di-, and oligosaccharides as well as polymers are potential stabilizers and a screening of various potential excipients is required. However, when the synthetic gene carrier is more sensitive against dehydration some excipients might be promising in freeze–thaw studies but might be not the optimal choice for lyophilization. Freeze–thaw studies have to be performed under the same conditions that will be applied during lyophilization. In a first step freeze–thaw studies should be performed only with few formulations: one without excipient and 3–4 samples with increasing concentrations (e.g., 5, 10, and 20% [m/V]) of well-established excipients such as sucrose.
2. Subsequently, freeze-thawing is performed once or several (5–10) times in a pilot scale freeze-drier. Samples are frozen at

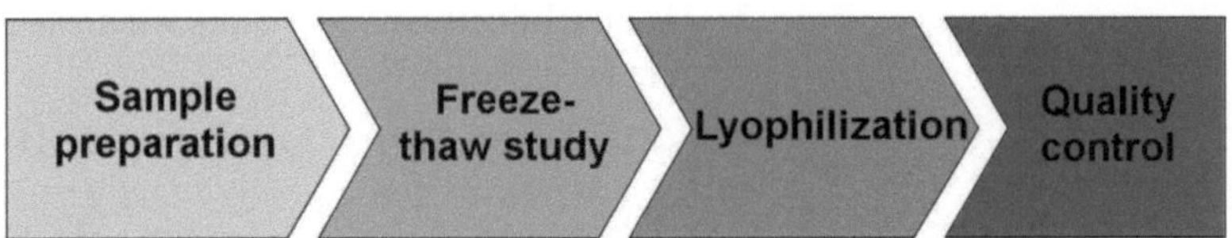

Fig. 1. Schematic work flow of a standard lyophilization process for synthetic gene carriers.

a typical freezing rate of −1°C/min down to −50°C. After 1 h at −50°C thawing is induced at a rate of 1°C/min up to 10°C. After 1 h at 10°C the next freeze–thaw cycle is started (see Note 7). The maximum/minimum number of those freeze–thaw cycles can be adjusted according to the respective purpose.

3. For data evaluation, quality attributes such as particle size, polydispersity, and transfection efficiency of the synthetic gene carriers after freeze-thawing need to be determined (see Subheading 3.4).
4. Based on these results the suitable excipient and the mass ratio of stabilizer to synthetic gene carriers can be assessed which is crucial for sufficient stabilization (see Note 8). If only low amounts of excipients (<10% [m/V]) are necessary for sufficient stabilization the use of disaccharides such as sucrose or trehalose is favored. In this case isotonicity of the formulations can be easily adjusted by varying the amount of stabilizer. If increased amounts (>10% [m/V]) of excipients are required, the use of disaccharides will lead to hypertonic formulations. In this case, oligo-/polysaccharides or polymers such as cyclodextrins or PVP can be used to obtain isotonic formulations (see Note 9). Moreover, if the synthetic gene carriers are highly susceptible for freeze-thawing-induced stresses the addition of low amounts of surfactants such as polysorbates or poloxamer can be considered.
5. Select the most promising compositions for lyophilization which need to withstand at least one freeze–thaw cycle.

3.2. Sample Preparation and Loading of the Freeze-Drier

1. Sample preparation: Mix the gene carrier stock solution with an adequate volume of stabilizer stock solution (evaluated in Subheading 3.1) in a reaction tube (see Note 6).
2. Preparation of a placebo formulation (same composition as samples but without any synthetic gene carriers): Mix the formulation buffer with an adequate volume of stabilizer stock solution in a reaction tube. The placebo formulation is used to fill the surrounding sample vials in order to avoid edge vial effects (see Note 10). For cost and materials saving reasons they can also be used as references for supplementary quality control parameters (see Subheading 3.3).
3. Placement (see Note 11) of empty lyophilization vials on the tray before filling helps to prevent the vials from tipping over (see Fig. 2).
4. Filling: Fill the sample (minimum 500 μL) into the vials in the center of the tray and fill the outer vials with the same volume of placebo formulation. Ensure that all sample vials are surrounded by other sample or placebo vials (see Fig. 2).

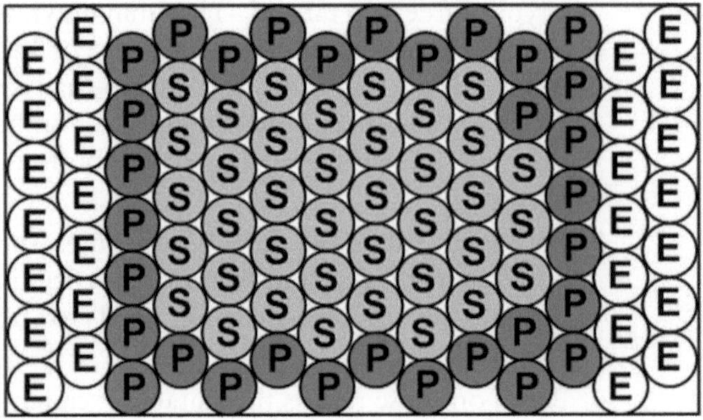

Fig. 2. Placement and filling of the vials on the lyophilization tray. First, empty vials (E) are placed on the whole tray in a hexagonal array. Then, sample vials (S) and placebo/"dummy" vials (P) are placed so that all sample vials are surrounded by "dummy" vials.

Fig. 3. Pictures of (**a**) a filled vial with inserted stopper in correct position prior to lyophilization, (**b**) a lyophilizate/cake after lyophilization, and (**c**) correct thermocouple placement in a filled vial.

5. Placement of stoppers on vials: Insert the shank of the stopper into the neck of the vial (see Fig. 3a). The bigger part of the cavity in the shank should be still outside the vial to enable the sublimation of the water (see Note 12).
6. Placement of thermocouples or resistance thermodetectors into selected samples (see Note 13): Ensure that the tip of the temperature probe is in the center bottom position of the vials as this is the position of latest drying (see Fig. 3c). If several temperature probes are available arrange them in vials at different locations throughout the tray.
7. Place the filled lyophilization tray onto the shelf of the freeze-drier.

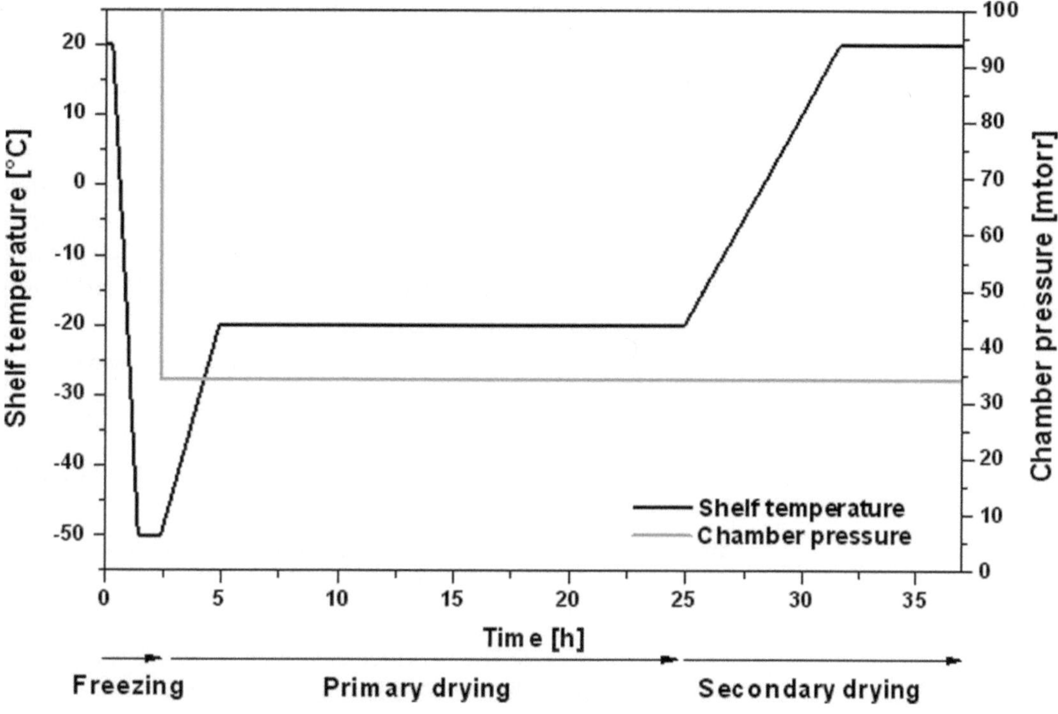

Fig. 4. Schematic depiction of a conservative lyophilization cycle. The lyophilization cycle can be divided into three stages: freezing, primary and secondary drying.

8. Connect thermocouples or resistance thermodetectors. Ensure that the temperature probes are not displaced after loading.
9. Remove the bottom of the tray if a tray with removable bottom is used.

When loading is completed, close the door (see Note 14) of the freeze-drier and ensure that the drain outlet and all other valves like the backfilling valve are properly closed.

3.3. Lyophilization

Start the following lyophilization cycle (Fig. 4. depicts the cycle scheme):

1. Freezing: Equilibrate the samples for 15 min at a shelf temperature of 20°C. Freezing is initiated by cooling of the shelves at a ramp of –1°C/min to –50°C (see Note 15). Hold this temperature for 1 h.
2. Primary drying: Apply vacuum to the chamber and reduce the pressure to, e.g., 34 mTorr. At the same time shelf temperature is increased at a ramp rate of 0.2°C/min to –20°C. These parameters determine a very conservative lyophilization cycle. The condenser temperature should always be –60°C or less.

For the theory behind primary drying and process variations see Note 16.

3. Endpoint determination of primary drying: Determine the endpoint of primary drying using product temperature data obtained by the temperature probes (see Note 17). The end of sublimation is indicated by an alignment of product and shelf temperature. As formulations in vials without thermocouples differ from vials with thermosensors, a "delay time" of approximately 15% of primary drying time should be included before secondary drying is initiated.
4. Secondary drying: Increase shelf temperature slowly at a ramp rate of 0.1°C/min to 20°C and hold there for 8 h (see Note 18).
5. Backfilling: Finally, backfill the product chamber with inert gas, e.g., nitrogen, to 600 mTorr.
6. Stoppering: Stopper the vials inside the chamber under the slight vacuum, e.g., 600 mTorr (see Note 19), using a manual or an automatic mechanism to converge the shelves.
7. Unloading: Aerate the product chamber to atmospheric pressure, disconnect thermocouples, and unload the tray.

3.4. Quality Control

In order to evaluate the suitability of the chosen formulations and the effect of the freeze-thawing or lyophilization process on the synthetic gene carriers quality control is required. After reconstitution of the samples' particle size, polydispersity and transfection efficiency should be performed. If aiming for a parenteral product further analytics such as determination of subvisible particles via light obscuration or microscopy and visual inspection supplemented by micro-flow imaging (MFI), nanoparticle tracking analysis (NTA), or asymmetrical flow field flow fractionation (AF4) should be assessed. Moreover, cake appearance and residual moisture content of the samples indicate the appropriateness of the lyophilization process.

1. Reconstitution: Reconstitute (see Note 20) the lyophilizate with the same amount of purified water compared to the original filling volume. Note the time that is required for the complete reconstitution and note whether any turbidity remains upon reconstitution. If necessary, reconstitution can be performed with less volume to increase the carrier concentration or with another liquid for adjustment of, e.g., isotonicity, pH, or ionic strength.
2. Particle size and polydispersity: Particle size and polydispersity are the most important quality parameters to investigate alterations upon freeze-thawing or lyophilization. These parameters can be determined by dynamic light scattering (DLS) and should be compared to the particle size and polydispersity of

freshly prepared samples. Additionally, compare the particle sizes with the freeze-thawed samples in order to assess if the synthetic gene carrier is more sensitive to freeze-thawing or dehydration-induced stresses.

3. Transfection efficacy: Maintenance of transfection efficiency upon freeze-thawing or lyophilization should be assessed in cell culture. For comparison, analyze freshly prepared samples and optionally freeze-thawed samples as reference.
4. Subvisible particles: When parenteral application is intended, particulate contamination with subvisible particles should be investigated which can be assessed via light obscuration or microscopy according to Ph. Eur. 2.9.19 (12).
5. Visual inspection: The presence of bigger particles should be assessed via visual inspection in accordance with the Ph. Eur. 2.9.20. (13). Additionally, other methods such as turbidimetry or nephelometry can be used.
6. Additional analytical methods: The aforementioned methods can be supplemented with further analytics such as MFI, NTA, or AF4.
7. Cake appearance: The lyophilizates should show an elegant cake appearance (see Fig. 3b). No traces of collapse or other structural alterations such as shrinkage or cracks on the surface should be visible. Moreover, inspect the batch for intra- and inter-batch homogeneity.
8. Moisture content: The final moisture content of the lyophilizate should be below 1% (see Note 21). The moisture content of the lyophilizates can be determined using Karl Fischer titration, thermal gravimetric analysis, or near-IR spectroscopy (14–16).
9. Storage stability: Long-term storage stability (12 months) at typically 2–8°C or potentially 25°C should be ensured. Accelerate stability studies should be performed at elevated temperature and relative humidity (RH) (40°C, 75% RH, 6 months) (17).

4. Notes

1. Depending on the manufacturer and type, freeze-driers considerably differ in their specifications. Regardless of the type, manufacturer's instructions should be followed.
2. In order to ensure direct contact between the vials and the shelf the use of a lyophilization tray with removable bottom is favored.

3. For lyophilization type I glass vials (18, 19) are preferred. Glass vials should be washed and heat sterilized (2 h at 180°C) or autoclaved (15 min, 121°C, 2 bar) if the absence of microbiological contamination is required. Filling volumes up to 1.5 mL in 2R vials are suitable. For larger volumes 6R or 10R vials should be used. In general, the filling height should range between 5 and 15 mm. Higher filling should be avoided as this results in significantly prolonged lyophilization processes. Vials must not be filled more than one-third as this may result in container cracking due to the thermal changes during freezing. In general, the properties of the glass vials such as glass type, bottom shape, and thickness as well as diameter will influence the heat transfer from the shelves to the product and, thus, affect the lyophilization process. Lyophilization stoppers (19) are made of a rubber elastic material such as butyl polymers. They are composed of a shank and a circular disc-shaped flange. The shank includes at least one cavity that provides a vapor port for the sublimed water. "Two-leg" stoppers with two and "three-leg" stoppers with three cavities are also available. The shank comprises a closed outer circumferential face whose diameter is slightly larger than the inner diameter of the vial neck and some blocking elements that allow for right positioning and closing of the stopper during lyophilization. Preferably siliconized stoppers should be used to allow easy stoppering. If required, stoppers can be autoclaved. Especially after steam sterilization stoppers should be dried prior to use to prevent the transfer of moisture into the lyophilized product during storage.
4. The effort necessary to select the optimal formulation for a synthetic gene carrier strongly depends on the sensitivity of the carrier system. Size, charge, and stability of the carriers are influenced by the ionic strength and the pH of the selected buffer (20). The ionic strength of the buffer should be low, about 10–20 mM. In order to keep the ionic strength low, but to provide sufficient buffer capacity, the pKa of the buffer salt should be close to the desired buffers' pH. Secondly, the ionic strength of the buffer should be as low as possible since cryo-concentration of the buffer salts will occur during freezing. High amounts of buffer salts also result in a decrease in the T_g' of the formulation and in an increased risk of cake collapse during lyophilization. Moreover, buffers that are based on two different buffer salts like sodium phosphate buffers should be avoided. In this case the disodium salt is less soluble during freeze-concentration and precipitates earlier compared to the monosodium salt resulting in a decrease in pH (21). Thus, buffers with no or slight freezing-induced pH shifts such as histidine, citrate, Tris, or Hepes should be chosen.
5. To ensure the reproducible preparation of large volumes of synthetic gene carrier stock solutions the use of an up-scaled and standardized preparation method can be beneficial (22, 23).

6. The preparation of a synthetic gene carrier stock solution at higher concentrations allows the easy mixing with stabilizers by simply adding the stabilizer stock solution. The concentration of the stabilizer stock solution depends on the targeted concentration of the stabilizer in the sample and on the mixing ratio with the gene carrier stock solution. If the preparation of a gene carrier stock solution at higher concentrations is not possible the gene carriers should be prepared directly in formulation buffer already containing the stabilizers.
7. The final freezing and thawing temperatures have to be held for 30 min (fill depth ≤5 mm), 1 h (fill depth ≤10 mm), and 2 h (fill depth ≥10 mm) in order to ensure complete freezing and thawing of the sample.
8. In general, the type of excipient and the mass ratio of excipient to synthetic gene carrier are of high importance for successful stabilization. The mass ratio highly depends on the type of gene carrier. For example, for polymer-based gene carriers such as DMAEMA-polyplexes a high mass ratio of about 1.250 (24), for transferrin–PEI complexes a mass ratio of 10.000 (1), and for PEI complexes a mass ratio of 7.500 (25) is necessary. For lipid-based complexes lower mass ratios are sufficient, e.g., 900 for DOTAP–DOPE complexes (25). However, high amounts of excipients can cause longer freeze-drying cycles and problems with the isotonicity.
9. In general, the use of crystallizing excipients such as mannitol or various amino acids should be avoided. Crystallization during freezing is additional stress and will reduce the stabilizing potential. Moreover, sugars with a very low T_g' should be handled cautiously as they tend to facilitate cake collapse. Furthermore, very low T_g' may require lower product temperatures which results in prolonged freezing and primary drying. In general, excipients with T_g' values above −30°C are preferred (3, 6, 26–28). Trehalose and sucrose are the preferred excipients.
10. In order to avoid edge vial effects (faster drying of the vials located along the periphery, specifically the edge of a tray, or edge vials) placebo vials should be placed along the periphery of the tray. These vials should be filled with the same volume of placebo formulation as that in the sample vials.
11. Placing the vials on the shelf in a hexagonal array allows for a uniform surrounding of each vial leading to improved sample uniformity, optimal utilization of the available space, and simplified removal of the tray bottom (see Fig. 2).
12. Stopper placement is critical. If the cavity left open is too small the rate at which water vapor leaves the container is decelerated which can result in a cake collapse.

13. If no special stoppers for the right positioning of the temperature probes are available, piercing of the stoppers and threading the temperature probe through the hole in the stopper allow easy positioning of the temperature probe. Small temperature probes should be used.
14. Aluminum foil should be placed on the inside of freeze-dryer doors consisting of acrylic glass in order to reduce radiation heat transfer from the chamber door to the sample vials.
15. The shelf temperature needs to be set below T_g' if the product is amorphous and below eutectic temperature (T_{eu}) for crystalline drugs during freezing. The cooling rate should be about 1°C/min. This cooling rate allows a moderate supercooling and formation of moderate ice surface areas and uniform ice structures and reduces inter- and intrabatch variety. The temperature should be held until the complete system is solidified. Consider the limited thermal conductivity between vials and shelf and, thus, a delay in cooling between shelf and product. In order to ensure homogeneous supercooling of the formulation and to avoid vial-to-vial variations two-step freezing can be applied. In this case cool down the shelf at −1°C/min to 5°C, hold for 15–30 min, further decrease the shelf temperature at −1°C/min to a temperature between −5 and −10°C, and hold for about 30 min before further reducing the shelf temperature. For crystalline drugs or bulking agents annealing can be required. After freezing is completed increase the product temperature to 10–20°C above T_g' but below T_{eu} to allow efficient crystallization of the drug/bulking agent. Hold this temperature for several hours. If no freeze-drier with refrigerable shelves is available the samples can be frozen in a suitable freezer or if necessary with liquid nitrogen or dry-ice acetone.
16. During primary drying the pressure is set below the saturated vapor pressure of ice. Sublimation takes place and continues as long as the chamber pressure is below the vapor pressure of ice. The product temperature should be maintained below T_g' to prevent collapse of the cake. T_g' of the formulation can be determined with differential scanning calorimetry (DSC), the collapse temperature via freeze-drying microscopy (29–31). The product temperature depends on several parameters like formulation properties, containers, shelf temperature, and chamber pressure but cannot be controlled directly. The two parameters that need to be controlled are shelf temperature and chamber pressure. Multiple combinations of those two parameters can lead to the target product temperature. In general, a higher sublimation rate is obtained at higher shelf temperature and lower chamber pressure. A chamber pressure which is about 10–30% of the vapor pressure of ice at the target

formulation temperature is often chosen. Subsequently, the shelf temperature is adjusted to ensure target product temperature. Sublimation rates should not result in an overload of the condenser which is typically not the case when smaller product volumes are used.

17. Besides the endpoint determination via product temperature measurement two other methods can be used: the pressure rise test or the manometric endpoint determination. For the pressure rise test the product chamber is isolated from the condenser by closing the connection valve for a few seconds. If the product is still wet ice will continue to sublime which cannot reach the condenser and, thus, results in an increased chamber pressure. If all ice has been removed the chamber pressure remains constant. For the manometric endpoint determination a Pirani gauge is needed in addition to the typically existing capacitance manometer. During sublimation, the chamber is filled with water vapor, at the end with nitrogen gas. The Pirani gauge which is sensitive to the chemical composition of the gas phase displays a different pressure compared to that measured by the capacitance manometer during primary drying. At the end both readings converge.
18. During secondary drying water (about 5–20%) which is adsorbed to the solute phase and has not sublimed during primary drying is removed. For this purpose the shelf temperature is increased to ambient or slightly higher temperatures. The higher the temperature is during secondary drying the lower the residual moisture level becomes at the end. In case of amorphous excipients the ramp rate should be low (~0.1°C/min) in order to avoid shrinkage of the cake. If crystalline excipients are used a faster ramp rate of 0.3 or 0.4°C/min can be applied. The typical time for secondary drying is about one-third of primary drying time. The use of an inert gas is preferable to minimize potential oxidation of the samples during storage which is important especially for lipid-based gene carriers.
19. Stoppering under slight vacuum (600 mTorr) helps to ensure that the vials are hermetically sealed, to avoid breakage of sample upon ventilation when opening the vials and to alleviate the addition of the reconstitution medium via needle/syringe systems through the stopper.
20. In order to minimize rehydration-induced stress, the reconstitution medium should be slowly injected against the vial wall and the vials should be gently rolled or swirled until complete dissolution. If rehydration-induced stress is a major problem the addition of surfactants like polysorbate 80 (e.g., 40 μM or 0.005% [m/V]) can be a useful option (32).

21. In general, residual moisture contents about ≤1% are aimed for. However, the drier is not always the better and the optimal residual moisture content has to be determined (33). During storage, lyophilizates with high residual moisture content may exhibit a reduced stability of the carrier and potentially collapse or melt.

References

1. Talsma H, Cherng J-Y, Lehrmann H, Kursa M, Ogris M, Hennink WE, Cotten M, Wagner E (1997) Stabilization of gene delivery systems by freeze-drying. Int J Pharm 157:233–238
2. Brus C, Kleemann E, Aigner A, Czubayko F, Kissel T (2004) Stabilization of oligonucleotide-polyethylenimine complexes by freeze-drying: physicochemical and biological characterization. J Control Release 95: 119–131
3. Kasper JC, Schaffert D, Ogris M, Wagner E, Friess W (2011) Development of a lyophilized plasmid/LPEI polyplex formulation with long-term stability—a step closer from promising technology to application. J Control Release 151:246–255
4. Tamura A, Nagasaki Y (2010) Smart siRNA delivery systems based on polymeric nanoassemblies and nanoparticles. Nanomedicine-Uk 5:1089–1102
5. Anchordoquy TJ, Koe GS (2000) Physical stability of nonviral plasmid-based therapeutics. J Pharm Sci 89:289–296
6. Abdelwahed W, Degobert G, Stainmesse S, Fessi H (2006) Freeze-drying of nanoparticles: formulation, process and storage considerations. Adv Drug Deliv Rev 58:1688–1713
7. Schwegman JJ, Hardwick LM, Akers MJ (2005) Practical formulation and process development of freeze-dried products. Pharm Dev Technol 10:151–173
8. Wang W (2000) Lyophilization and development of solid protein pharmaceuticals. Int J Pharm 203:1–60
9. Anchordoquy TJ, Armstrong TK, MdC M, Allison SD, Zhang Y, Patel MM, Lentz YK, Koe GS (2004) Physical stabilization of plasmid DNA-based therapeutics during freezing and drying. In: Constantino HR, Pikal MJ (eds) Lyophilization of biopharmaceuticals. AAPS Press, Arlington, TX
10. Tang X, Pikal MJ (2004) Design of freeze-drying processes for pharmaceuticals: practical advice. Pharm Res 21:191–200
11. Franks F (1998) Freeze-drying of bioproducts: putting principles into practice. Eur J Pharm Biopharm 45:221–229
12. Palmer CNA, Irvine AD, Terron-Kwiatkowski A, Zhao YW, Liao HH, Lee SP, Goudie DR, Sandilands A, Campbell LE, Smith FJD, O'Regan GM, Watson RM, Cecil JE, Bale SJ, Compton JG, DiGiovanna JJ, Fleckman P, Lewis-Jones S, Arseculeratne G, Sergeant A, Munro CS, El Houate B, McElreavey K, Halkjaer LB, Bisgaard H, Mukhopadhyay S, McLean WHI (2006) Common loss-of-function variants of the epidermal barrier protein filaggrin are a major predisposing factor for atopic dermatitis. Nat Genet 38:441–446
13. Weidinger S, Illig T, Baurecht H, Irvine AD, Rodriguez E, Diaz-Lacava A, Klopp N, Wagenpfeil S, Zhao YW, Liao HH, Lee SP, Palmer CNA, Jenneck C, Maintz L, Hagemann T, Behrendt M, Ring J, Nothen MM, McLean WHI, Novak N (2006) Loss-of-function variations within the filaggrin gene predispose for atopic dermatitis with allergic sensitizations. J Allergy Clin Immunol 118:214–219
14. May JC, Wheeler RM, Etz N, Del Grosso A (1992) Measurement of final container residual moisture in freeze-dried biological products. Dev Biol Stand 74:153–164
15. May JC (2010) Regulatory control of freeze-dried products: importance and evaluation of residual moisture. In: Rey L, May JC (eds) Freeze drying/lyophilization of pharmaceutical and biological products, 3rd edn. Informa Healthcare, New York, NY
16. Oetjen G-W, Haseley P (2004) Measurement of the residual moisture content (RM). In: Oetjen G-W, Haseley P (eds) Freeze-drying, 2nd edn. Viley-VHC, Weinheim
17. ICH Guidance for Industry (2003) Q1A(R2) stability testing of new drug substances and products
18. Dietrich C, Maurer F, Roehl H, Friess W (2010) Pharmaceutical packaging for lyophilization applications. In: Rey L, May JC (eds) Freeze drying/lyophilization of pharmaceutical and biological products, 3rd edn. Informa Healthcare, New York, NY
19. DeGrazio FL (2010) Closure and container considerations in lyophilization. In: Rey L, May JC (eds) Freeze drying/lyophilization of

pharmaceutical and biological products, 3rd edn. Informa Healthcare, New York, NY

20. Ogris M (2005) Gene delivery using polyethylenimine and copolymers. In: Amiji MM (ed) Polymeric gene delivery, principles and applications. CRC Press LLC, Boca Raton, FL, pp 97–106
21. Pikal-Cleland KA, Rodríguez-Hornedo N, Amidon GL, Carpenter JF (2000) Protein denaturation during freezing and thawing in phosphate buffer systems: monomeric and tetrameric [beta]-galactosidase. Arch Biochem Biophys 384:398–406
22. Kasper JC, Schaffert D, Ogris M, Wagner E, Friess W (2011) The establishment of an up-scaled micro-mixer method allows the standardized and reproducible preparation of well-defined plasmid/LPEI polyplexes. Eur J Pharm Biopharm 77:182–185
23. Davies LA, Nunez-Alonso GA, Hebel HL, Scheule RK, Cheng SH, Hyde SC, Gill DR (2010) A novel mixing device for the reproducible generation of nonviral gene therapy formulations. BioTech 49:666–668
24. Cherng J-Y, van de Wetering P, Talsma H, Crommelin DJA, Hennink WE (1997) Freeze-drying of poly((2-dimethylamino)ethyl methacrylate)-based gene delivery systems. Pharm Res 14:1838–1841
25. Molina MC, Allison SD, Anchordoquy TJ (2001) Maintenance of nonviral vector particle size during the freezing step of the lyophilization process is insufficient for preservation of activity: insight from other structural indicators. J Pharm Sci 90:1445–1455
26. Hinrichs WLJ, Manceñido FA, Sanders NN, Braeckmans K, De Smedt SC, Demeester J, Frijlink HW (2006) The choice of a suitable oligosaccharide to prevent aggregation of PEGylated nanoparticles during freeze thawing and freeze drying. Int J Pharm 311:237–244
27. Anchordoquy TJ, Armstrong TK, Molina MC (2005) Low molecular weight dextrans stabilize nonviral vectors during lyophilization at low osmolalities: concentrating suspensions by rehydration to reduced volumes. J Pharm Sci 94:1226–1236
28. Anchordoquy TJ, Armstrong TK, Molina MDC, Allison SD, Zhang Y, Patel MM, Lentz YK, Koe GS (2004) Formulation considerations for DNA-based therapeutics. In: Lu DR, Oeie S (eds) Cellular drug delivery. Humana Press Inc., Totowa, NJ, pp 237–263
29. Kett V, McMahon D, Ward K (2005) Thermoanalytical techniques for the investigation of the freeze drying process and freeze-dried products. Curr Pharm Biotechnol 6:239–250
30. Ward KR, Mateijtschuk P (2010) The use of microscopy, thermal analysis, and impedance measurements to establish critical formulation parameters for freeze-drying cycle development. In: Rey L, May JC (eds) Freeze drying/lyophilization of pharmaceutical and biological products, 3rd edn. Informa Healthcare, New York, NY
31. Passot S, Tréléa IC, Marin M, Fonseca F (2010) The relevance of thermal properties for improving formulation and cycle development: application to freeze-drying of proteins. In: Rey L, May JC (eds) Freeze drying/lyophilization of pharmaceutical and biological products, 2nd edn. Informa Healthcare, New York, NY
32. Yu J, Anchordoquy TJ (2009) Synergistic effects of surfactants and sugars on lipoplex stability during freeze-drying and rehydration. J Pharm Sci 98:3319–3328
33. Hsu CC, Ward CA, Pearlman R, Nguyen HM, Yeung DA, Curley JG (1992) Determining the optimum residual moisture in lyophilized protein pharmaceuticals. Dev Biol Stand 74: 255–270

Chapter 11

Surface- and Hydrogel-Mediated Delivery of Nucleic Acid Nanoparticles

Angela K. Pannier and Tatiana Segura

Abstract

Gene expression within a cell population can be directly altered through gene delivery approaches. Traditionally for nonviral delivery, plasmids or siRNA molecules, encoding or targeting the gene of interest, are packaged within nanoparticles. These nanoparticles are then delivered to the media surrounding cells seeded onto tissue culture plastic; this technique is termed bolus delivery. Although bolus delivery is widely utilized to screen for efficient delivery vehicles and to study gene function in vitro, this delivery strategy may not result in efficient gene transfer for all cell types or may not identify those delivery vehicles that will be efficient in vivo. Furthermore, bolus delivery cannot be used in applications where patterning of gene expression is needed. In this chapter, we describe methods that incorporate material surfaces (i.e., surface-mediated delivery) or hydrogel scaffolds (i.e., hydrogel-mediated delivery) to efficiently deliver genes. This chapter includes protocols for surface-mediated DNA delivery focusing on the simplest and most effective methods, which include nonspecific immobilization of DNA complexes (both polymer and lipid vectors) onto serum-coated cell culture polystyrene and self-assembled monolayers of alkanethiols on gold. Also, protocols for the encapsulation of DNA/cationic polymer nanoparticles into hydrogel scaffolds are described, including methods for the encapsulation of low amounts of DNA (<0.2 μg/μL) and high amounts of DNA (>0.2 μg/μL) since incorporation of high amounts of DNA poses significant challenges due to aggregation.

Key words: Gene delivery, Hydrogel, Surface-mediated, Transfection, Nonviral

1. Introduction

Gene expression within a cell population can be directly altered through gene delivery approaches, which have tremendous potential for therapeutic applications, such as gene therapy to correct genetic deficiencies or tissue engineering scaffolds, where gene delivery can present chemical factors to guide tissue formation in

Manfred Ogris and David Oupicky (eds.), *Nanotechnology for Nucleic Acid Delivery: Methods and Protocols*,
Methods in Molecular Biology, vol. 948, DOI 10.1007/978-1-62703-140-0_11,

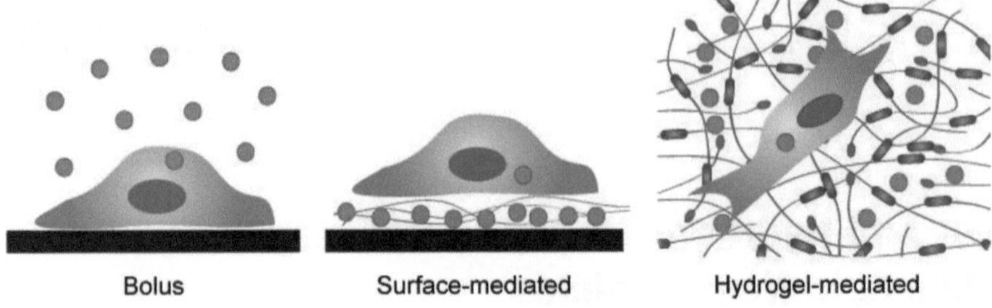

Fig. 1. Bolus, surface- and hydrogel-mediated gene delivery.

regeneration matrices for treatment of organ loss and failure. Furthermore, gene delivery is often critical to research applications, such as functional genomics, cell culture studies, and biotechnological assays. Gene delivery can be performed with both viral and nonviral vectors, the latter of which are the focus of the techniques described in this chapter. Traditionally for nonviral delivery, plasmids or siRNA molecules, encoding or targeting the gene of interest, are packaged within nanoparticles. These nanoparticles are then delivered to the media surrounding cells seeded onto tissue culture plastic; this technique is termed bolus delivery (Fig. 1). Although bolus delivery is widely utilized to screen for efficient delivery vehicles and to study gene function in vitro, this delivery strategy may not result in efficient gene transfer for all cell types or may not identify those delivery vehicles that will be efficient in vivo. Furthermore, bolus delivery cannot be used in applications where patterning of gene expression is needed. In this chapter, we describe methods that incorporate material surfaces (i.e., surface-mediated delivery) or hydrogel scaffolds (i.e., hydrogel-mediated delivery) to efficiently deliver genes (Fig. 1).

Surface-mediated delivery, also termed solid-phase delivery or reverse transfection, refers to the delivery of nucleic acid nanoparticles from a surface or biomaterial that supports cell adhesion. DNA nanoparticles are immobilized to the surface or biomaterial and cells are plated directly on top of the immobilized nanoparticles (Fig. 1), which enhances the extent of transgene expression, along with increasing cell viability (1). This method of delivery was first described in 2002 using a specific avidin–biotin bond to tether nanoparticles to surfaces (2, 3) and was later extended to nonspecific absorption of the nanoparticles to a variety of biomaterial surfaces (1, 4–10). Nonspecific adsorption of nanoparticles is accomplished through noncovalent mechanisms (1, 4–14), including hydrophobic, electrostatic, and van der Waals interactions. Nonspecific binding depends upon the molecular composition of the vector (e.g., lipid versus polymer) and the relative quantity of each (e.g., nitrogen-to-phosphate ratio or N/P), as well as properties of the surface (e.g., hydrophilicity, charge, presence of serum proteins).

In surface-mediated delivery, DNA is concentrated at the delivery site and targeted to cells adhered to the substrate (1–3, 15). The major advantages of surface-mediated DNA delivery are the following: (1) decreased nucleic acid nanoparticle aggregation and degradation, since nanoparticles can be produced in the most favorable buffer for their formation and then immobilized to the surface, which preserves the size observed in solution and avoids exposure of the particles in solution to the cell culture media, which often induces aggregation and decomplexation; (2) cells cultured on the substrate are exposed to elevated DNA concentrations within the local microenvironment (but typically overall concentration of DNA is minimized when compared to bolus techniques), which enhances transfection and minimizes cytotoxicity (1–10, 16); (3) elimination of mass transport limitations for the nanoparticles to reach the cell; and (4) immobilization of DNA complexes to substrates allows for the ability to pattern transfection (9, 17), allowing for spatial control of gene delivery, which is critical for biotechnological and tissue engineering applications. Below, the reader will find protocols for surface-mediated DNA delivery focusing on the simplest and most effective methods, which include nonspecific immobilization of DNA complexes (both polymer and lipid vectors) onto serum-coated cell culture polystyrene and self-assembled monolayers (SAMs) of alkanethiols on gold.

Hydrogel-mediated delivery has been studied for over a decade primarily through the encapsulation of naked DNA during hydrogel formation. Naked DNA has been successfully incorporated inside hydrogels composed of collagen (18), pluronic-hyaluronic acid (19), dimethacrylated poly(lactic acid)-b-poly(ethylene glycol)-b-poly(lactic acid) triblock copolymers (20), alginate (21), oligo(polyethylene glycol) fumarate (22), and engineered silk elastin (23). Although naked DNA has shown gene expression and ability to guide regeneration in vivo (18, 24), limitations with low gene transfer efficiency and rapid diffusion of the DNA from the hydrogel scaffold motivated the use of DNA nanoparticles instead of naked DNA. DNA condensed either with cationic peptides, lipids, or polymers has been introduced into fibrin hydrogels (25–29), enzymatically degradable PEG hydrogels (30), and PEG-hyaluronic acid hydrogels (26). The delivery of genes from hydrogel scaffolds is becoming an attractive route to introduce foreign genes to cells for several reasons. First, in the biotechnology area, where new delivery agents are being investigated for in vivo gene transfer, the delivery of genes inside a matrix may be a better model for in vivo gene transfer. For example, hydrogels can be used to recreate disease models in vitro, including multiple cell types and complex extracellular matrices (e.g., cancer and organoid models). Second, in tissue engineering and regenerative medicine applications, it is often desired to transfect cells either infiltrating the scaffold (endogenous) or that are transplanted. In the case of endogenous cell transfection, the hydrogel scaffold serves as a depot that contains genes or siRNA that can transfect cells as the cells infiltrate the scaffold or the matrix is degraded and the genes or siRNA are released. In the case of transplanted cell transfection, the matrix and cells are encapsulated together. The goal here is to transfect the transplanted cells at a later time to induce their differentiation or the differentiation of nearby cells. Below, the reader will find protocols for the encapsulation of DNA/cationic polymer nanoparticles into hydrogel scaffolds. We divide the encapsulation into low amounts of DNA (<0.2 μg/μL) and high amounts of DNA (>0.2 μg/μL) since incorporation of high amounts of DNA poses significant challenges due to aggregation.

2. Materials

For surface-mediated delivery, the following reagents are necessary.

2.1. Dulbecco's Phosphate-Buffered Saline (PBS): 1× Solution containing 2.67 mM KCl, 1.47 mM KH_2PO_4, 137.93 mM NaCl, and 8.1 mM Na_2HPO_4 with no calcium or magnesium, prepared in MilliQ water, pH 7.4. Stored at room temperature. Sterile filtered.

2.2. 10% Fetal Bovine Serum (FBS): 10% FBS prepared in 1× PBS (prepared as described above). Prepare sterilely and aliquots may be stored at 4°C for a short term (<2 weeks) or −20°C for longer time periods.

2.3. TBS Buffer: 100 mM tris(hydroxymethyl)aminomethane–HCl, pH 7.4, 150 mM NaCl.

2.4. Opti-MEM® I Reduced Serum Media (1×) (Invitrogen™ by Life Technologies) (see Note 1).

2.5. Tissue Culture Polystyrene (TCPS): Delivery can be performed on any TCPS substrate (microscope slides, 6-, 12-, 24-, 48-, or 96-well plates or dishes, see Note 2), with the requirement that it is "tissue culture" polystyrene (indicating that it has been surface modified, typically through corona discharge or gas-plasma, to promote protein and cell adhesion (31, 32)).

These reagents are needed for the preparation of SAMs:

2.6. Ethanol: 200 Proof, filtered through a 0.22 μm filter, degassed with N_2 gas for at least 15 min.

2.7. Argon Gas: It is recommended to store the alkanethiol solutions and gold substrates under argon gas. Do not store SAMs under argon as it can cause toxicity issues.

2.8. 11-Mercaptoundecanoic Acid (MUA): 2 mM solution in ethanol, described above.

2.9. Gold Substrates: Gold-coated glass slides, composed of a titanium adhesion layer and 100–500 Å of gold, can be purchased commercially or prepared using e-beam evaporation. Gold-coated slides should be cut into smaller pieces with a diamond-tipped glass cutter so that pieces fit into standard 48-well tissue culture plates (for this protocol, may be adjusted based on experimental setup).

The following reagents are used for DNA nanoparticle preparation:

2.10. Reporter Plasmid DNA: Transfection is typically assessed by either fluorescent reporter genes (e.g., GFP or DS-red) or luminescent reporter genes (e.g., Firefly or Renilla Luciferase), or both. Plasmids are available commercially with these reporter genes, typically driven by CMV promoters. These plasmids can be expanded using commercially available prep kits such as the endotoxin free Giga Prep kit from Qiagen (see Note 3) and their expression assayed using fluorescent microscopy or commercially available kits. DNA is typically stored in TE buffer (10 mM Tris–Cl, pH 7.4–8.0, 1 mM EDTA, pH 8.0) at −20°C. Prepare stock solution in TE of 1 μg/μL for complexation protocol below.

2.11. Branched Poly(Ethylene Imine) (PEI) Stocks:

1. Prepare 10 mg/mL solution of PEI (Sigma, branched, 25 kDa) in 0.1 M $NaHCO_3$ (sodium bicarbonate) (pH 8.2). Branched PEI is a liquid.
2. Using 6,000–8,000 MWCO tubing, dialyze the PEI against MilliQ water for 24 h, changing water at least three times over the course of dialysis.
3. After dialysis, lyophilize the PEI solution.
4. When lyophilization is complete, weigh the lyophilized PEI and resuspend in TBS buffer at 1 mg/mL. Store stocks at −20°C.

2.12. Linear PEI Stocks: Linear PEI (Polysciences, 25 kDa) comes as a very basic polymer with Cl^- as a counterion. To remove excess Cl^- ions, which would be present at neutral pH, the PEI is precipitated and dialyzed.

1. Dissolve 100 mg of PEI in 10 mL of DI water at room temperature. While stirring the solution bring the pH down to 7.4 using 1 N HCl to help dissolve it.
2. Precipitate the PEI by raising the pH to 12 with 1 N NaOH. Keep stirring the solution. The solution will turn white and blurry, but still stay in suspension. The PEI will

be neutral since it will be deprotonated and it will become hydrophobic and insoluble in water.

3. Using 6,000–8,000 MWCO tubing, dialyze the precipitated PEI against DI water, pH 12 for 1 day to remove the chlorine ion (change water 2× every 2–3 h using 3–4 L DI water, pH 12 each time) and for 2–3 days against DI water (pH 7) to re-protonate.
4. The PEI will become partially soluble in water again. After dialysis lyophilize the PEI solution.
5. Redissolve the entire 100 mg in 10 mL water or, alternatively, weigh out the desired amount of powder and dissolve it in water for a 10 mg/mL solution. About 10 μL of HCl per 3 mg of PEI will be required to help it dissolve. After the PEI is dissolved adjust the pH to 7.4.
6. Aliquot 1 mg/tube and lyophilize the PEI solution a second time. Store the lyophilized stocks at -20°C and resuspend in water or the desired buffer/saline solution. A typical final concentration is 1 μg/μL. Adjust final pH to 7.4.

2.13. Lipofectamine™ 2000 (Invitrogen™ by Life Technologies) is used (see Note 4).

Materials for hydrogel-mediated delivery are the following:

2.14. Triethanol amine (TEOA) Buffer: 0.3 M TEOA (stored under argon). Pipette 19.2 mL of ultrapure water into a beaker, add 0.8 mL of TEOA, and mix. Set pH to 8.6–8.7 with 37% HCl (the initial pH is 9–10 and requires 20–40 droplets). Sterile filter the final buffer. Use immediately or alternatively the buffer can be placed into aliquots and stored at -20°C under argon until use.

2.15. PEG-VS:

1. Dissolve four-arm PEG-OH in tetrahydrofuran (THF) under inert atmosphere and heat to reflux in a Soxhlet apparatus filled with molecular sieves for 3–4 h.
2. Allow the solution to cool to the touch and add sodium hydride (NaH; please note that NaH is a highly flammable material when in contact with water), at fivefold molar excess over OH groups, followed by the addition of divinyl sulfone, at 50-fold molar excess over OH groups.
3. Allow the reaction to continue at room temperature (RT) under argon atmosphere with constant stirring for 3 days.
4. Neutralize the reaction solution with acetic acid, filter it, concentrate by roto-evaporation, and precipitate it in ice-cold diethyl ether.

5. Precipitation should be repeated three times to remove unreacted divinyl sulfone.
6. Dry the final product under vacuum and store under argon at −20°C.
7. The degree of functionalization is confirmed with 1H NMR (in $CDCl_3$): 3.6 ppm (PEG backbone), 6.1 ppm (d, 1H, CH_2), 6.4 ppm (d, 1H, CH_2), and 6.8 ppm (dd, 1H, $-SO_2CH$).

2.16. HA-AC:

1. Dissolve 2 g hyaluronic acid (HA) (60 kDa) in 400 mL DI water.
2. Add 36.3 g adipic acid dihydrazide (ADH, 21.8 molar excess over carboxylic acid groups), dissolve completely, and adjust the pH to 4.75 with 1 N HCl.
3. Add 4 g 1-ethyl-3-[3-dimethylaminopropyl] carbodiimide hydrochloride (EDC, 3.95 molar excess over carboxylic acid groups) as a catalyst.
4. Maintain the pH at 4.75 until it becomes stable (~2 h) and let it react overnight.
5. Purify HA–ADH through dialysis (6,000–8,000 MWCO) against DI water for 3 days, changing the water at least 2× per day.
6. Collect the purified HA–ADH and lyophilize it until completely dry (~2 days). Dry HA–ADH can be stored at −20°C until the next step.

Using these reaction conditions, ~40% of the carboxyl groups are modified with ADH based on the trinitrobenzene sulfonic acid (TNBSA, Pierce) assay.

7. Make reaction buffer: 10 mM HEPES with 150 mM NaCl and 10 mM EDTA at pH 7.2.
8. Dissolve HA–ADH in reaction buffer for a final concentration of ~5.5 mg/mL. Mix using a stir bar and plate for about 2 h to completely dissolve.
9. Weigh out 1 g *N*-acryloxysuccinimide (NHS-Ac, ~3–4 molar excess over amine groups) and dissolve in 10 mL DMSO. Keep on ice.
10. React HA–ADH with NHS-Ac by adding 1 mL NHS-Ac in DMSO every 15–20 min and let it react overnight.
11. Purify HA-AC through dialysis (6,000–8,000 MWCO) against DI water for 3 days, changing the water at least 2× per day.
12. Collect the purified HA-Ac and lyophilize it until completely dry (~2 days). Dry HA-Ac can be stored at −20°C until used.

The degree of functionalization should be confirmed with 1H NMR (in D2O).

2.17. Peptide Crosslinker Aliquots: Michael addition chemistry can be used to cross-link PEG-VS or HA-AC into a hydrogel. Any di-thiol-containing molecule would work; however, to make the hydrogel degradable by cells protease degradable peptides are generally used. The peptide Ac-GCRDGPQGIWGQDRCG-NH_2 (synthesized through solid-phase peptide synthesis or commercially purchased) has been used by several groups (for reviews see refs. 33, 34). Aliquots of the peptide are made to prevent its exposure to oxygen and disulfide bond formation. The peptide crosslinker powder is dissolved in 0.1% TFA, aliquoted into microcentrifuge tubes, and lyophilized. Because TFA is a volatile salt the resulting lyophilized powder contains only the peptide. The amount to be aliquoted depends on the number of cross-links desired or *r* ratio, which is the molar ratio of –SH groups to –VS or –AC groups. For example, for 100% modified PEG-VS hydrogels with 6% PEG an *r* ratio close to 1 is used, while for a 40% modified HA-AC with 3% gel an *r* ratio close to 0.3 is used. Typically enough material is aliquoted to be used in a single use. No freeze–thaw cycles are recommended.

2.18. RGD Peptide Aliquots: Michael addition is used to modify PEG-VS or HA-AC with cell adhesion peptides such as RGD. Similar to the peptide crosslinker aliquots, RGD peptide aliquots are made to prevent disulfide bond formation. RGD peptide powder (Ac-GCGYGRGDSPG-NH_2, synthesized through solid-phase peptide synthesis or commercially purchased) is dissolved in 0.1% TFA, aliquoted into microcentrifuge tubes, and lyophilized. The amount to be aliquoted depends on the amount of RGD desired inside the hydrogel material (typically 30–500 μM final concentration for HA hydrogels and 30–200 μM for PEG-VS gels). Typically enough material is aliquoted to be used in a single use. No freeze–thaw cycles are recommended.

2.19. Reporter Plasmid DNA: To assess transfection from cells seeded inside hydrogel scaffolds, secreted proteins are recommended. The two most commonly used are secreted alkaline phosphatase mammalian expression vector (pSEAP, Genlantis, San Diego, CA) and Gaussia luciferase (pGLUC, New England Biolabs, Ipswich, MA). Both plasmids can be expanded using commercially available prep kits such as the endotoxin free Giga Prep kit from Qiagen and their expression assayed using commercially available kits.

2.20. Sigmacote-Coated Slides:

1. Prepare Sigmacote®-coated glass slides by immersion of slides in Sigmacote® solution using a 50 mL conical tube for 1 min and air drying for 3 min. Repeat this 3× for each slide (see Note 5).
2. After the third coating, place slides into oven above 100°C for 2 h to overnight to allow for complete drying.
3. Wash the slides with ethanol to sterilize before use. You should notice that the water slides off the slide easily.

2.21. Agarose: Dissolve 670 mg low-melting-point agarose in 1 L MQ H_2O, allow 1–2 h to mix using a stir bar and stir plate, and filter solution using a 0.22 μm filter. Store stock solution at RT.

2.22. Sucrose: Dissolve 3.5 g sucrose in 10 mL Millipore water and filter solution using a 0.22 μm filter. Store stock solution at 4°C.

3. Methods

3.1. Surface-Mediated DNA Delivery

For substrate-mediated gene delivery, the properties of the surface are critical to both immobilization strategies and transfection efficiencies, as are the properties of the DNA complexes. Surface-mediated delivery has been reported on a variety of surfaces, including SAMs (9, 10), and standard tissue culture polystyrene (PS) plastic (1, 4–6, 35). As the highest transfection levels were reported on carboxylic acid-terminated SAMs (9) and serum-coated PS (1), methods for these two surfaces are reported in this chapter. Methods are included for both polymer- and lipid–DNA complexes (Fig. 2).

3.1.1. For Surface Preparation of Serum-Coated Polystyrene Proceed as Follows (See Also Notes 6–8)

1. Add 100 μL of 10% FBS per cm^2 (e.g., add 100 μL per well of a 48-well plate).
2. Let plate incubate at room temperature for 2 h (see Note 9).
3. After incubation, remove 10% FBS solution from wells and wash with 1× PBS (see Note 10).

3.1.2. For Surface Preparation of COO^--Terminated SAMs Proceed as Follows (See Note 11)

1. Prepare 2 mM solution of MUA in filtered, degassed ethanol (see Note 12).
2. Wash gold substrates with copious amounts of acetone followed by ethanol. Dry gold substrates under a stream of N_2 gas.
3. Place gold surfaces into 2 mM MUA solution and allow monolayer formation to proceed for 45 min (see Note 13).

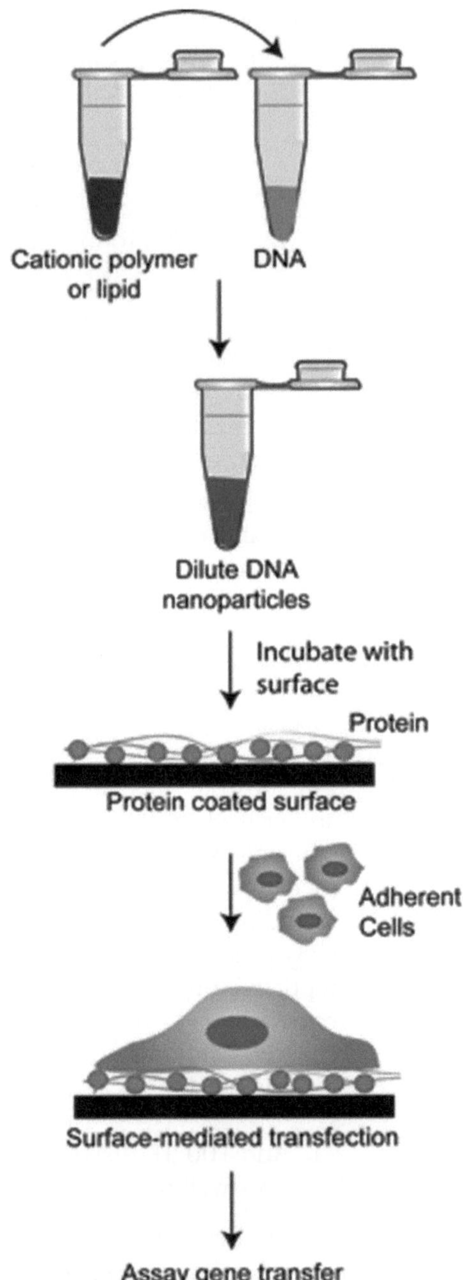

Fig. 2. Visual protocol for surface-mediated gene transfer.

4. After monolayer formation, rinse SAMs with ethanol and dry surfaces under stream of N_2 gas.
5. Allow surfaces to equilibrate in PBS for 15 min to ensure deprotonation of terminal functional groups at physiological pH.

3.1.3. PEI–DNA Complexes Are Prepared with Branched PEI, 25 kDA with 3 μg of DNA in 300 μL Per Well at an N/P of 20 (See Notes 14 and 15)

1. Add 177 μL of TBS (see Note 16) to a microfuge tube labeled "DNA." Add 3 μL of plasmid DNA stock solution (1 μg/μL) and mix by pipetting up and down (see Note 17).
2. Add 111.8 μL of TBS to a microfuge tube labeled "PEI." Add 8.2 μL of PEI stock solution (1 mg/mL) and mix by pipetting up and down (see Note 18).
3. Add the contents of the PEI tube to the DNA solution (DNA tube) (see Note 19), vortex gently for 10 s, and let complexes form for 15 min at room temperature (see Note 20).

3.1.4. Lipofectamine 2000–DNA Complexes Are Generated with 1 μg of DNA in 300 μL Per Well, and a DNA:Lipofectamine 2000 Ratio of 1:2 (See Notes 14 and 15)

1. Add 147 μL of Opti-MEM media to a microfuge tube labeled "DNA." Add 3 μL of plasmid DNA stock solution (1 μg/μL) and mix by pipetting up and down.
2. Add 144 μL of Opti-MEM media to a microfuge tube labeled "LF2000." Add 6 μL of Lipofectamine 2000 stock solution and mix by pipetting up and down.
3. Add the contents of the LF2000 tube to the DNA solution (DNA tube) (see Note 21), pipette up and down, and let complexes form for 20 min at room temperature (see Note 20).

3.1.5. After Formation, Complexes are Immobilized

1. Once complex formation is complete (for either PEI–DNA complexes OR Lipofectamine 2000–DNA Complexes), add 300 μL of complexes to surfaces (for either FBS-coated PS or COO^--terminated SAMs) and let deposit for 2 h at room temperature.
2. After 2 h of complex deposition, remove unbound complexes by washing surfaces twice with TBS (for PEI–DNA complexes) or Opti-MEM (for Lipofectamine 2000–DNA complexes) (see Notes 22 and 23).

3.1.6. After Complex Immobilization to Surfaces, Cells Are Seeded onto the Complexes and the Transfection Efficiency Analyzed

1. Seed 300 μL of cells in each well (15,000 cells/well) (see Note 24).
2. Incubate cells at 37°C and 5% CO_2 for 24–48 h.
3. Assay for transfection using fluorescence microscopy or luminescence assay (depending on the reporter gene present on plasmid) after 24–48 h (see Notes 25–27).

3.2. Hydrogel-Mediated Gene Delivery: Direct Encapsulation (<0.2 μg DNA/μL Hydrogel)

The encapsulation of DNA into hydrogel scaffolds will require different protocols depending on the quantity of DNA loaded. Low amounts of DNA, <0.2 μg/μL, may be loaded into hydrogel scaffolds without significant aggregation. The following protocol is for the synthesis of a 100 μL 3% HA or 100 μL 6% PEG hydrogel scaffold with 0.05 μg DNA/μL of hydrogel. We will be incorporating DNA complexed with linear PEI at an N/P ratio of 7 (see Note 28).

3.2.1. HA-RGD/PEG-RGD Is Synthesized as Follows

1. Weight out and dissolve HA-AC (3 mg) or PEG-VS (6 mg) into 38 μL or 27 μL TEOA buffer. For the HA polymer, incubate for 20 min at 37°C to speed up the dissolution process. Ensure that polymer solution pH is 8.2–8.3 at this point.
2. Take an RGD peptide aliquot out from the −30°C freezer and place it on ice.
3. Use the dissolved HA-AC or PEG-VS to dissolve the RGD aliquot. Vortex thoroughly and allow to incubate for 30 min at 37°C.
4. Add 5 μL of 0.3 M TEOA to HA-RGD (see Note 29).
5. Keep the newly reacted HA-RGD or PEG-RGD on ice.

3.2.2. Thereafter, DNA Nanoparticles (Polyplexes) Are Formed at an N/P Ratio of 7

1. Place 5 μg of DNA into a sterile microcentrifuge tube and dilute to 10 μL in Millipore water.
2. Place 4.57 μg of PEI into a sterile microcentrifuge tube and dilute to 8 μL in Millipore water.
3. Add the PEI solution to the DNA solution, vortex gently for 15 s, and incubate for 10 min at room temperature. After the 10-min incubation place the DNA nanoparticles (polyplexes) on ice for at least 5 min.

3.2.3. The Hydrogel Is Synthesized as Follows

1. Take a crosslinker aliquot out of the freezer and place in −20°C cooler.
2. Subculture cells and dilute to 25,000 cells/μL or 20,000 cells/μL. The final desired cell concentration in hydrogel is 1,000–5,000 cells/μL of gel. Place cells on ice.

The next steps must be done quickly to avoid premature gelation in the tube!

3. Bring the polyplex solution to 0.3 M TEOA by adding 2 μL of 3 M TEOA and immediately add the entire 20 μL polyplex solution to the HA-RGD or the PEG-RGD. Place on ice.
4. Add 20 μL or 31 μL of the 25,000 cells/μL or 20,000 cells/μL cell stock to the HA-RGD/DNA nanoparticle or PEG-RGD/DNA nanoparticle solution and mix thoroughly but gently. Place on ice (see Note 30).
5. Dissolve the crosslinker in 17 μL or 22 μL of ice-cold TEOA buffer and quickly add to the HA-RGD/Cell/polyplex suspension or the PEG-RGD/Cell/polyplex suspension. Mix thoroughly using a positive displacement pipette or a wide orifice tip (see Note 31). The final gel solution should have a pH of 8.0–8.1.
6. Quickly cast the desired gels either as one large 100 μL gel or several small gels into the gel caster (Fig. 3). Place gel caster at 37°C for 30 min.

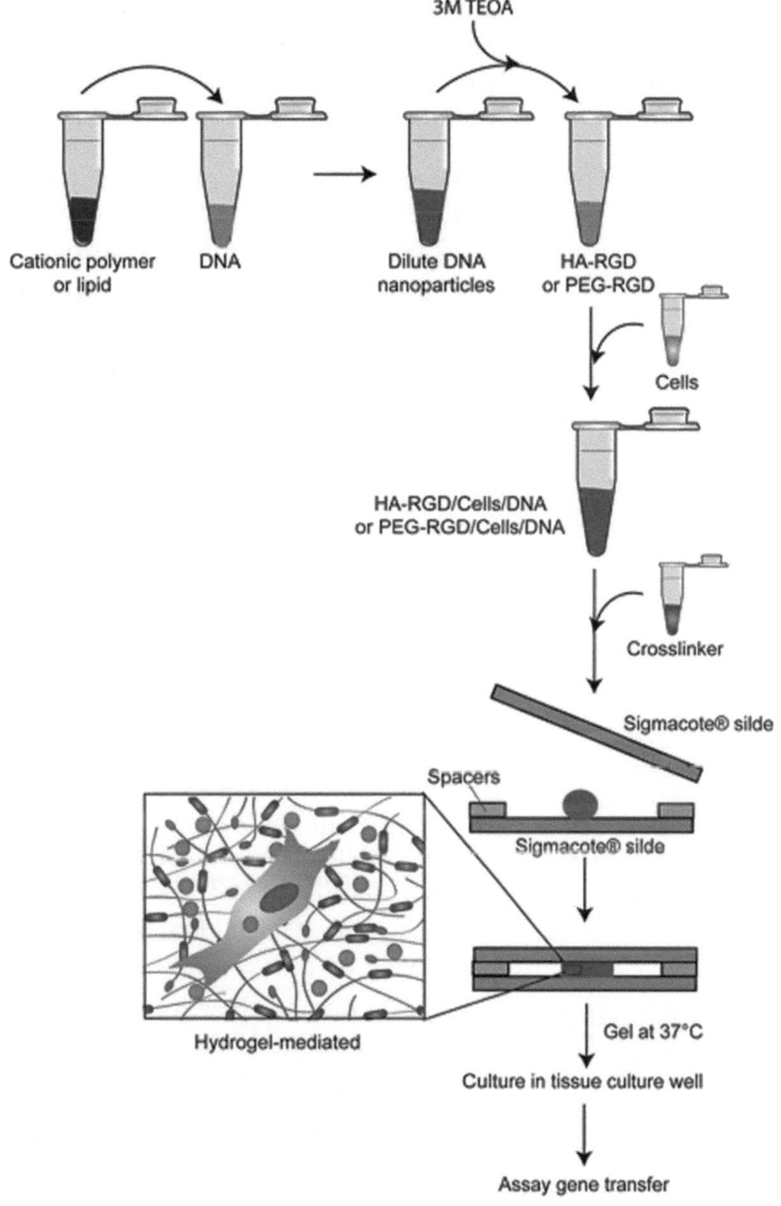

Fig. 3. Visual protocol for hydrogel-mediated gene transfer for low concentrations of DNA.

7. Place the gel(s) into the desired tissue culture plate size. The large gel can be cut with a biopsy punch to result in several small gels. 96- or 48-well plates are recommended for culturing to avoid diluting out the reporter protein.
8. By collecting the media of the cells and performing the appropriate assay, Phospha-light SEAP reporter gene assay system or Biolux gaussia luciferase assay kit, transfection can be monitored. Gene transfer has been observed as early as 2 days and as late as 7 days depending on the cell type and how well the cells migrate through the material.

3.3. Hydrogel-Mediated Gene Delivery: Caged Nanoparticle Encapsulation

For high concentrations of DNA polyplexes direct encapsulation results in severe aggregation. The Segura laboratory has developed a process that can be used to encapsulate high concentrations, up to 5 μg/μL, into hydrogel scaffolds (36). As an example, the following protocol is for the synthesis of 100 μL HA-AC/MMP hydrogels with 100 μg of DNA or 100 μL PEG-VS/MMP hydrogels with 100 μg of DNA.

First, polyplex powder is prepared:

1. Dilute pDNA (100 μg) with 1,000 μL MilliPore water H_2O in a 15 mL conical tube.
2. In a separate 15 mL tube, add 100 μL (350 mg/mL) sucrose and 2,200 μL MilliPore water.
3. Add 91.3 μL of L-PEI to the diluted sucrose solution.
4. Add the dilute PEI/sucrose solution to the pDNA solution, vortex for 15 s, and incubate at room temperature for 15 min.
5. Add 1.5 mL of (.67 mg/mL) agarose to DNA/PEI solution, vortex for 10 s, and immediately dip into liquid nitrogen.
6. Allow 5 min to freeze completely and lyophilize.
7. Remove tubes from lyophilizer and place tubes at −20°C until ready for use.

3.3.1. HA-RGD/PEG-RGD Synthesis Is Carried Out Slightly Different as Described in Subheading 3.2

1. Weight out and dissolve HA-AC (3 mg) or PEG-VS (6 mg) into 38 μL or 27 μL TEOA buffer. For the HA polymer incubate for 20 min at 37°C to speed up the dissolution process.
2. Take an RGD peptide aliquot out from the −20°C freezer and place it on ice.
3. Use the dissolved HA-AC or PEG-VS to dissolve the RGD aliquot. Vortex thoroughly and allow to incubate for 30 min at 37°C.
4. Add 25 μL of 0.3 M TEOA to HA-RGD (see Note 29).
5. Keep the newly reacted HA-RGD or PEG-RGD on ice.

3.3.2. Hydrogels Are Synthesized in the Following Way

1. Take a crosslinker aliquot out of the freezer and place in −20°C cooler.
2. Subculture cells and dilute to 25,000 cells/μL or 20,000 cells/μL. The final desired cell concentration in hydrogel is 1,000–5,000 cells/μL of gel. Place cells on ice.

The next steps must be done quickly to avoid premature gelation!

3. Place polyplex powder in a 1.5 mL microcentrifuge tube and mix with HA-RGD/cell or PEG-RGD/cell suspension.
4. Add 20 μL or 31 μL of the 25,000 cells/μL or 20,000 cells/μL cell stock to the HA-RGD/DNA nanoparticle or PEG-RGD/DNA nanoparticle solution and mix thoroughly but gently. Place on ice.

5. Dissolve the crosslinker in 17 μL or 22 μL of ice-cold TEOA buffer and quickly add to the HA-RGD/Cell/polyplex suspension or the PEG-RGD/Cell/polyplex suspension. Mix thoroughly using a positive displacement pipette or a wide orifice tip (see Note 30). The final gel solution should have a pH of 8.0–8.1.
6. Quickly cast the desired gels either as one large 100 μL gel or several small gels into the gel caster (Fig. 4). Place gel caster at 37°C for 30 min.

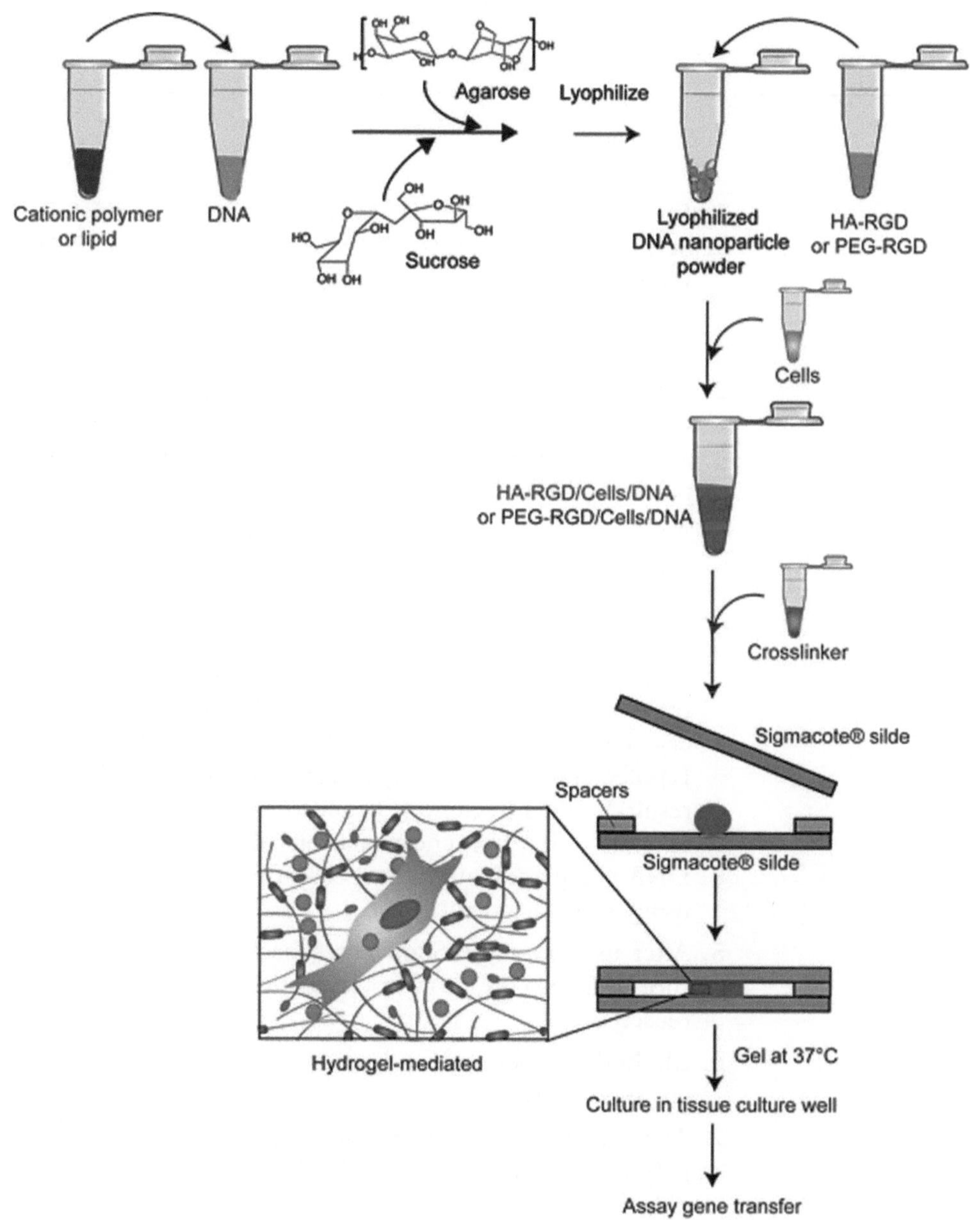

Fig. 4. Visual protocol for hydrogel-mediated gene transfer for high concentrations of DNA.

7. Place the gel(s) into the desired tissue culture plate size. The large gel can be cut with a biopsy punch to result in several small gels. 96- or 48-well plates are recommended for culturing to avoid diluting out the reporter protein.
8. By collecting the media of the cells and performing the appropriate assay, Phospha-light SEAP reporter gene assay system or Biolux gaussia luciferase assay kit, transfection can be monitored. Gene transfer has been observed as early as 2 days and as late as 7 days depending on the cell type and how well the cells migrate through the material.

4. Notes

1. Invitrogen recommends the use of Opti-MEM® I Reduced Serum Media for transfection with Lipofectamine 2000 (37). We have found that its use does enhance transfection, but for many cell lines other serum-free media can be used for formation of the lipoplexes.
2. Most of our protocols have been optimized with and utilize 48-well plate format. The least effective well for surface-mediated delivery is the 96-well plate, as those wells are often not perfectly flat and thus complexes deposit irregularly.
3. Any protocol for DNA preparation from bacterial cultures is sufficient, but "endotoxin-free" buffers are essential for transfection-grade DNA.
4. This chapter includes protocols for one lipid-based transfection reagent (Lipofectamine 2000) and one polymer-based transfection reagent (PEI). However, surface-mediated delivery has also been accomplished with jet-PEI (3, 16), Lipofectamine LTX (35), FugeneHD (Rocher) (unpublished results), and Effectene (Qiagen) (35), as well as viral vectors (15, 38–43). The encapsulation of lipoplexes and jetPEI/DNA polyplexes into hydrogel scaffolds has also been possible using a similar protocol (36).
5. After use, pour the Sigmacote® back into a dark glass bottle (not back into the clean stock solution). This Sigmacote can be reused 3–4×.
6. Methods are per cm^2, and can be scaled accordingly for different surface areas.
7. These methods can be modified to other surfaces including poly(lactide-co-glycolide) (PLGA) (7, 8, 11, 42) and collagen (11, 43).
8. This protocol can also be modified to use purified extracellular matrix/serum proteins, including fibronectin, collagen, and

laminin (5, 7), with fibronectin typically resulting in the highest transfection levels.

9. Serum incubation can proceed for longer periods of time, even overnight at 4°C, without negatively affecting transfection levels.
10. Perform this step right before adding complexes, as described under "Subheading 3.1.6."
11. Incorporation of poly(ethylene glycol) groups into carboxylic acid-terminated SAMs may increase transfection (10). In this case, prepare SAMs by immersing gold substrates into a mixed alkanethiol solution containing both MUA as well as $HS(CH_2)_{11}(OCH_2CH_2)_6OH$, in ratios identical to what is desired in the final SAM (typically 40% or less of the PEGylated thiol).
12. Prepare alkanethiol solution fresh for each experiment and use amber bottles or those covered with aluminum as thiols are light sensitive. Cover with argon if not used immediately.
13. Typically we use gold substrates of ~0.5 cm^2 cut from a gold-coated microscope slide, as described above. These small pieces of gold are best handled with three-prong forceps. We typically form SAMs in foil-lined glass scintillation vials that contain 7–8 mL of MUA solution.
14. Both DNA amount and N/P ratio (or DNA:lipid ratio) need to be optimized for each cell type; the conditions used in this chapter represent good starting optimization points for NIH/3T3s; also these amounts are for a single well, so scale accordingly for multiple wells. For Lipofectamine™ 2000, Invitrogen recommends a Lipofectamine™/DNA ratio (μL/μg) between 0.5 and 5 for optimum transfection (37). For branched PEI, N/Ps are recommended between 10 and 25.
15. Adsorption is typically 10–20% of total DNA added to surfaces (1, 9); thus use approximately 5–10 times as much DNA as would typically be used in bolus conditions.
16. Can also use 150 mM NaCl to form polyplexes.
17. 60% of total complex volume should be with DNA.
18. 40% of total complex volume should be with PEI.
19. This order of addition is critical to transfection success.
20. During this time prepare surfaces for complex immobilization; for FBS-coated PS, remove the FBS and rinse with PBS as described above or for SAMs, incubate with PBS to ensure deprotonation.
21. We have found that even for lipids this order of addition is critical to transfection success.

22. Longer adsorption times result in larger mass of DNA complexes immobilized, but can also be associated with loss of activity; thus adsorption time is another variable that can be manipulated.
23. Be sure that cell suspension is prepared and cells are ready to seed immediately after this rinsing step.
24. Cell density is for NIH/3T3s; this value will need to be optimized for each cell type.
25. Time after delivery to assay transfection may also need to be optimized.
26. For transfection studies on SAMs, move each SAM to a new well before proceeding with luminescence or protein assay to ensure that transfection is only measured in cells adhered to SAM and not on surrounding PS.
27. For transfection studies on SAMs, gold quenching of fluorescence may limit observation of transfection by fluorescent reporter genes. To reduce quenching, be sure to use upright microscope or fix cells and then place SAMs face down into well before imaging.
28. Encapsulation of polyplexes into fibrin, collagen, and UV cross-linked hydrogel scaffolds is also possible using the same protocols (36).
29. This 5 μL or 25 μL volume could be used to introduce other factors to the hydrogel such as growth factors. Further, this volume could be added to the cell volume to make the cell stock less concentrated.
30. The cell concentration in the gel and the cell concentration used in the stock solution should be optimized per cell type. For example, we have found that for MSCs 25,000 cells/μL is the highest concentration that can be used before severe cell death is observed post encapsulation. It is recommended to make different cell stock concentrations and make hydrogels without DNA/siRNA nanoparticles to determine the ideal cell concentration.
31. The amount of crosslinker added should be ~0.3 moles of thiol/mole of acrylate for the HA-AC hydrogel and 0.9 thiol/mole of vinyl sulfone for the PEG hydrogel. This is assuming a 40% modified HA molecule and a 100% modified PEG-VS polymer. This ratio can be varied to make hydrogels with different mechanical properties. It is recommended to use a spreadsheet program such as Microsoft Excel to calculate the ratio of thiols to acrylate or vinyl sulfone. Further the amount of crosslinker added will also be dependent on the activity of the crosslinker. Since the crosslinkers are bi-thiol molecules they are easily oxidized into disulfides. To determine the

activity of the crosslinker, a thiol quantification assay such as the Ellman's assay should be performed. The moles of the crosslinker used to achieve a desired thiol/acrylate or vinyl sulfone ratio should be adjusted accordingly.

Acknowledgements

We would like to thank Talar Tokatlian, Shiva Gojgini, Tim Martin, Sarah Plautz, and Tadas Kasputis for helpful comments and suggestions. We also like to thank the National Institutes of Health (R21EB009516, TS), the National Science Foundation (CAREER 0747539, TS), the American Heart Association (10SDG2640217), the University of Nebraska Foundation (Layman Funds), the Nebraska Research Initiative, NSF EPSCoR First Award, USDA CSREES-Nebraska (NEB-21-146), University of Nebraska-Lincoln IANR Strategic Investments Funds, and the J.A. Woollam Co. for financial support.

References

1. Bengali Z, Pannier AK, Segura T, Anderson BC, Jang JH, Mustoe TA, Shea LD (2005) Gene delivery through cell culture substrate adsorbed DNA complexes. Biotechnol Bioeng 90:290–302
2. Segura T, Shea LD (2002) Surface-tethered DNA complexes for enhanced gene delivery. Bioconjug Chem 13:621–629
3. Segura T, Volk MJ, Shea LD (2003) Substrate-mediated DNA delivery: role of the cationic polymer structure and extent of modification. J Control Release 93:69–84
4. Bengali Z, Rea JC, Gibly RF, Shea LD (2009) Efficacy of immobilized polyplexes and lipoplexes for substrate-mediated gene delivery. Biotechnol Bioeng 102:1679–1691
5. Bengali Z, Rea JC, Shea LD (2007) Gene expression and internalization following vector adsorption to immobilized proteins: dependence on protein identity and density. J Gene Med 9:668–678
6. Bengali Z, Shea LD (2005) Gene delivery by immobilization to cell-adhesive substrates. MRS Bull 30:659–662
7. De Laporte L, Yan AL, Shea LD (2009) Local gene delivery from ECM-coated poly(lactide-co-glycolide) multiple channel bridges after spinal cord injury. Biomaterials 30:2361–2368
8. Jang JH, Bengali Z, Houchin TL, Shea LD (2006) Surface adsorption of DNA to tissue engineering scaffolds for efficient gene delivery. J Biomed Mater Res A 77:50–58
9. Pannier AK, Anderson BC, Shea LD (2005) Substrate-mediated delivery from self-assembled monolayers: effect of surface ionization, hydrophilicity, and patterning. Acta Biomater 1:511–522
10. Pannier AK, Wieland JA, Shea LD (2008) Surface polyethylene glycol enhances substrate-mediated gene delivery by nonspecifically immobilized complexes. Acta Biomater 4:26–39
11. Bielinska AU, Yen A, Wu HL, Zahos KM, Sun R, Weiner ND, Baker JR Jr, Roessler BJ (2000) Application of membrane-based dendrimer/DNA complexes for solid phase transfection in vitro and in vivo. Biomaterials 21:877–887
12. Kneuer C, Sameti M, Bakowsky U, Schiestel T, Schirra H, Schmidt H, Lehr CM (2000) A nonviral DNA delivery system based on surface modified silica-nanoparticles can efficiently transfect cells in vitro. Bioconjug Chem 11: 926–932
13. Manuel WS, Zheng JI, Hornsby PJ (2001) Transfection by polyethyleneimine-coated microspheres. J Drug Target 9:15–22
14. Zhang JT, Chua LS, Lynn DM (2004) Multilayered thin films that sustain the release of functional DNA under physiological conditions. Langmuir 20:8015–8021

15. Levy RJ, Song C, Tallapragada S, DeFelice S, Hinson JT, Vyavahare N, Connolly J, Ryan K, Li Q (2001) Localized adenovirus gene delivery using antiviral IgG complexation. Gene Ther 8:659–667
16. Segura T, Chung PH, Shea LD (2005) DNA delivery from hyaluronic acid-collagen hydrogels via a substrate-mediated approach. Biomaterials 26:1575–1584
17. Houchin-Ray T, Whittlesey KJ, Shea LD (2007) Spatially patterned gene delivery for localized neuron survival and neurite extension. Mol Ther 15:705–712
18. Bonadio J, Smiley E, Patil P, Goldstein S (1999) Localized, direct plasmid gene delivery in vivo: prolonged therapy results in reproducible tissue regeneration. Nat Med 5: 753–759
19. Chun KW, Lee JB, Kim SH, Park TG (2005) Controlled release of plasmid DNA from photo-cross-linked pluronic hydrogels. Biomaterials 26:3319–3326
20. Quick DJ, Anseth KS (2004) DNA delivery from photocrosslinked PEG hydrogels: encapsulation efficiency, release profiles, and DNA quality. J Control Release 96:341–351
21. Kong HJ, Kim ES, Huang YC, Mooney DJ (2008) Design of biodegradable hydrogel for the local and sustained delivery of angiogenic plasmid DNA. Pharm Res 25:1230–1238
22. Kasper FK, Jerkins E, Tanahashi K, Barry MA, Tabata Y, Mikos AG (2006) Characterization of DNA release from composites of oligo(poly(ethylene glycol) fumarate) and cationized gelatin microspheres in vitro. J Biomed Mater Res Part A 78A:823–835
23. Megeed Z, Haider M, Li D, O'Malley BW Jr, Cappello J, Ghandehari H (2004) In vitro and in vivo evaluation of recombinant silk-elastin like hydrogels for cancer gene therapy. J Control Release 94:433–445
24. Jang JH, Rives CB, Shea LD (2005) Plasmid delivery in vivo from porous tissue-engineering scaffolds: transgene expression and cellular transfection. Mol Ther 12:475–483
25. Lei P, Padmashali RM, Andreadis ST (2009) Cell-controlled and spatially arrayed gene delivery from fibrin hydrogels. Biomaterials 30:3790–3799
26. Wieland JA, Houchin-Ray TL, Shea LD (2007) Non-viral vector delivery from PEG-hyaluronic acid hydrogels. J Control Release 120:233–241
27. Saul JM, Linnes MP, Ratner BD, Giachelli CM, Pun SH (2007) Delivery of non-viral gene carriers from sphere-templated fibrin scaffolds for sustained transgene expression. Biomaterials 28:4705–4716
28. Trentin D, Hall H, Wechsler S, Hubbell JA (2006) Peptide-matrix-mediated gene transfer of an oxygen-insensitive hypoxia-inducible factor-1alpha variant for local induction of angiogenesis. Proc Natl Acad Sci USA 103: 2506–2511
29. Trentin D, Hubbell J, Hall H (2005) Non-viral gene delivery for local and controlled DNA release. J Control Release 102:263–275
30. Lei YG, Segura T (2009) DNA delivery from matrix metal lop roteinase degradable poly(ethylene glycol) hydrogels to mouse cloned mesenchymal stem cells. Biomaterials 30:254–265
31. Amstein CF, Hartman PA (1975) Adaption of plastic surfaces for tissue culture by glow discharge. J Clin Microbiol 2:46–54
32. Ramsey WS, Hertl W, Nowlan ED, Binkowski NJ (1984) Surface treatments and cell attachment. In Vitro 20:802–808
33. West JL, Hubbell JA (1999) Polymeric biomaterials with degradation sites for proteases involved in cell migration. Macromolecules 32:241–244
34. Lutolf MP, Hubbell JA (2005) Synthetic biomaterials as instructive extracellular microenvironments for morphogenesis in tissue engineering. Nat Biotechnol 23:47–55
35. Pannier AK, Ariazi EA, Bellis AD, Bengali Z, Jordan VC, Shea LD (2007) Bioluminescence imaging for assessment and normalization in transfected cell arrays. Biotechnol Bioeng 98:486–497
36. Lei Y, Huang S, Sharif-Kashani P, Chen Y, Kavehpour P, Segura T (2010) Incorporation of active DNA/cationic polymer polyplexes into hydrogel scaffolds. Biomaterials 31:9106–9116
37. www.invitrogen.com (2011) Lipofectamine 2000
38. Abrahams JM, et al. (2002) Endovascular microcoil gene delivery using immobilized anti-adenovirus antibody for vector tethering. Stroke 33(5): 1376–82
39. Klugherz BD, et al. (2002) Gene delivery to pig coronary arteries from stents carrying antibody-tethered adenovirus. Hum Gene Ther, 13(3):443–54.
40. Pandori M, Hobson D, and Sano T (2002) Adenovirus-microbead conjugates possess enhanced infectivity: a new strategy for localized gene delivery. Virology. 299(2):204–12.
41. Hobson DA, Pandori MW, and Sano T (2003) In situ transduction of target cells on solid surfaces by immobilized viral vectors. BMC Biotechnol 3(1):4

42. Kofron MD and Laurencin CT (2004) Development of a calcium phosphate co-precipitate/poly(lactide-co-glycolide) DNA delivery system: release kinetics and cellular transfection studies. Biomaterials 25(13):2637–43.

43. Katz JM, Roth CM, and Dunn MG, Factors that influence transgene expression and cell viability on DNA-PEI-seeded collagen films. Tissue Eng 11(9-10):1398–406.

Chapter 12

Layer-by-Layer Assembled Gold Nanoparticles for the Delivery of Nucleic Acids

Eva-Christina Wurster, Asmaa Elbakry, Achim Göpferich, and Miriam Breunig

Abstract

The delivery of nucleic acids to mammalian cells requires a potent particulate carrier system. The physicochemical properties of the used particles, such as size and surface charge, strongly influence the cellular uptake and thereby the extent of the subsequent biological effect. However the knowledge of this process is still fragmentary because heterogeneous particle collectives are applied. Therefore we present a strategy to synthesize carriers with a highly specific appearance on the basis of gold nanoparticles (AuNPs) and the Layer-by-Layer (LbL) technique. The LbL method is based on the alternate deposition of oppositely charged (bio-)polymers, in our case poly(ethylenimine) and nucleic acids. The size and surface charge of those particles can be easily modified and accordingly systematic studies on cellular uptake are accessible.

Key words: Delivery of nucleic acids, Layer-by-Layer, Gold nanoparticles, Poly(ethylenimine)

1. Introduction

The introduction of nucleic acids into mammalian cells is a promising approach for the treatment of genetic diseases. Those strategies need a potent carrier system, the so-called vector, to transport nucleic acids (DNA or RNA). But there are three major challenges a gene delivery system has to overcome: First, the nucleic acids have to be protected against the degradation by nucleases. Second, the cellular barriers, membranes and endosome, have to be crossed. And finally, the carrier itself should be toxicologically and immunologically inert (1, 2).

A promising alternative to viral gene vectors are complexes of positively charged polymers or lipids and negatively charged nucleic

Manfred Ogris and David Oupicky (eds.), *Nanotechnology for Nucleic Acid Delivery: Methods and Protocols*, Methods in Molecular Biology, vol. 948, DOI 10.1007/978-1-62703-140-0_12,

acids. A big disadvantage of those particle collectives is that they are usually heterogeneous in size and surface charge, which are the main properties influencing the cellular uptake. Consequently, systematic studies to investigate the relationship between physicochemical particle properties and the endocytotic pathways into the cells as well as the intracellular fate of the gene vectors are nearly impossible. That is why we decided to engineer a novel gene carrier system with a highly specific appearance: The size and the shape of the particles are defined by a solid core of gold nanoparticles (AuNPs). The charge of the particles is affected by charged polymers which are deposited on the particles' surface in a highly ordered manner.

In this methodology chapter we would like to introduce a novel gene delivery strategy on the basis of nanoparticles which are specific in size, shape, and surface charge.

1.1. Layer-by-Layer

The basic principle of the Layer-by-Layer (LbL) idea is the self-assembly that means the autonomous organization (3) of oppositely charged polyelectrolytes and was introduced by Gero Decher and colleagues 20 years ago (4). Since this time LbL assemblies have entered various scientific fields, from materials science (5) and physical chemistry (6) to electrochemistry (7) and biomedical engineering (8). The fundamental concept is the electrostatic interaction of positively and negatively charged polymers, namely, polycations and polyanions. The alternation of oppositely charged polymers results in a stable nanoscale film coating on a flat or a curved template (9). The characteristics of those films can be tailored with nanometer precision with each layer. The popularity of LbL approaches in biomedical research is also due to the fact that biological components can be integrated into LbL systems under physiological conditions only by electrostatic interactions that means under conservation of the native structure.

In the field of drug delivery research there are two major approaches for LbL-based particles: On one hand there are hollow microcapsules, constructed of a dissolvable core with a shell of LbL-assembled polyelectrolytes. Here, the drug is encapsulated into the interior of the hollow sphere before or after destruction of the core material. Those microcapsules are in the size range of several micrometers in diameter (10, 11). They can be characterized by light microscopy and imaging but their biological applications are limited, because only a few cell types are able to process microparticles. On the other hand, there are core/shell particles, consisting of a solid core, coated with a shell of multilayers. In this case, the active substance can be part of the shell-multilayer (12). LbL-assembled nanoparticles are much more relevant as drug carriers because they can be applied into the bloodstream and taken up by various cell types via endocytosis.

1.2. Gold Nanoparticles

AuNPs are widely used metal nanoparticles for biological and biomedical purposes. Due to the easy synthesis and outstanding physical properties of colloidal gold, the applications reach from molecular imaging and diagnostics, to functional bioconjugates, treatment of cancer, and up to drug delivery strategies (13, 14). Of course, gold atoms carry a high electron density and therefore give a high contrast in Transmission Electron Microscopy (TEM) (15). Even more important is the phenomenon of the surface plasmon resonance: Electromagnetic waves are absorbed and scattered effectively, if the wavelength hits the resonance frequency of the electron clouds surrounding the gold atoms. According to the Mie Theory, the plasmon resonance frequency is influenced by the size and the shape of the particle and by the dielectric constant of the medium (16). Therefore, suspensions of spherical AuNPs show absorption of light in the region from 510 to 540 nm. The absorption spectra of colloidal gold can be tuned to the red and near-infrared (NIR) spectra, the so-called "water window," where no absorption of biological materials occurs. This is the basis for most imaging and diagnosis applications using AuNPs. Plasmon resonance is also the physical background of the hyperthermia treatment of cancer, where AuNPs emit absorbed light as heat energy (17).

Chemical features are also important for the wide distribution of AuNPs. Gold surfaces can easily be modified because of a high affinity to soft bases, such as thiol groups (18). Alkanethiols form a stable self-assembled monolayer around the gold core. The mechanism of this interaction is still under discussion, but probably relies on the interaction of thiolate ions and oxidized Au^{+} ions on the particle surface. In the case of alkanethiols with further reactive head groups, e.g., carboxylic acids, the particle surface can easily be functionalized. This opportunity opens the door for a large number of bioconjugation strategies (19).

1.3. Layer-by-Layer Modified Gold Nanoparticles

The versatility of the LbL surface modifications is applied to AuNPs in our studies. Those strategies have been published earlier (20), but not in combination with nucleic acids in the shell of LbL-AuNPs.

The core–shell AuNPs for nucleic acid delivery consist of four polymer layers deposited on a gold core (Fig. 1). Gold was chosen as the size template for several reasons: The physical properties of gold allow the detection of the deposition of polymer layers by vis-spectroscopy. The coating process can be monitored because of a red shift of the plasmon absorption band. It is therefore also possible to identify particle aggregation during polymer deposition triggered by the change of salt concentrations or purification conditions. Additionally AuNPs are visible by TEM with a high contrast because of their large electron density. TEM is therefore a useful tool to determine the amount and location of particles which have been endocytosed by cells.

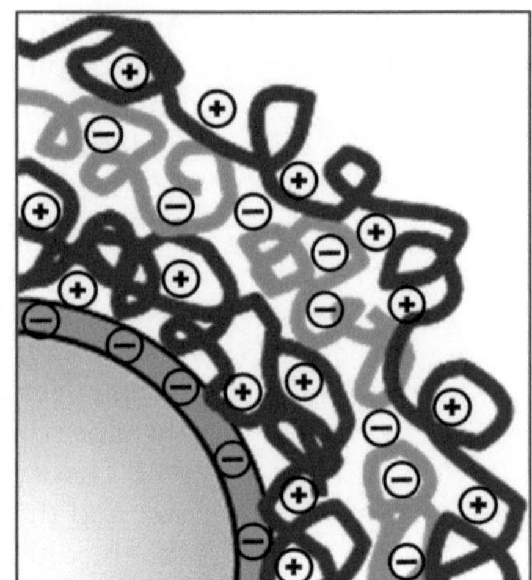
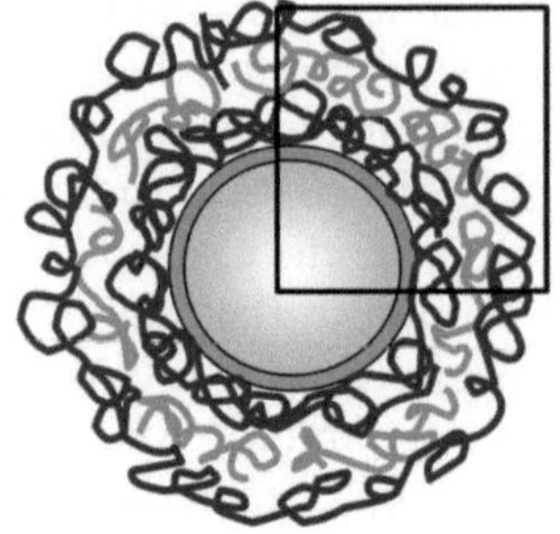

Fig. 1. Core–shell gold nanoparticles consist of a solid gold core and four layers. The first layer is the stabilizing agent MUA which conserves a negative charge on the particle surface. The positively charged layer of PEI is followed by the negatively charged nucleic acid. The nucleic acids are protected by a last layer of PEI. (Adapted with permission from Elbakry et. al (25). Copyright 2009 American Chemical Society).

The first layer which is deposited on the gold surface is the stabilizing agent mercaptoundecanoic acid (MUA). The thiol group binds fairly stable to the gold core and the deprotonated carboxylic group introduces a negative charge on the particle surface (21). This step is important to avoid aggregation of the particles and to get a stable adherence of the following polyelectrolyte layers. The basis for the nucleic acid delivery with our particles is poly(ethylenimine) (PEI) which is known to be an effective transfection agent (22, 23). PEI consists of primary, secondary, and tertiary amino groups which can be protonated and therefore carry a positive charge. This is important for the deposition of negatively charged nucleic acids and may also be responsible for the endosomal escape after cellular uptake (24). Accordingly PEI forms the second layer of the LbL core/shell particles, followed by the layer of negatively charged nucleic acids and a second layer of PEI as the outer shell. We could show that the last layer of PEI is essential to get a biological effect, in our case a transfection effect, with these particles and assume that it is necessary to avoid degradation of the nucleic acids (25).

In this methodology chapter we describe a strategy to get size- and charge-specific particles for the cellular delivery of nucleic acids. It is noteworthy that this strategy is not a general protocol which can easily be conferred to other polymers and nucleic acid sequences. The coating of nanoparticles is challenging because of aggregation of small particles and coating conditions have to be adjusted for each case.

2. Materials

2.1. Synthesis of Gold Nanoparticles

1. 100 ml three-neck round-bottom flask and elliptic stir bar, thoroughly rinsed with aqua regia and distilled water (see Note 1).
2. Oil bath and condenser.
3. $HAuCl_4$-solution (stock solution 1% w/v).
4. Sodium citrate solution (stock solution 1% w/v).

2.2. Stabilization of AuNPs with MUA (AuNP–MUA)

1. Colloidal gold suspension.
2. MUA (stock solution: 20 mg/ml).
3. 1 mM Sodium hydroxide solution, pH-Meter.
4. Snap cap vials, stir plate, and stir bars.
5. 2 ml Eppendorf tubes and centrifuge for purification.
6. Distilled water for resuspension of the nanoparticles.

2.3. Coating with PEI (AuNP–MUA–PEI)

1. Stabilized AuNPs.
2. PEI (molecular weight: 25 kDa, stock solution: 10 mg/ml).
3. Snap cap vials, stir plate, and stir bars.
4. 2 ml Eppendorf tubes and centrifuge for purification.
5. 1 mM Sodium chloride solution.

2.4. Coating with Nucleic Acids (AuNP–MUA–PEI–DNA/RNA)

1. AuNPs coated with PEI.
2. DNA or RNA stock solution (100 μM).
3. Snap cap vials, stir plate, and stir bars.
4. Eppendorf tubes and centrifuge for purification.
5. 1 mM Sodium chloride solution.

2.5. Characterization by Dynamic Light Scattering, Electrophoretic Mobility, and UV–Vis Spectrometry

1. Characterization of the particles by dynamic light scattering and electrophoretic mobility (Zeta-Potential) was performed on a Zetasizer Nano ZS (Malvern Instruments).
2. UV–vis-Spectra were measured with a Uvikon 900 double beam photometer (Kontron Instruments).

2.6. Characterization by TEM

1. Transmission electron micrographs were taken on a Philips CM 12 microscope (FEI, Eindhoven, The Netherlands).
2. Carbon-coated copper grids, pretreated in a plasma beam.
3. Image J software for statistical analysis of TEM micrographs.

3. Methods

3.1. Preparation of AuNPs

1. AuNPs are prepared according to the method published by Frens (26). The size of the particles can be adjusted by variation of the $HAuCl_4$-to-sodium citrate ratio. The preparation described here resulted in particles of about 50 nm in diameter. 25 ml of a 0.01% HAuCl4 solution is prepared in a three-neck round-bottom flask with a condenser. The solution was heated to 100°C in an oil bath and stirred vigorously. 180 μl of the sodium citrate solution (stock solution 1% w/v) was added to the reaction mixture and heating is continued until the color changed from pale yellow to dark red. This indicates the formation of AuNPs. The AuNPs obtained from this synthesis are stabilized by a shell of citrate ions preventing aggregation of the particles. The suspension can be stored over months if they are not purified by dialysis or centrifugation.

3.2. Stabilization with MUA (AuNP–MUA)

1. The AuNPs are further stabilized by MUA. Gold surfaces show a stable affinity to sulfur groups, such as thiols and disulfides. MUA consists of a thiol group and a deprotonated carboxylic acid group. This conserves a stable negative charge on the particle surface, which cannot be displaced like the citrate ligands (Fig. 2).

   This negative charge is important for the stability of the particle suspension and avoids aggregation and is the basis for the deposition of positively charged polymer layers on the surface.

2. The pH of the nanoparticle suspension is adjusted to pH 11 with 1 mM NaOH.

   MUA is added under stirring to give a final concentration of 0.1 mg/ml. Stirring is continued over 3 days to give a stable surface modification. The particles are purified by centrifugation to remove the excess of MUA, which interferes with the charged polymers in the next layer deposition steps.

3.3. Purification of AuNPs

1. The gold suspensions are purified after each modification to remove excess reagents. The most convenient method is the purification by centrifugation. The centrifugation conditions depend strongly on the particle size (see Note 2). An Eppendorf centrifuge and 2 ml Eppendorf tubes are used (see Note 3).
2. For 50 nm particles the suspensions are centrifuged at 5,000 × *g* for 10 min at 4°C. The supernatant is removed and the pellet is washed with water, twice (see Note 4). For the following LbL deposition the last resuspension is carried out in 1 mM sodium chloride.

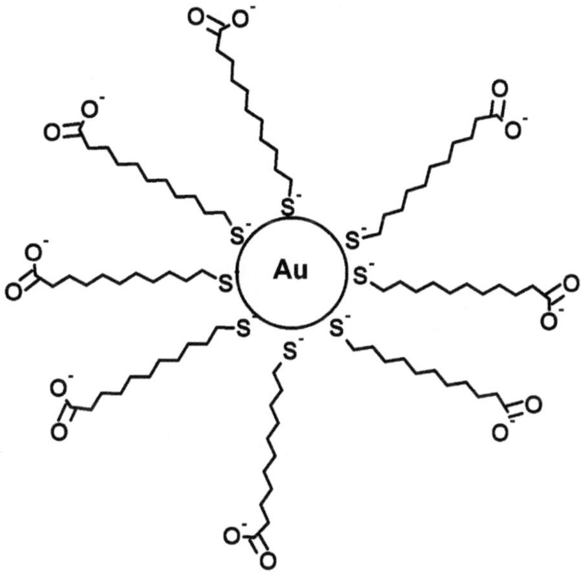

Fig. 2. Schematic figure of a gold nanoparticle core stabilized by MUA. The thiol groups displace the citrate ligands due to their high affinity to the gold surface (binding mechanism still under discussion). The negative charge of the deprotonated carboxylic acid group conserves a negative charge on the particle surface which is the basis for the further deposition of positively charged polymers and prevents aggregation by electrostatic repulsion.

3.4. Coating with PEI: Deposition of the First and Third Layer (AuNP–MUA–PEI and AuNP–MUA–PEI–DNA/RNA–PEI)

1. The stabilized particles are purified by centrifugation as described above and the last resuspension is performed in 1 mM NaCl.
2. The PEI stock solution (10 mg/ml in 1 mM NaCl) is placed in a small snap cap vial and stirred. The final concentration of PEI is 1 mg/ml. The AuNP solution is added dropwise and stirring is continued for 30 min at room temperature (Fig. 3). The suspension is purified afterwards as described above.

3.5. Coating with Nucleic Acids: Deposition of the Second Layer (AuNP–MUA–PEI–DNA/RNA)

1. The coating of the nanoparticles with nucleic acids is performed according to the coating with PEI. The nanoparticles are resuspended in 1 mM NaCl after purification. Double-stranded DNA or RNA is placed into a snap cap vial at a final concentration of 1.5 μM for DNA or 2.0 μM for RNA. The nanoparticle suspension is added dropwise under stirring. After 30 min the particles are purified and resuspended in 1 mM NaCl for the deposition of the last layer of PEI (see Notes 5–7).

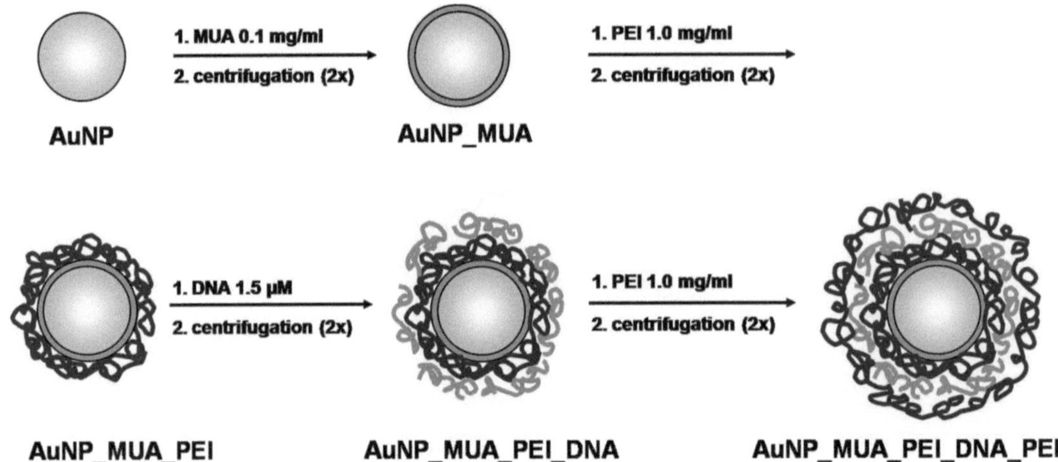

Fig. 3. Schematic illustration of the LbL-process. The stabilized gold nanoparticles are alternatively covered with PEI and DNA to give the LbL-assembled core–shell particles. Reprinted (adapted) with permission from Elbakry et al. (25). Copyright 2009 American Chemical Society.

3.6. Characterization of Core–Shell AuNPs by Vis-Spectroscopy, Dynamic Light Scattering, and Zeta Potential

1. The deposition of each polymer should be followed by measuring three important parameters: particle size, zeta potential, and Vis-absorbance (see Note 8).
2. The particle size should increase with each polymer layer but the extent should not be more than 10 nm per layer and the size distribution should be as narrow as possible (Fig. 4a).
3. The zeta potential is a parameter for the surface charge of the particles. The polymers have opposite charges that means that the surface charge of the particles should turn from negative (AuNP–MUA) to positive (AuNP–MUA–PEI) and so on (Fig. 4b).
4. A special characteristic of AuNP is the surface plasmon resonance, leading to a red color of the particle suspensions. The absorbance peak of the plasmon resonance depends on the particle size and the electrostatic environment of the particles, among other parameters. The size increase during the LbL process and particle aggregation can be followed by measuring the absorbance maximum of the particles. The absorbance peak should show a slight red shift of about 2 nm per polymer layer (Fig. 4c). A shift of more than 10 nm or a broadening of the absorbance spectra is a sign for particle aggregation.

3.7. Characterization by TEM

1. Samples for TEM images are prepared by depositing the gold suspension onto carbon-coated copper grids. The grids are treated in a plasma beam before use to get a more hydrophilic surface. The grids are air dried before use.

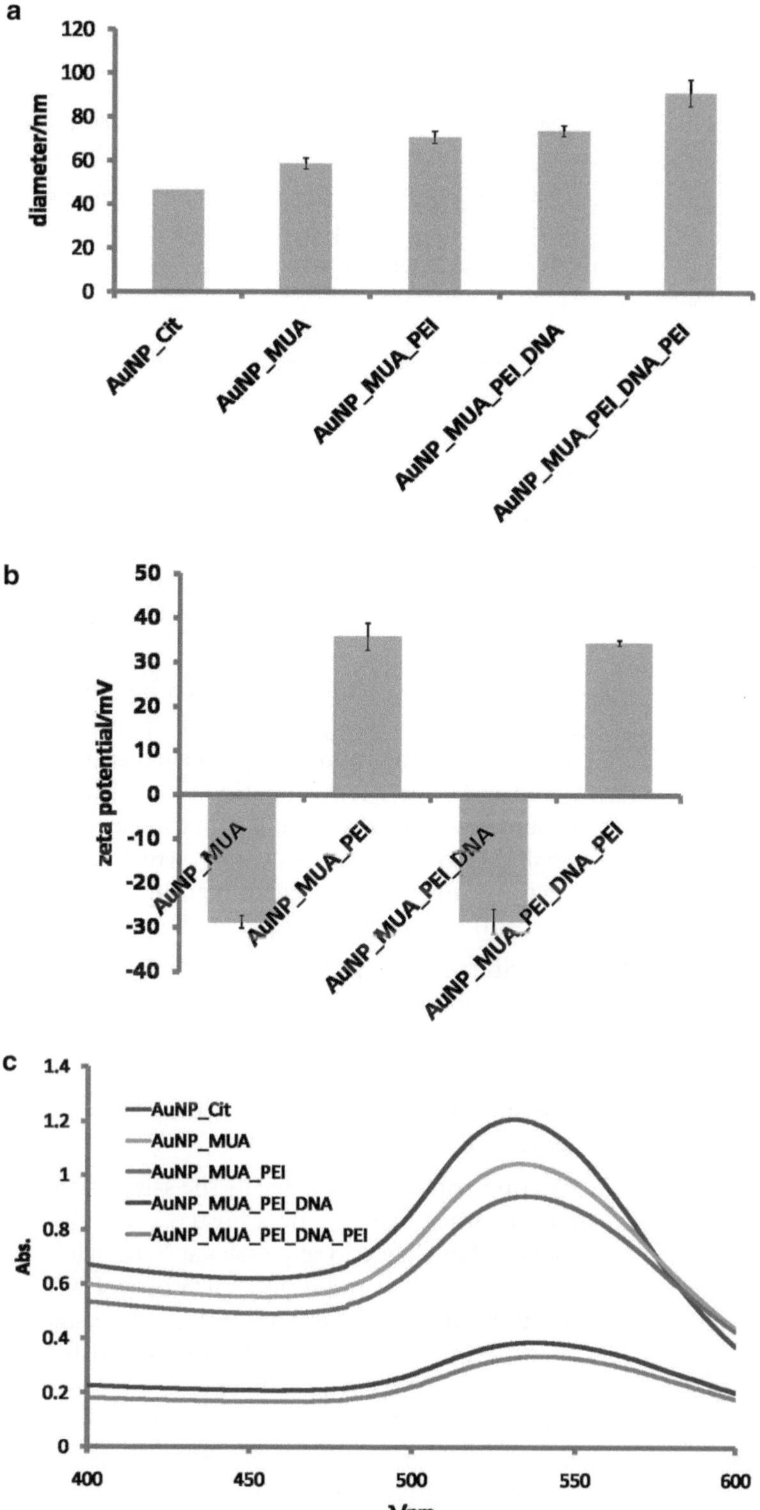

Fig. 4. (**a**) The particle size increases with the deposition of the polymer layers. The size of the particles was determined by dynamic light scattering of three independent coating batches. The extent of size growth is less than 10 nm per layer. (**b**) The zeta potential of the core–shell assembled particles inverts its sign with the deposition of the oppositely charged polymers. (**c**) The Plasmon resonance wavelength of the gold particles depends on the particle size. That means that the absorbance maximum shows a shift of 2 nm per particle layer.

2. The size distribution of gold suspension is analyzed by TEM images in addition to the light scattering measurements to ensure a homogenous particle size distribution.
3. TEM images can be analyzed using the free of charge available software Image J. A detailed introduction into this software can be found on the Internet.

4. Notes

1. The glassware and stir bars for all preparation steps should be thoroughly cleaned. Especially the flask and stir bar for the AuNP synthesis should be rinsed with aqua regia and afterwards with Millipore water. This avoids contaminations during the reduction step and enhances the quality of the nanoparticles.
2. Centrifugation conditions have to be adjusted to the particle size to avoid aggregation. Larger particles need slower and shorter centrifugation steps, and smaller particles need a higher speed and longer centrifugation times. The pellet of the AuNPs should have a red color and move loosely in the vessel. A dark, small spot on the wall of the cup, which cannot be resuspended easily, means that the centrifugation conditions have been too hard and the particles aggregated.
3. Centrifugation is performed in an Eppendorf centrifuge with 2 ml Eppendorf cups filled only with 1 ml of the particle suspension. The use of larger centrifugation vessels might lead to particle aggregation.
4. The supernatant should be centrifuged as well to achieve a better recovery rate.
5. Coating conditions, especially the salt concentrations and incubation times, have to be adjusted carefully to the used polymers and nucleic acid sequences and particle sizes. The protocol described here is not a general protocol which may not work for all kinds of nucleic acids and particle sizes. Before starting an experiment a concentration series with different salt concentrations (for example between 1 mM and 10 mM NaCl) and different polymer concentrations (for example between 0.5 and 5 mg/ml polymer) should be performed to establish the ideal coating conditions.
6. The nanoparticles need special storing conditions. In general purified particles should not be stored. The AuNP–MUA are stable over weeks at room temperature (the excess of MUA may precipitate at lower temperatures). The particles covered with PEI are stable for weeks as well, as long as the excess of

the polymer is still present. Storing the particles with the nucleic acids as the outer layer should be avoided, because nucleic acids are sensitive to nucleases. So particles should be stored either with the first or the last layer of PEI on the particle surface.

7. The coating steps of polymer layers needed at least 30 min, but the time can be extended largely.
8. It is recommended to follow each deposition step by measuring the size, zeta potential, and Vis absorbance of the particles. The increase of the particle size should be less than 10 nm per polymer layer and the shift of the absorbance maximum should be 1 or 2 nm per layer. If the particle suspension turns to a purple or a violet color the particles are aggregated.

References

1. Gao X et al (2007) Nonviral gene delivery: what we know and what is next. AAPS J 9:92–104
2. Mintzer MA, Simanek EE (2009) Nonviral vectors for gene delivery. Chem Rev 109:259–302
3. Whitesides GM, Grzybowski B (2002) Self-assembly at all scales. Science 295:2418–2421
4. Decher G (1997) Fuzzy nanoassemblies: toward layered polymeric multicomposites. Science 277:1232–1237
5. Hammond PT (2004) Form and function in multilayer assembly: new applications at the nanoscale. Adv Mater 16:1271–1293
6. Ariga K et al (2007) Layer-by-layer assembly as a versatile bottom-up nanofabrication technique for exploratory research and realistic application. Phys Chem Chem Phys 9:2319–2340
7. Lutkenhaus JL, Hammond PT (2007) Electrochemically enabled polyelectrolyte multilayer devices: from fuel cells to sensors. Soft Matter 3:804–816
8. Boudou T et al (2010) Multiple functionalities of polyelectrolyte multilayer films: new biomedical applications. Adv Mater 22:441–467
9. Decher G, Schlenoff JB (2002) Multilayer thin films: sequential assembly of nanocomposite materials, vol 1. Wiley VCH, Weinheim
10. Becker AL et al (2010) Layer-by-layer-assembled capsules and films for therapeutic delivery. Small 6:1836–1852
11. De Cock LJ et al (2010) Polymeric multilayer capsules in drug delivery. Angew Chem Int Ed Engl 49:6954–6973
12. Ai H et al (2003) Biomedical applications of electrostatic layer-by-layer nano-assembly of polymers, enzymes, and nanoparticles. Cell Biochem Biophys 39:23–43
13. Daniel MC, Astruc D (2004) Gold nanoparticles: assembly, supramolecular chemistry, quantum-size-related properties, and applications toward biology, catalysis, and nanotechnology. Chem Rev 104:293–346
14. Boisselier E, Astruc D (2009) Gold nanoparticles in nanomedicine: preparations, imaging, diagnostics, therapies and toxicity. Chem Soc Rev 38:1759–1782
15. Mayhew TM et al (2009) A review of recent methods for efficiently quantifying immunogold and other nanoparticles using TEM sections through cells, tissues and organs. Ann Anat 191:153–170
16. Kelly KL et al (2002) The optical properties of metal nanoparticles: The influence of size, shape, and dielectric environment. J Phys Chem B 107:668–677
17. Cademartiri L, Ozin GA (2009) Concepts of nanochemistry. Wiley-VCH-Verl, Weinheim
18. Bain CD et al (1989) Formation of monolayer films by the spontaneous assembly of organic thiols from solution onto gold. J Am Chem Soc 111:321–335
19. Giljohann DA et al (2010) Gold nanoparticles for biology and medicine. Angew Chem Int Ed Engl 49:3280–3294
20. Schneider G, Decher G (2008) Functional core/shell nanoparticles via layer-by-layer assembly. Investigation of the experimental parameters for controlling particle aggregation and for enhancing dispersion stability. Langmuir 24:1778–1789

21. Gittins DI, Caruso F (2001) Tailoring the polyelectrolyte coating of metal nanoparticles. J Phys Chem B 105:6846–6852
22. Breunig M et al (2007) Breaking up the correlation between efficacy and toxicity for nonviral gene delivery. Proc Natl Acad Sci USA 104:14454–14459
23. Boussif O et al (1995) A versatile vector for gene and oligonucleotide transfer into cells in culture and in vivo: polyethylenimine. Proc Natl Acad Sci USA 92:7297–7301
24. Sonawane ND et al (2003) Chloride accumulation and swelling in endosomes enhances DNA transfer by polyamine-DNA polyplexes. J Biol Chem 278:44826–44831
25. Elbakry A et al (2009) Layer-by-layer assembled gold nanoparticles for siRNA delivery. Nano Lett 9:2059–2064
26. Frens G (1973) Controlled nucleation for regulation of particle-size in monodisperse gold suspensions. Nat Phys Sci 241:20–22

Chapter 13

In Situ AFM Analysis Investigating Disassembly of DNA Nanoparticles and Nano-Films

Yi Zou, Lei Wan, Jenifer Blacklock, David Oupicky, and Guangzhao Mao

Abstract

Synthetic vector-based gene delivery systems continue to gain strength as viable alternatives to viral vectors due to safety and other concerns. DNA release dynamics is key to the understanding and control of gene delivery from nano-systems. Here we describe atomic force microscope application to the understanding of DNA release dynamics from bioreducible polycation-based nano-systems. The two nano-systems are polyplex nanoparticles and layer-by-layer films.

Key words: In situ AFM, Gene delivery, DNA release, Polyplex, Layer-by-layer film

1. Introduction

The promise of gene therapy to treat a variety of genetic and acquired diseases has fueled research on gene delivery systems. Nonviral gene carriers, especially polycations, are attractive alternatives to viral carriers because they show lower safety risks and can be tailored to specific therapeutic needs. A major problem of nonviral gene carriers is their low transfection efficiency. Currently the field is dominated by "black-box" strategies that test reporter gene expression levels of various formulations either in vitro or in vivo without detailed knowledge of carrier behavior along the gene delivery pathways. A critical question to the overall gene delivery efficiency is how and when plasmid DNA is dissociated from its complexes with polycations (polyplexes) (1–6). Several intracellular barriers have been identified that include endosomal escape, cytoplasmic transport, and nuclear entry (7–9). Extracellular barriers have also been identified including extracellular matrix, serum

Manfred Ogris and David Oupicky (eds.), *Nanotechnology for Nucleic Acid Delivery: Methods and Protocols*, Methods in Molecular Biology, vol. 948, DOI 10.1007/978-1-62703-140-0_13,

proteins, and soluble polyelectrolytes such as glycosaminoglycans (10, 11). The poor correlation between in vitro and in vivo transfection results suggests that extracellular barriers can also hinder DNA delivery.

Atomic force microscope (AFM) has two basic operational modes—the contact mode and the tapping mode (sometimes referred to as the AC mode). In the contact mode, AFM is operated at a constant cantilever deflection so that the interaction force between sample and probe is kept constant during imaging. In the tapping mode, the probe oscillates near its resonance frequency with constant, damped amplitude when scanning the surface. Both modes can be operated in air or solution (with a fluid tip holder). The tapping mode can provide clearer images of soft biological samples in solution by lowering frictional force between probe and sample (12, 13). In addition to the study of morphology, adsorption, and condensation of nucleic acids, (14–17) AFM has been used to study DNA condensation dynamics in solution with potential applications in gene delivery (18). Our research employs tapping mode in solution to monitor disassembly of polymeric gene delivery nano-systems in real time and at the single DNA molecular scale. Figure 1 shows in situ AFM setup for the DNA release study (Dimension 3100, VEECO). In situ AFM was used to study DNA release mechanisms from their polyelectrolyte assemblies with polycations.

For targeted gene delivery, polyplex disassembly can be triggered by the redox potential gradient between extracellular environment and various subcellular organelles in states compatible with the physiologic conditions (19–24). DNA release is triggered by depolymerization of high-molecular-weight polycations into low-molecular-weight oligocations via thiol and disulfide exchange reaction. The number of charges per reduced oligomer fragment plays an important role in regulating DNA release from polyplexes (5).

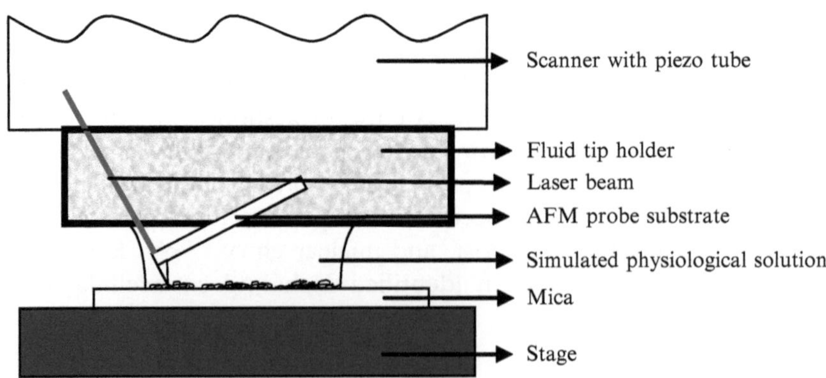

Fig. 1. A typical AFM setup for liquid operation. The AFM scanning can be either tip scanning or sample scanning. In this figure (Dimension 3100), the scanning is tip scanning.

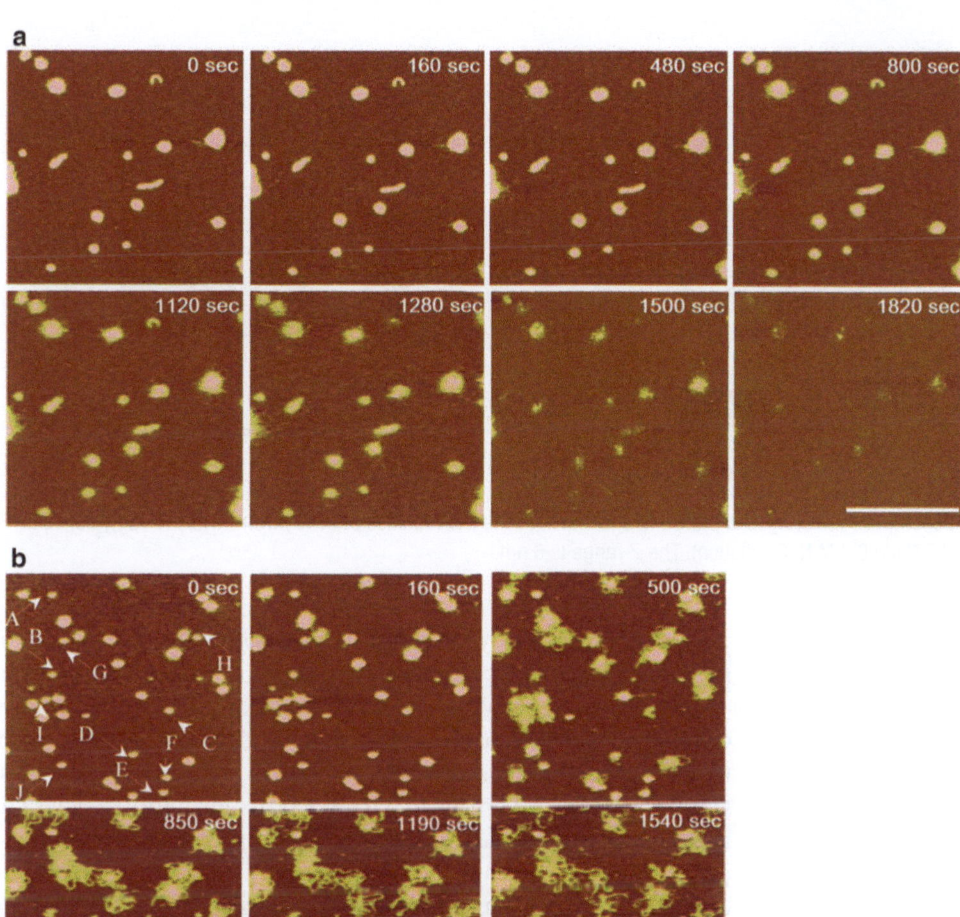

Fig. 2. In situ AFM sequence of DNA release from (**a**) NLS^{6+} polyplexes and (**b**) HRP^{10+} polyplexes in 20 mM DTT and 0.4 M NaCl solution. Time zero corresponds to the addition of DTT. The *Z*-range is 10 nm. Scan size is $2.6 \times 2.6\ \mu m^2$ for (**a**) and $2.0 \times 2.0\ \mu m^2$ for (**b**). Reproduced with permission from (5). Copyright 2008 American Chemical Society.

In Fig. 2a, DNA release from nuclear localization signal peptide (NLS^{6+}, CGAGPKKKRKVC) polyplexes displays an abrupt and size-independent disassembly route. In contrast, Fig. 2b shows a cooperative and size-dependent disassembly route from histidine-rich peptide (HRP^{10+}, CKHHHKHHHKC) polyplexes. This result is consistent with relatively low in vitro transfection efficiency of NLS^{6+} polyplexes. The results suggest that the abrupt disassembly of NLS^{6+} polyplexes and DNA release prevents intact polyplexes from reaching the nucleus where the role of NLS is manifested. On the other hand, the higher stability of HRP^{10+} polyplexes allows histidyl residues to carry out the buffering function in the endo/lysosomal pH. The AFM study provides a potential strategy to control DNA release timing by subtly varying the charge density of oligocations.

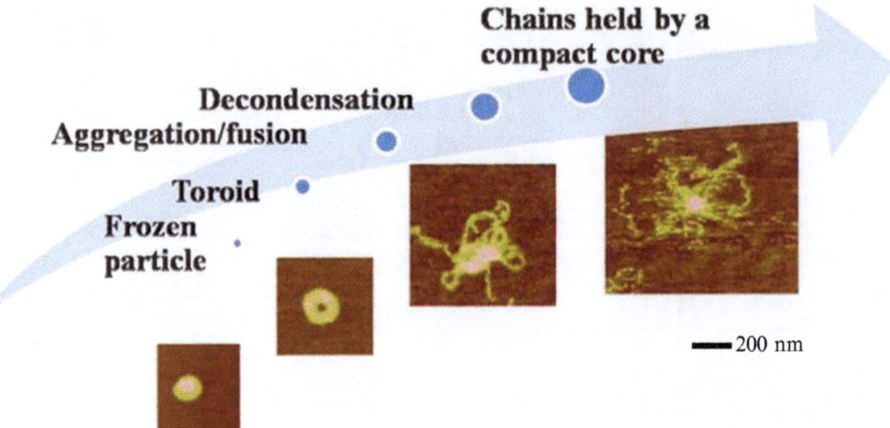

Fig. 3. Schematic AFM images captured in situ showing the DNA release pathway from RHB/DNA polyplexes (N/P = 4 in 20 mM DTT and 0.2 M NaCl solution). The *Z*-range is 6 nm.

In another study of bioreducible disassembly (25), common nanostructures were identified among bioreducible poly(amido amine) (PAA) polyplexes with different polymer structures, e.g., disulfide content, molecular weight, and molecular architecture. Figure 3 shows the disassembly to roughly consist of three stages. In stage 1, various metastable polyplex nanoparticles change into the toroid structure. In stage 2, the toroids interact with each other by aggregation and fusion. In the final stage, DNA wormlike chains gradually unravel from the polyplex resulting in loose loops/tails that are held by a central compact core. In situ AFM is capable of capturing transient polyplex nanostructures at the single polyplex level. It can be a powerful tool to evaluate polyplex transfection efficiency at the nanoscale.

In the disassembly of layer-by-layer (LbL) films containing bioreducible PAAs and DNA (26), AFM was used to capture disassembly, rearrangement, and release of molecules from the surface due to thiol–disulfide exchange reaction. Salt was found to accelerate the overall rate of film disassembly. Additionally, it was found that the LbL films disassemble much slower than polyplex nanoparticles. The predominant intermediate structure is the fiber bundle structure (Fig. 4) for the DNA layer during film disassembly in contrast to the toroid structure for the polyplex layer. This study offers a simple means to modulate DNA release from LbL films by utilizing both condensed and uncondensed DNA in different layers.

The disassembly of polyplexes can be induced by interpolyelectrolyte exchange reaction with various endogenous polyanions including heparin and albumin. This mechanism is responsible for DNA clearance in blood circulation and extracellular and intracellular transport. Figure 5 shows the polyplex disassembly process by

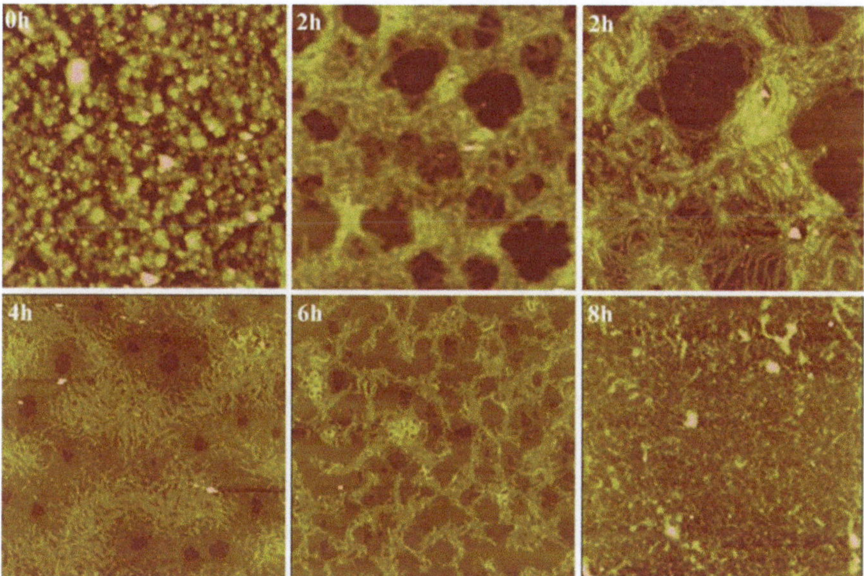

Fig. 4. Time elapse AFM images of the RHB/DNA/RHB film in 20 mM DTT and 0.1 M NaCl. Scan size = 1 μm for the third image and 4 μm for the rest. *Z*-range = 15 nm. Reproduced with permission from Blacklock et al. (26). Copyright 2008 American Chemical Society.

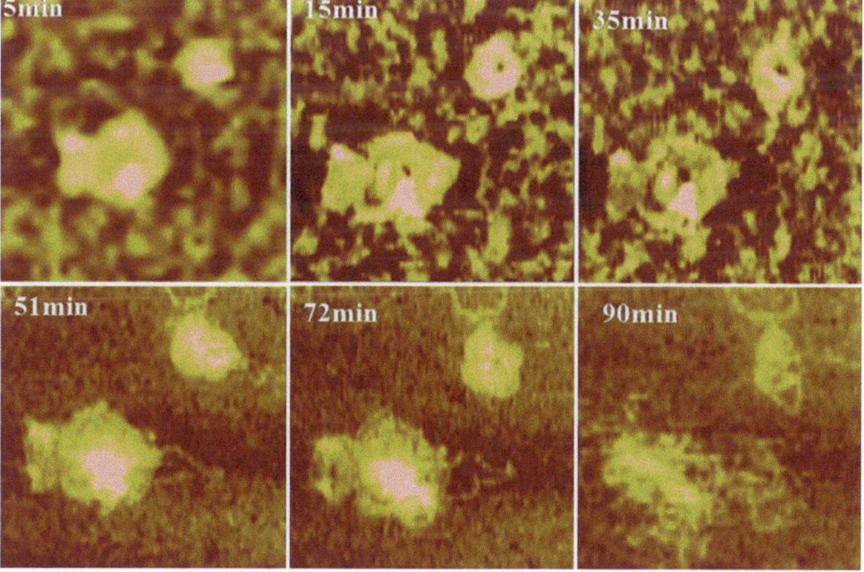

Fig. 5. In situ AFM image sequence for LPAA/DNA polyplexes (N/P = 12) treated by 500 μg/mL heparin for 5 s and then imaged in 30 mM pH 4.5 acetate buffer. Scan size is 2.2 μm; *Z*-range is 8 nm.

polyelectrolyte exchange with heparin at the single polyplex level using in situ AFM imaging. AFM captured distinct polyplex structures with partial DNA release that include the core–shell, toroid, nanoparticle-decorated toroid, and loose chains held by a compact core structure. This study contributes to the basic understanding of interpolyelectrolyte interaction at the solid/liquid interface and its effect on polyplex stability.

2. Materials

1. Plasmid DNA vectors, gWiz High-Expression GFP plasmid (6.7 kb) and gWiz High-Expression Secreted Alkaline Phosphatase (SEAP) plasmid (5.8 kb) (Aldevron). The contour length of DNA with 6.7 kb is estimated to be 2.3 μm.
2. Dithiothreitol (DTT, Sigma).
3. 1-(2-Aminoethyl)piperazine (AEPZ, Aldrich).
4. 1-Methylpiperazine (Aldrich).
5. *N,N′*-Methylenebisacrylamide (MBA, Aldrich).
6. *N,N′*-Cystaminebisacrylamide (CBA, Polysciences).
7. 1,5-Diiodopentane (DIP) (Acros Organics).
8. Heparin sodium salt (Sigma, H4784).
9. Stainless steel T304 mesh with 120 mesh woven wire with diameter 94.0 mm (TWP Inc., Berkeley, CA).
10. Custom-synthesized peptides (Sigma-Genosys): NLS (CGAGPKKKRKVC, Mr 1274) and HRP (CKHHHKHHHKC, Mr 1431).
11. Water deionized to 18 MΩ × cm resistivity using the Nanopure system from Barnstead.
12. Grade V5 muscovite mica was purchased from Ted Pella. Hand cleaved just before use.

3. Methods

3.1. Synthesis of Reducible Polypeptides

1. Polypeptides are synthesized at 30 mM concentration in phosphate-buffered saline containing 30 vol% DMSO via oxidation of the terminal cysteinyl thiol groups. Molecular weight is controlled by introducing compounds containing single thiol functionality, such as 2-aminoethanethiol.
2. Reaction product is purified from DMSO and cyclic by-products using centrifugal ultrafilters with molecular weight cutoff of 10 kDa.
3. Titration and size exclusion chromatography (SEC) were used to determine the concentration, conversion, and molecular weight.

3.2. Synthesis of Reducible Poly(Amido Amine)s

1. Reducible PAAs are synthesized by Michael addition copolymerization of a triamine (AEPZ) and bisacrylamide monomers (CBA and MBA). The chain architecture is controlled by feed ratio, i.e., half molar ratio for hyperbranched PAA (RHB) and equal molar ratio for linear PAA (LPAA). For example, CBA (0.260 g, 1.0 mmol) and MBA (0.308 g, 2.0 mmol) are added to a small vial containing AEPZ (0.193 g, 1.5 mmol) in methanol/water mixture (5 mL, 7/3 v/v) in order to obtain reducible RHB with 33 % disulfide content.
2. Reaction product is precipitated in cold acetone followed by dialysis in water with semipermeable membrane cutoff molecular weight of 10 kDa. Polymer powder is obtained by lyophilization.

3.3. Polyplex Preparation

1. Prepare 30 mM acetate buffer at desired pH (4.5 or 5.0) with or without sodium chloride according to different experiment design (see Note 1). All reagents such as DTT and heparin should be dissolved in this buffer (see Note 2).
2. Thaw DNA stock and dilute it with buffer to desired concentration. Apply gentle vortexing and let it stay at room temperature for 10 min.
3. Add the polymer solution to the DNA solution and mix by vortexing (Fisher Scientific Vortex Mixer Model 231) for 10 s followed by incubating at room temperature for 30 min (see Note 3).
4. Vary the feed ratio to control the N/P ratio (amine-to-DNA phosphate molar ratio) of the polyplexes.

3.4. LbL Film Preparation

1. Prepare substrates: In order to characterize the nanostructure of the LbL films, the films should be deposited on uniformly charged substrates. Different substrates were used following appropriate cleaning procedures. Silicon wafers are cleaned in piranha solution (volume ratio 3:1 of concentrated sulfuric acid to 30 % aqueous hydrogen peroxide solution). Quartz substrates are cleaned using sodium hydroxide followed by sulfuric acid. Stainless steel mesh is sterilized. Mica is freshly cleaved.
2. To assemble the films, immerse the substrate into the polycation solution for 15 min, and then rinse three times with deionized water for 2 min each. Repeat deposition alternatively in the DNA and polycation solution (DNA concentration 0.25 g/L, polymer 1–2 g/L) until a desired number of layers is obtained.
3. The LbL films can be cross-linked to improve film mechanical integrity and cell attachment. Place the film in 1 M DIP solution in water with 5 % hexane at 50–55 °C for 1 h, then rinse with deionized water, and dry with nitrogen.

3.5. In Situ AFM for Depolymerization-Induced DNA Release

1. Mount 1 cm^2 mica on a steel disk using epoxy glue. Double-sided tapes, often used for AFM imaging in air, are not suitable for fluid operation.
2. Place 20 μL of polyplexes solution on 1 cm^2 freshly cleaved mica. Avoid the exposure time of fresh mica surface to air in order to minimize contamination.
3. After 5 min, remove excess solution and rinse the surface with deionized water three times. Dry the surface using nitrogen.
4. Image sample using the tapping mode in air at first to characterize polyplexes' morphology. Silicon tip is used in the tapping mode with a factory-specified spring constant of 40 N/m.
5. Switch to the contact mode in air using a silicon nitride tip. Usually the tip has a nominal radius of curvature of 20 nm and cantilever spring constant of 0.38 N/m as provided by the manufacturer (see Note 4). Find an imaging area with appropriate polyplex density and distribution. Also confirm the probe condition, and clean or replace the tip if necessary.
6. Inject 40 μl buffer. Adjust operational parameters to achieve best image quality. Typically, the surface was imaged continuously at an average rate of 1–2 Hz on a 2×2 or 5×5 μm^2 area. The ranges of frequency, amplitude, integral gain, and proportional gain used are 8 kHz, 0.5–1.0 V, 0.2–1, and 0.4–2.0, respectively (see Note 5).
7. Inject 10 μL DTT solution and maintain the DTT concentration at 20 mM. The surface is scanned with AFM until no significant changes are observed.

3.6. In Situ AFM for Polyelectrolyte Exchange-Induced DNA Release

1. The procedures are similar with the above case except the following differences. After depositing polyplexes on mica surface and drying the surface, coat a layer of heparin on top of the polyplexes by placing a droplet of heparin solution for a short time. For example, 50 μL 500 μg/mL heparin solution is kept for 5 s.
2. The polyelectrolyte exchange reaction is followed in buffer solution. Image quality is improved by imaging in buffer instead of the polyanion solution.

3.7. LbL Film Disassembly and AFM Imaging

1. Conduct film disassembly in 20 mM DTT solution (pH 5–7, salt concentration 0–0.2 M). Complete film disassembly can take many hours.
2. Rinse substrates three times with deionized water for 2 min each and dry with nitrogen. Then mount substrates on steel disks for AFM imaging.
3. Use the tapping mode in air (see Note 6). Adjust scan rate according to scan size. Larger scan size requires lower scan rate

and smaller gains. The ranges of frequency, amplitude, integral, and proportional gains used are 150 kHz, 1.0–1.5 V, 0.1–0.4, and 0.2–0.8, respectively.

4. Notes

1. DNA molecules are fragile. Always avoid shear force such as high-speed vortexing and violent pipetting. Freeze and thaw cycles also should be avoided. Because AFM study is dependent on the DNA molecular morphology, it is essential to keep circular plasmid DNA structure intact and fully untangled. It is recommended to distribute DNA stock to aliquots at desired concentration and volume in order to avoid thaw and freeze cycle. Immediately before polyplex preparation, thaw DNA stock completely and gently vortex it.
2. For uniform polyplex preparation, a droplet of polymer solution is placed on the vial wall with the vial held horizontally. The vial is turned vertically to proceed immediately with mixing and vortexing. If the polymer concentration is nonuniform, polyplexes tend to aggregate at low polymer concentration and form spheroids at higher concentration.
3. Fresh DTT and GSH solutions should be used to avoid oxidation in order to obtain more consistent degradation rate results.
4. AFM probe selection is critical. For biological applications, low-stiffness cantilevers are recommended due to the low applied forces. V-shaped cantilevers (spring constant ~0.38 N/m) with oxide-sharpened silicon nitride tips were found suitable for the above disassembly studies for in situ tapping.
5. The tapping mode frequency choice is also important when frequency tuning is conducted manually in fluid tapping operation. The peak around 8 kHz provides best images while larger peaks may not yield the best results. Drive amplitudes are set at RMS signal 0.5 V. Large RMS may not be suitable for soft biomaterials. During in situ AFM imaging, it is necessary to keep track of the drive amplitude and set point in order to remain at minimal force. Scan rate should be adjusted with the scan area to remain at 2–5 μm/s. Integral and proportional gains can be optimized to improve contrast and decrease background noise.
6. It is more challenging to follow the disassembly of the LbL film in situ because of more film materials coming off the surface and film swelling. In addition, the disassembly of the LbL film takes longer time than individual polyplexes made of the same bioreducible polymers in similar conditions.

References

1. Chen J, Wu C, Oupicky D. Bioreducible hyperbranched poly(amido amine)s for gene delivery. Biomacromolecules. 2009;10:2921–7.
2. Honore I, Grosse S, Frison N, Favatier F, Monsigny M, Fajac I. Transcription of plasmid DNA: influence of plasmid DNA/polyethylenimine complex formation. J Control Release. 2005;107:537–46.
3. Pollard H, Remy JS, Loussouarn G, Demolombe S, Behr JP, Escande D. Polyethylenimine but not cationic lipids promotes transgene delivery to the nucleus in mammalian cells. J Biol Chem. 1998;273:7507–11.
4. Schaffer DV, Fidelman NA, Dan N, Lauffenburger DA. Vector unpacking as a potential barrier for receptor-mediated polyplex gene delivery. Biotechnol Bioeng. 2000;67:598–606.
5. Wan L, Manickam DS, Oupicky D, Mao GZ. DNA release dynamics from reducible polyplexes by atomic force microscopy. Langmuir. 2008;24:12474–82.
6. Zabner J, Fasbender AJ, Moninger T, Poellinger KA, Welsh MJ. Cellular and molecular barriers to gene-transfer by a cationic lipid. J Biol Chem. 1995;270:18997–9007.
7. Schaffert D, Wagner E. Gene therapy progress and prospects: synthetic polymer-based systems. Gene Ther. 2008;15:1131–8.
8. Wong SY, Pelet JM, Putnam D. Polymer systems for gene delivery-past, present, and future. Prog Polym Sci. 2007;32:799–837.
9. Burke RS, Pun SH. Extracellular barriers to in Vivo PEI and PEGylated PEI polyplex-mediated gene delivery to the liver. Bioconjug Chem. 2008;19:693–704.
10. Ruponen M, Ylä-Herttuala S, Urtti A. Interactions of polymeric and liposomal gene delivery systems with extracellular glycosaminoglycans: physicochemical and transfection studies. Biochim Biophys Acta. 1999;1415:331–41.
11. Zhou Q-H, Wu C, Manickam DS, Oupicky D. Evaluation of pharmacokinetics of bioreducible gene delivery vectors by real-time PCR. Pharm Res. 2009;26:1581–9.
12. Putman CAJ, Vanderwerf KO, Degrooth BG, Vanhulst NF, Greve J. Tapping mode atomic-force microscopy in liquid. Appl Phys Lett. 1994;64:2454–6.
13. Schabert FA, Engel A. Reproducible acquisition of Escherichia coli porin surface topographs by atomic force microscopy. Biophys J. 1994;67:2394–403.
14. Lindsay SM, Nagahara LA, Thundat T, Knipping U, Rill RL, Drake B, Prater CB, Weisenhorn AL, Gould SAC, Hansma PK. STM and AFM images of nucleosome DNA under water. J Biomol Struct Dyn. 1989;7:279–87.
15. Weisenhorn AL, Gaub HE, Hansma HG, Sinsheimer RL, Kelderman GL, Hansma PK. Imaging single-stranded-DNA, antigen-antibody reaction and polymerized Langmuir-Blodgett-films with an atomic force microscope. Scanning Microsc. 1990;4:511–6.
16. Hansma HG, Weisenhorn AL, Gould SAC, Sinsheimer RL, Gaub HE, Stucky GD, Zaremba CM, Hansma PK. Progress in sequencing deoxyribonucleic-acid with an atomic force microscope. J Vac Sci Technol B. 1991;9:1282–4.
17. Bustamante C, Vesenka J, Tang CL, Rees W, Guthold M, Keller R. Circular DNA molecules imaged in air by scanning force microscopy. Biochemistry. 1992;31:22–6.
18. Martin AL, Davies MC, Rackstraw BJ, Roberts CJ, Stolnik S, Tendler SJB, Williams PM. Observation of DNA-polymer condensate formation in real time at a molecular level. FEBS Lett. 2000;480:106–12.
19. Tang FX, Hughes JA. Use of dithiodiglycolic acid as a tether for cationic lipids decreases the cytotoxicity and increases transgene expression of plasmid DNA in vitro. Bioconjug Chem. 1999;10:791–6.
20. Byk G, Wetzer B, Frederic M, Dubertret C, Pitard B, Jaslin G, Scherman D. Reduction-sensitive lipopolyamines as a novel nonviral gene delivery system for modulated release of DNA with improved transgene expression. J Med Chem. 2000;43:4377–87.
21. Saito G, Swanson JA, Lee KD. Drug delivery strategy utilizing conjugation via reversible disulfide linkages: role and site of cellular reducing activities. Adv Drug Deliv Rev. 2003;55:199–215.
22. Read ML, Bremner KH, Oupicky D, Green NK, Searle PF, Seymour LW. Vectors based on reducible polycations facilitate intracellular release of nucleic acids. J Gene Med. 2003;5:232–45.
23. Chittimalla C, Zammut-Italiano L, Zuber G, Behr JP. Monomolecular DNA nanoparticles for intravenous delivery of genes. J Am Chem Soc. 2005;127:11436–41.
24. Oishi M, Hayama T, Akiyama Y, Takae S, Harada A, Yamasaki Y, Nagatsugi F, Sasaki S, Nagasaki Y, Kataoka K. Supramolecular assemblies for the cytoplasmic delivery of antisense oligodeoxynucleotide: polyion complex (PIC)

micelles based on poly(ethylene glycol)-SS-oligodeoxynucleotide conjugate. Biomacromolecules. 2005;6:2449–54.

25. Wan L, You Y, Zou Y, Oupicky D, Mao G. DNA release dynamics from bioreducible poly(amido amine) polyplexes. J Phys Chem B. 2009;113:13735–41.

26. Blacklock J, Mao G, Oupicky D, Mohwald H. DNA release dynamics from bioreducible layer-by-layer films. Langmuir. 2010;26(11):8597–605.

Chapter 14

Enhancing Nucleic Acid Delivery with Ultrasound and Microbubbles

Steven K. Cool, Bart Geers, Ine Lentacker, Stefaan C. De Smedt, and Niek N. Sanders

Abstract

For gene therapy to work in vivo, nucleic acids need to reach the target cells without causing major side effects to the patient. In many cases the gene only has to reach a subset of cells in the body. Therefore, targeted delivery of genes to the desired tissue is a major issue in gene delivery. Many different possibilities of targeted gene delivery have been studied. A relatively novel approach to target nucleic acids and other drugs to specific regions in the body is the use of ultrasound and microbubbles. Microbubbles are gas-filled spheres with a stabilizing lipid, protein, or polymer shell. When these microbubbles enter an ultrasonic field, they start to oscillate. The bubble expansion and compression are inversely related to the pressure phases in the ultrasonic field. When microbubbles are exposed to high-intensity ultrasound they will eventually implode and fragment. This generates shockwaves and microjets which can temporarily permeate cell membranes and blood vessels. Nucleic acids or (non)-viral vectors can extravasate through these pores to gain access to the cell's cytoplasm or the surrounding tissue. The nucleic acids can either be mixed with the microbubbles or loaded on the microbubbles. Nucleic acid-loaded microbubbles can be obtained by coupling nucleic acid-containing particles (i.e., lipoplexes) to the microbubbles. Upon ultrasound-mediated implosion of the microbubbles, the nucleic acid-containing particles will be released and will deliver their nucleic acids in the ultrasound-targeted region.

Keywords: Ultrasound, Microbubbles, Lipoplex, Biotin–avidin coupling, Targeted delivery

1. Introduction

The enhancement of naked nucleic acid delivery in combination with microbubbles and ultrasound (US) has been widely investigated in the last decade (1, 2). The bio-effects have been attributed

Steven K. Cool and Bart Geers contributed equally to this work.

Manfred Ogris and David Oupicky (eds.), *Nanotechnology for Nucleic Acid Delivery: Methods and Protocols*, Methods in Molecular Biology, vol. 948, DOI 10.1007/978-1-62703-140-0_14, © Springer Science+Business Media, LLC 2013

to the cavitation of microbubbles. The addition of synthetic microbubbles can therefore drastically increase the efficacy of ultrasound-assisted delivery of nucleic acids. However, in most studies, a high nucleic acid concentration was needed to obtain a biological effect (2–4) because of enzymatic degradation and short life span of US-induced pores. Degradation of nucleic acids can be prevented by complexing them with a cationic carrier. Several studies have shown that ultrasound and microbubbles can increase the efficacy of both viral and nonviral vectors and target them to specific tissues in vivo (5, 6). To obtain a high concentration of nucleic acids near the pores in the vasculature or cell membrane, different research groups designed microbubbles that were loaded with nucleic acids (7, 8). Direct loading of nucleic acids to cationic microbubbles has been studied by several groups. Alternatively, preformed nonviral vectors have been attached to microbubbles. We showed that an efficient and local gene delivery can be obtained by loading PEGylated lipoplexes on microbubbles and exposing them to ultrasound (9). To ascertain that the lipoplexes remain attached to the microbubbles in the presence of anionic serum components we loaded the lipoplexes via a biotin–(strept)avidin linkage to the microbubbles. In the following paragraph we discuss the production of microbubbles, the loading of lipoplexes on the microbubbles, and the evaluation of ultrasound-assisted gene delivery using microbubbles.

Microbubbles are spherical particles, composed of a lipid or a polymer shell, encapsulating and stabilizing a gaseous core. The gaseous core consists of hydrophobic, physiologically inert gas molecules, like perfluorcarbons (10–12). These microbubbles are often prepared just before injection and they are stable in the bloodstream for a few minutes. The gas is exhaled unaltered and the lipids enter the normal lipid metabolite pathways (12–15).

Lipid microbubbles behave differently in low- and high-intensity US fields. Low-intensity US will lead to the so-called stable cavitation state (Fig. 1a). The microbubbles will oscillate through different cycles of compression and expansion. The oscillation will give rise to microstreams in the vicinity of the microbubbles which cause intense friction and shear-stress with the surrounding structures. When microbubbles encounter high-intensity US (more than 1 MPa; 1 MHz), the amplitude of microbubble oscillation rises promptly, so that the surrounding liquid cannot follow the oscillation movement of the microbubbles (Fig. 1b). Surrounding liquid will rush towards the microbubbles, while the microbubbles already enter their expansion phase, which causes them to implode. This process is called inertial cavitation and results in shockwaves and sometimes microjets (16). Microjets can be described as a powerful stream of liquid caused by asymmetric implosion of microbubbles (10). The microstreams, formed during stable cavitation, and especially the shockwaves and microjets, resulting from

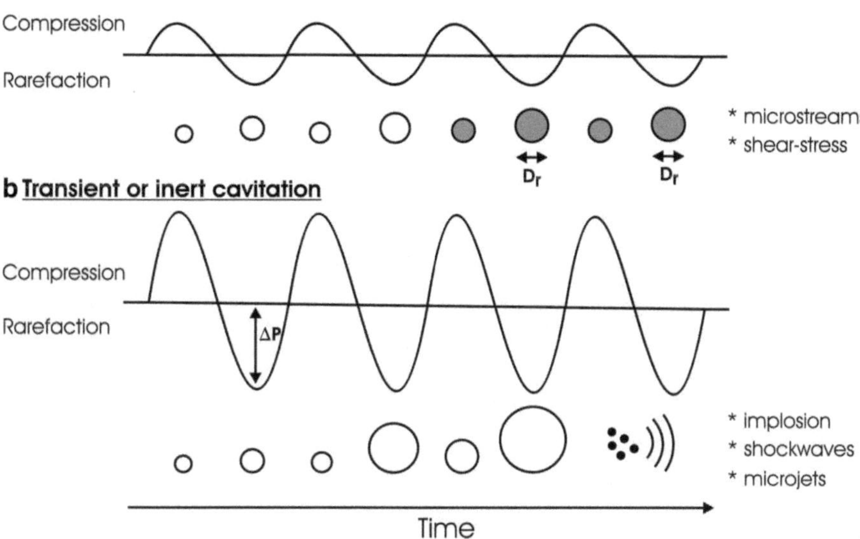

Fig. 1. Representation of the influence of ultrasound on a microbubble. (**a**) The influence of low-intensity ultrasound, shown as pressure change (ΔP) in function of time, on a microbubble. The microbubble will oscillate with a certain resonance diameter (D_r). (**b**) The influence of high-intensity ultrasound on a microbubble, causing microjets and shockwaves. Image adapted from (16).

inertial cavitation, will give rise to temporary pores in vessel walls and the membrane of surrounding cells (1, 10, 17–21).

The microbubbles we use are lipid-shelled microbubbles, based on the commercially available Definity® contrast agents. The Definity®-microbubbles are safe to use in vivo and are approved by the FDA. They can be administered i.v. through a bolus or an infusion. US is applied immediately after administration of the microbubbles.

Coupling lipoplexes or other components to microbubbles can be achieved by incorporating a biotinylated lipid in the microbubble shell and in the lipoplexes. For example, microbubbles containing a certain amount of biotinylated lipids are incubated with avidin. The excess of avidin is washed off before adding the lipoplexes which also contain lipid-PEG-biotin. This creates a biotin–avidin–biotin link between the two components (Fig. 2).

It is easy to see how these microbubbles can be used as a universal drug-carrier, where not only lipoplexes but also viruses and other nanoparticles containing nucleic acids and/or regular drugs can be coupled to the microbubbles.

The most frequently used liposomes for nucleic acid delivery consist of the cationic lipid *N*-(1-(2,3-dioleoyloxy)propyl)-*N*,*N*,*N*-trimethyl-ammoniumchloride (DOTAP) and the neutral lipid dioleoyl phosphatidylethanolamine (DOPE). Lipoplexes are created by controlled mixing of nucleic acids (NA) with lipids or preformed

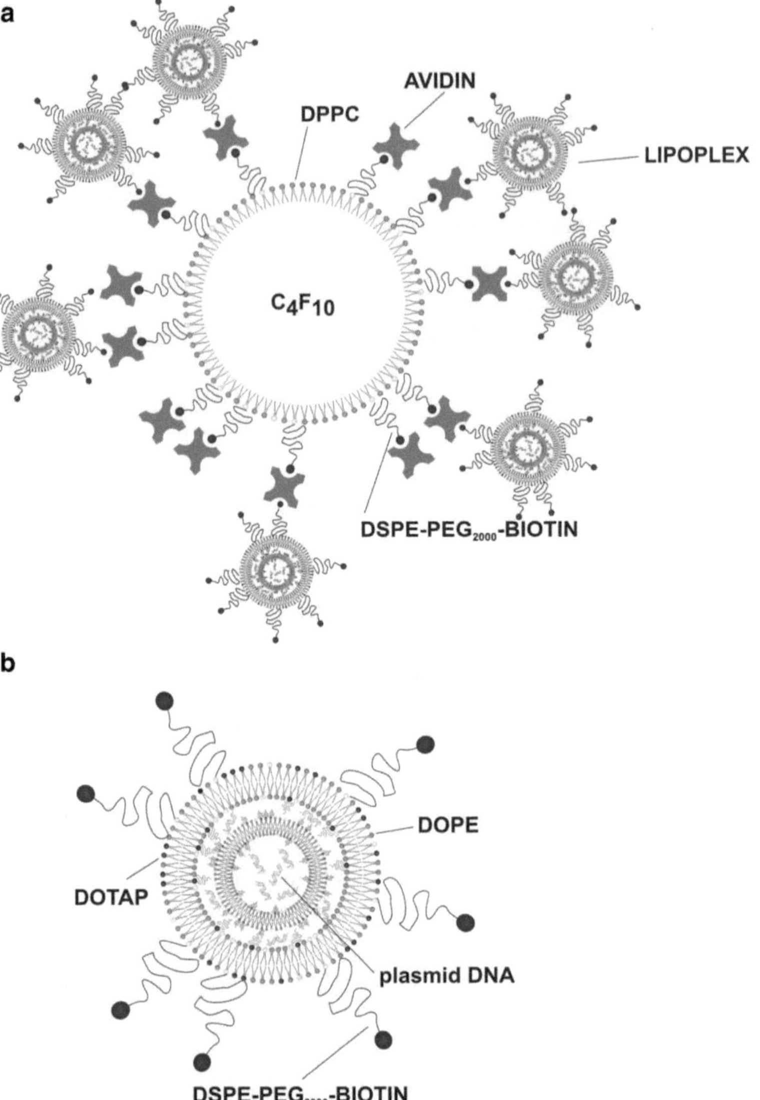

Fig. 2. (**a**) Schematic depiction of a lipoplex-loaded microbubble. The *white* disk surrounded by the lipids, dipalmitoylphosphatidylcholine (DPPC) and 1,2-distearoyl-*sn*-glycero-3-phosphoethanolamine-*N*-[biotinyl-PEG-2000] (DSPE–PEG–biotin), represents an avidinylated lipid microbubble with a perfluorobutane (C_4F_{10}) gas core. Lipoplexes with a certain amount of DSPE–PEG–biotin are attached to these avidinylated microbubbles via biotin–avidin–biotin bridges. (**b**) Detailed illustration of a single biotinylated lipoplex. *DOTAP* *N*-(1-(2,3-dioleoyloxy)propyl)-*N*,*N*,*N*-trimethylammonium chloride, *DOPE* dioleoyl phosphatidyl ethanolamine (image reproduced from (9)).

liposomes. The nucleic acids bind to the cationic lipids through electrostatic interactions. By adding a lipid with a PEG chain attached to it, the lipoplexes are more resistant to aggregation and have a lowered toxicity and a higher circulation time. However, the created PEG-shell interferes with the delivery process of the NA.

This problem is overcome by coupling them to microbubbles and subsequent exposure to US (9, 22–24).

To assess the transfection-efficiency of lipoplex-loaded microbubble in vitro, cells need to be grown in ultrasound transparent cell culture plates. This is to prevent unwanted reflections of the ultrasound waves during ultrasound treatment. OptiCells™ are a cell-culture system composed of two parallel, respiratory active membranes (RAM). These are gas-permeable polystyrene membranes, which reflect very little of the US waves. Cells are grown in the OptiCell™ and microbubble-solution can simply be injected in the OptiCell™ and thus added to the cells. The OptiCell™ is placed in a water tank with absorbing material at the bottom to prevent US reflection. The US transducer is placed on the OptiCell™ and evenly moved across the surface (Fig. 3). Transfection efficiency can be assayed following a standard luciferase protocol.

In vivo, the lipoplex-loaded microbubbles can be injected intravenously, intratumorally, or intramuscularly before applying US. The US-irradiated region can subsequently be imaged with an in vivo optical imaging system like the IVIS Lumina II (Caliper LS, USA).

Multiple US systems can be used in combination with microbubbles. The systems we use are the Sonidel™ SP100 (Sonidel™Ltd, UK) and the similar Sonitron2000 (Rich-Mar, Chattanooga, TN, USA).

Some groups also use commercially available ultrasound imaging devices for US-microbubble-mediated gene and drug delivery (25). Finally, many groups design their own ultrasound device and setup (26). An ultrasound device basically consists of a waveform generator, which generates electrical waveforms, coupled to a transducer probe, which is a piezoelectric crystal that converts the electrical waves into physical vibrations, creating the sound waves.

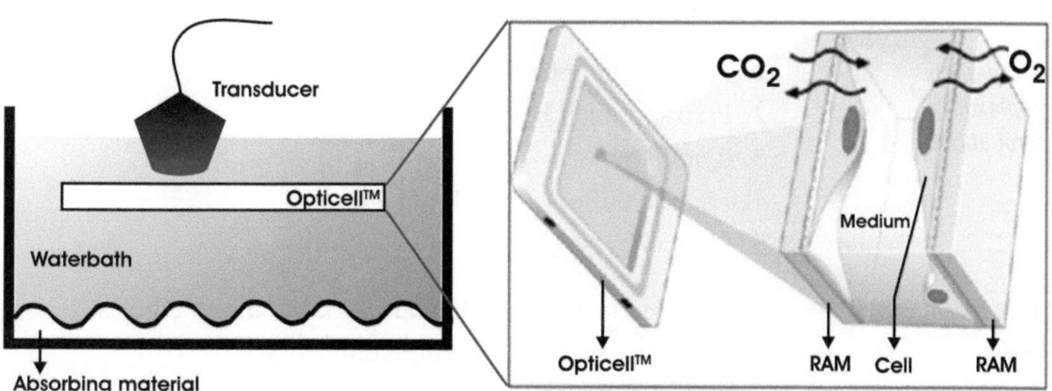

Fig. 3. Setup for US-application on an Opticell™ culture system with a cross-section of an Opticell™ (image adapted from www.nuncbrand.com).

The protocols for the preparation, loading, and administration of nucleic acid-loaded microbubbles and the assessment of the transfection efficiency, using a luciferase readout, are discussed below.

2. Materials

2.1. Lipoplex Preparation

1. Purified nucleic acids (pDNA, siRNA). In this example the pGL3 pDNA was used (Promega, Leiden, The Netherlands).
2. HEPES buffer (20 mM, pH 7.4).
3. DOTAP (Avanti polar lipids, Alabaster, AL).
4. DOPE (Avanti polar lipids, Alabaster, AL).
5. 1,2-Distearoyl-*sn*-glycero-3-phosphoethanolamine[biotin (polyethylene glycol)-2000] (DSPE–PEG–biotin) (Avanti polar lipids, Alabaster, AL).
6. Rotavapor.
7. Mini-extruder (Avanti polar lipids, Alabaster, AL).
8. Polycarbonate filters (200 nm, Whatman, Belgium).

2.2. Microbubble Preparation

1. 1,2-Dipalmitoyl-*sn*-glycero-3-phosphocholine (DPPC) (Avanti polar lipids, Alabaster, AL).
2. 1,2-Distearoyl-*sn*-glycero-3-phosphoethanolamine-*N*-[biotin(polyethylene glycol)-2000] (DSPE–PEG–biotin) (Avanti polar lipids, Alabaster, AL).
3. Round-bottom flask.
4. Rotavapor.
5. HEPES buffer (20 mM, pH 7.4).
6. C_4F_{10} gas (F2 chemicals, Preston, UK).
7. Tip sonifier (Branson, UK).

2.3. Coupling Lipoplexes to Microbubbles

1. Prepared lipoplexes.
2. Prepared microbubbles.
3. Avidin.
4. HEPES buffer (20 mM, pH 7.4).
5. Centrifuge.

2.4. Transfection In Vitro

1. Prepared and activated lipoplex-loaded microbubbles.
2. Opticell™ with cells at 90 % confluency.
3. Cell growth medium.
4. Phosphate-buffered saline.

5. Cell Culture Lyse Reagent (CCLR) (Promega, Leiden, The Netherlands).
6. Ultrasound device.
7. Luciferase assay tools.

3. Methods

3.1. Lipoplex Preparation

1. Mix in a round-bottom flask DOTAP and DOPE, dissolved in $CHCl_3$, in a 1:1 molar ratio with 5 % DSPE–PEG–biotin.
2. Evaporate the $CHCl_3$ in a rotavapor (Note 1).
3. When the $CHCl_3$ is evaporated, dissolve the lipid film in HEPES buffer at a final concentration of 5 mg lipids/ml (Note 2).
4. Extrude the resulting liposomes through a polycarbonate membrane (pore size 200 nm) using a mini-extruder (Avanti polar lipids, Alabaster, AL) (Fig. 4) (Note 3).

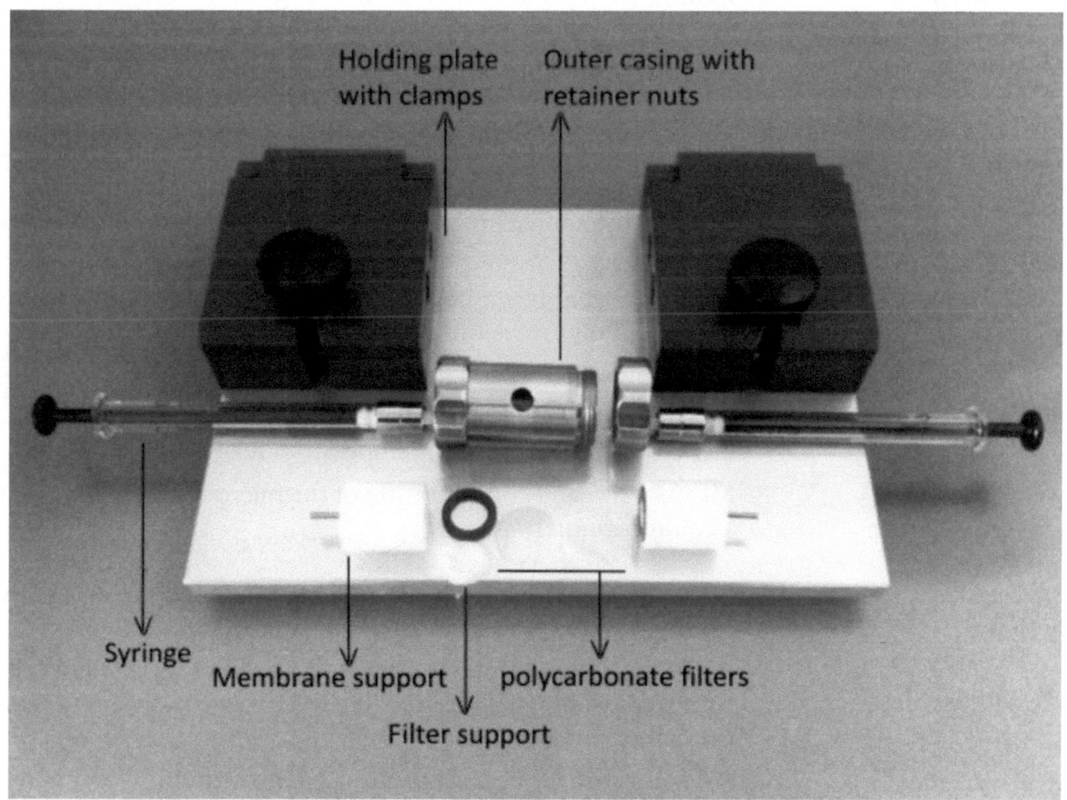

Fig. 4. Picture showing the different mini-extruder parts (Avanti polar lipids, Alabaster, AL). The system consists of two syringes, each connected to a membrane support unit which tightly holds two polycarbonate filters in place.

5. Dissolve pDNA in HEPES buffer to a concentration of 0.41 mg/ml.
6. Add the pDNA to an equal volume of cationic liposomes (5 mM DOTAP). Immediately add HEPES to make a final pDNA concentration of 0.126 mg/ml.
7. Vortex and incubate at room temperature for 30 min.
8. The resulting lipoplexes have a positive/negative charge ratio of 4.

3.2. Microbubble Preparation

1. Mix in a round-bottom flask DPPC and DSPE–PEG–biotin dissolved in $CHCl_3$ in a 85:15 molar ratio.
2. Evaporate the $CHCl_3$ in a rotavapor (Note 1).
3. When the $CHCl_3$ is evaporated, dissolve the lipid film in HEPES buffer at a final concentration of 5 mg lipid/ml (Note 2).
4. Add perfluorobutane (C_4F_{10}) gas on top of the liposome dispersion and sonicate the solution with a tip sonicator for about 2 min. It is important to hold the tip of the tip sonicator just below the liquid surface.

3.3. Coupling Lipoplexes to Microbubbles

1. Centrifuge the microbubbles at 118 g for 10 min to remove liposomes that did not form microbubbles.
2. Add avidin (final concentration 10 mg/mL) and incubate for 10 min.
3. Remove excess avidin by centrifuging microbubbles, as described above.
4. Remove the subnatant below the floating microbubbles with a syringe.
5. Wash microbubbles two times more with 3 ml fresh HEPES buffer and remove subnatant by centrifugation.
6. Dissolve microbubbles in 5 ml fresh HEPES buffer.
7. Add desired amount of lipoplexes to the avidinylated microbubbles and incubate at room temperature for 5 min.
8. Measure size and concentration of the microbubbles with, e.g., a coulter counter.

3.4. Transfection In Vitro

1. Fill an Opticell™ with 2×10^6 cells in 10 ml medium and grow overnight to 90 % confluency in cell-optimal conditions.
2. Add 1 ml of the lipoplex-loaded microbubbles to 10 ml of cell-culture medium and add this to the Opticell™ after removing the culture medium.
3. Submerge the Opticells™ in a water tank with 37 °C water and with a US absorbing material at the bottom.

4. Deliver US by evenly moving the US transducer over the whole plate during 1–2 min (Note 4).
5. US conditions: 1–3 MHz, 20–100 % duty cycle, and intensity 2–3 W/cm^2. This should be optimized for each cell type.
6. Incubate the cells for 2 h at 37 °C in the cell incubator.
7. Remove transfection medium and wash with PBS.
8. Add 10 ml of culture medium and incubate for 24 h.
9. Remove culture medium and wash with PBS.
10. Cut out the regions on the Opticell that have been exposed to ultrasound and that contain the attached cells. Subsequently, transfer the pieces of membrane to a 24-well plate. Alternatively, the substrate with d-luciferin can also be added to the Opticell and the bioluminescent light can be quantified using an in vivo optical imaging system.
11. Add 80 μl of CCLR buffer to each well.
12. Incubate at room temperature for at least 20 min.
13. Transfer 20 μl of the cell lysate from the 24-well plate to a 96-well plate. This step is needed if your luminometer only works with 96-well plates.
14. Luciferase expression can be assayed in a luminometer using a standard luminescence protocol.

4. Notes

1. Should it occur that the created lipid film does not look homogenous, one can simply add a few ml of $CHCl_3$ to dissolve the film again and restart the evaporation process.
2. Adding one or two small glass beads can help to suspend the lipid film.
3. Extruding the liposomes requires some force to be applied. If the extrusion suddenly becomes easier, it is an indication that the filter has ripped. One should use a new filter and start over.
4. Test the US transducer regularly to see if it is working. One can submerge the tip of the transducer into a volume in water and check for waves created by the transducer. Also when moving the transducer over an Opticells™, one should note the movement of bubbles away from the irradiated region.

References

1. Hernot S, Klibanov AL (2008) Microbubbles in ultrasound-triggered drug and gene delivery. Adv Drug Deliv Rev 60:1153–1166
2. Duvshani-Eshet M, Machluf M (2005) Therapeutic ultrasound optimization for gene delivery: a key factor achieving nuclear DNA localization. J Control Release 108:513–528
3. Kinoshita M, Hynynen K (2005) Intracellular delivery of Bak BH3 peptide by microbubble-enhanced ultrasound. Pharm Res 22:716–720
4. Manome Y, Nakayama N, Nakayama K, Furuhata H (2005) Insonation facilitates plasmid DNA transfection into the central nervous system and microbubbles enhance the effect. Ultrasound Med Biol 31:693–702
5. Koch S, Pohl P, Cobet U, Rainov NG (2000) Ultrasound enhancement of liposome-mediated cell transfection is caused by cavitation effects. Ultrasound Med Biol 26:897–903
6. Lawrie A, Brisken AF, Francis SE, Cumberland DC, Crossman DC, Newman CM (2000) Microbubble-enhanced ultrasound for vascular gene delivery. Gene Ther 7:2023–2027
7. Frenkel PA, Chen S, Thai T, Shohet RV, Grayburn PA (2002) DNA-loaded albumin microbubbles enhance ultrasound-mediated transfection in vitro. Ultrasound Med Biol 28:817–822
8. Teupe C, Richter S, Fisslthaler B, Randriamboavonjy V, Ihling C, Fleming I, Busse R, Zeiher AM, Dimmeler S (2002) Vascular gene transfer of phosphomimetic endothelial nitric oxide synthase (S1177D) using ultrasound-enhanced destruction of plasmid-loaded microbubbles improves vasoreactivity. Circulation 105:1104–1109
9. Lentacker I, De Smedt SC, Demeester J, Van Marck V, Bracke M, Sanders NN (2007) Lipoplex-loaded microbubbles for gene delivery: A Trojan horse controlled by ultrasound. Adv Funct Mater 17:1910–1916
10. Pitt WG, Husseini GA, Staples BJ (2004) Ultrasonic drug delivery–a general review. Expert Opin Drug Deliv 1:37–56
11. Tinkov S, Bekeredjian R, Winter G, Coester C (2009) Microbubbles as ultrasound triggered drug carriers. J Pharm Sci 98:1935–1961
12. Toft KG, Hustvedt SO, Hals PA, Oulie I, Uran S, Landmark K, Normann PT, Skotland T (2006) Disposition of perfluorobutane in rats after intravenous injection of Sonazoid. Ultrasound Med Biol 32:107–114
13. Grayburn PA (2002) Current and future contrast agents. Echocardiography 19:259–265
14. Klibanov AL (1999) Targeted delivery of gas-filled microspheres, contrast agents for ultrasound imaging. Adv Drug Deliv Rev 37:139–157
15. Unger EC, Porter T, Culp W, Labell R, Matsunaga T, Zutshi R (2004) Therapeutic applications of lipid-coated microbubbles. Adv Drug Deliv Rev 56:1291–1314
16. Newman CM, Bettinger T (2007) Gene therapy progress and prospects: ultrasound for gene transfer. Gene Ther 14:465–475
17. Brujan EA (2004) The role of cavitation microjets in the therapeutic applications of ultrasound. Ultrasound Med Biol 30:381–387
18. Collis J, Manasseh R, Liovic P, Tho P, Ooi A, Petkovic-Duran K, Zhu Y (2010) Cavitation microstreaming and stress fields created by microbubbles. Ultrasonics 50:273–279
19. Marmottant P, Hilgenfeldt S (2003) Controlled vesicle deformation and lysis by single oscillating bubbles. Nature 423:153–156
20. Postema M, van Wamel A, Lancee CT, de Jong N (2004) Ultrasound-induced encapsulated microbubble phenomena. Ultrasound Med Biol 30:827–840
21. Sundaram J, Mellein BR, Mitragotri S (2003) An experimental and theoretical analysis of ultrasound-induced permeabilization of cell membranes. Biophys J 84:3087–3101
22. Lentacker I, Wang N, Vandenbroucke RE, Demeester J, De Smedt SC, Sander NN (2009) Ultrasound exposure of lipoplex loaded microbubbles facilitates direct cytoplasmic entry of the lipoplexes. Mol Pharm 6:457–467
23. Remaut K, Sanders NN, De Geest BG, Braeckmans K, Demeester J, De Smedt SC (2007) Nucleic acid delivery: where material sciences and bio-sciences meet. Mat Sci Eng R 58:117–161
24. Vandenbroucke RE, Lentacker I, Demeester J, De Smedt SC, Sanders NN (2008) Ultrasound assisted siRNA delivery using PEG-siPlex loaded microbubbles. J Control Release 126:265–273
25. Bekeredjian R, Kroll RD, Fein E, Tinkov S, Coester C, Winter G, Katus HA, Kulaksiz H (2007) Ultrasound targeted microbubble destruction increases capillary permeability in hepatomas. Ultrasound Med Biol 33:1592–1598
26. Palussiere J, Salomir R, Le Bail B, Fawaz R, Quesson B, Grenier N, Moonen CT (2003) Feasibility of MR-guided focused ultrasound with real-time temperature mapping and continuous sonication for ablation of VX2 carcinoma in rabbit thigh. Magn Reson Med 49:89–98

Chapter 15

Magnetic and Acoustically Active Microbubbles Loaded with Nucleic Acids for Gene Delivery

Dialechti Vlaskou, Christian Plank, and Olga Mykhaylyk

Abstract

Targeted gene or drug delivery aims to locally accumulate the active agent and achieve the maximum local therapeutic effect at the target-site while reducing unwanted effects at nontarget sites. A further development of the magnetic drug-targeting concept is combining it with an ultrasound-triggered delivery using magnetic microbubbles as a carrier for gene or drug delivery. For this purpose, selected magnetic nanoparticles (MNPs), phospholipids, and nucleic acid are assembled in the presence of perfluorocarbon gas into flexible formulations of magnetic lipospheres or microbubbles. This manuscript describes the protocols for preparation of magnetic lipospheres and microbubbles for nucleic acid delivery, and it also describes the procedures for labeling the components of the bubbles (lipids, MNPs, and nucleic acids) for the visualization of the vectors and their characterization, such as magnetic responsiveness and ultrasound contrast effects. Protocols are given for the transfection procedure in adherent cells, evaluation of the association of the magnetic vectors with the cells, reporter gene expression analysis, and cell viability assessment.

Key words: Gene delivery, Magnetofection, Microbubble, Magnetic nanoparticles, Nonviral vectors, Sonoporation, Ultrasound contrast

1. Introduction

The delivery of nucleic acids to cells is required for both gene therapy and numerous research applications. Despite substantial progress in the development of gene delivery vectors, efficient gene vector delivery is a bottleneck in gene therapy applications. Viral vectors are able to mediate gene transfer with high efficiency and achieve long-term gene expression. However, the acute immune response, immunogenicity, and insertion mutagenesis demonstrated in gene therapy clinical trials have highlighted serious safety issues. Nonviral vectors formulated by nucleic acids assembled

Manfred Ogris and David Oupicky (eds.), *Nanotechnology for Nucleic Acid Delivery: Methods and Protocols*, Methods in Molecular Biology, vol. 948, DOI 10.1007/978-1-62703-140-0_15,

with synthetic or naturally occurring compounds to compact and stabilize DNA for delivery of the transgene into cells present certain advantages over viral ones with regard to the safety, ease, and cost of fabrication. Physical methods are also employed to further improve delivery efficacy, including the use of gene guns, hydrodynamic delivery, electroporation, and ultrasound radiation. These methods employ a physical force that permeates the cell membrane and facilitates intracellular gene transfer. During the last decade, the concept of magnetic drug targeting in nucleic acid delivery was developed (magnetofection) and is defined as nucleic acid delivery under the influence of a magnetic field that acts on nucleic acid vectors that are associated with magnetic (nano)particles (1, 2).

These days, medical ultrasound imaging is a well-established technique for clinical diagnostics. Ultrasound can improve drug delivery into tissues and cells (3–7). Although ultrasound has been proven to increase cell membrane permeability on its own (8), the use of microbubbles has a significant additive effect. Perturbation of the cell membrane and the vessel wall by insonated microbubbles increases the permeability and interstitial delivery as described (9). The microjets cause shear stress on the cell membrane and create transient, nonlethal holes in the plasma membrane through which a drug or a gene is able to diffuse (10–12). Additionally, it is believed that the generation of intracellular reactive oxygen species following the application of ultrasound might contribute to permeabilization of the cell membrane without affecting the cell viability (13). The main advantage of ultrasound-mediated microbubble-assisted transfection lies in the ability to achieve nucleic acid delivery and expression in only the insonated areas and not in the non-targeted organs.

Microbubbles comprise small cavities filled with a gas that has a low solubility in blood and that are stabilized by a layer of lipids or proteins (albumin), a polymer, or a mixture thereof. There are ultrasonic contrast agents that are used clinically (e.g., Sonovue®, Bracco; Optison®, Mallinckrodt). They are suitable carriers for drugs or nucleic acids because their outer layer provides an ideal surface for the incorporation of targeting ligands, imaging agents, and drugs and are "activated" by the local application of ultrasound at the target site (9, 14–19). Microbubbles are considered to be advantageous compared to other gene delivery methods because they can be applied for both therapeutic and diagnostic purposes. Based on our experience in the field of magnetofection, we developed magnetic microbubble (MMB) formulations that can be remotely controlled by both magnetic fields and ultrasound (20, 21) (Fig. 1). Magnetic nanomaterials, primarily iron oxides of various compositions, are tightly associated with lipid cavities known as microbubbles. Different genes, antisense oligonucleotides, synthetic siRNA, and drugs are associated with the magnetically responsive materials in the nano- to micrometer-size range

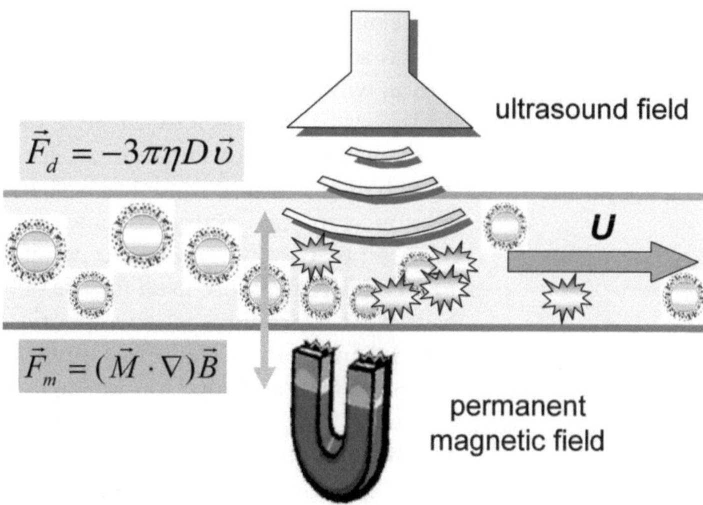

Fig. 1. Schematic presentation of magnetic microbubbles and the principle of function. Magnetic lipospheres (MAALs) and microbubbles (MMBs) are assembled from lipids, nucleic acids, and magnetic nanoparticles in a perfluoropropane atmosphere. As carriers of nucleic acids, MAALs/MMBs can be used for combined magnetic and ultrasound-targeted delivery. This figure demonstrates the MAALs'/MMBs' assembly, the magnetic attraction of MAALs/MMBs loaded with nucleic acid in the bloodstream by a suitable external magnetic field, and nucleic acid deposition triggered by ultrasound radiation.

and can be navigated by magnetic force. Our first generation of MMBs, the so-called magnetic acoustically active lipospheres, or MAALs (20, 22–26), is based on a composition developed for the delivery of hydrophobic drugs and is made up of soybean oil and amphipathic lipids (27). Additionally, the composition is loaded with magnetic nanoparticles (MNPs) and comprises a cationic lipid transfection reagent to bind nucleic acids (20). This formulation has a high binding capacity for nucleic acids and favorable magnetic properties but has a lower responsiveness to ultrasound compared to ultrasound contrast agents that have been approved for clinical use. Our second-generation MMBs (21), which are similar to the composition reported by Stride et al. and Lentacker et al. (28, 29), are highly responsive to both ultrasound and magnetic fields and give excellent contrast in ultrasound imaging.

The protocols provided here are for the fabrication of MMBs and enable the loading of the magnetic vesicles with plasmid DNA, small interfering RNA (siRNA), and antisense oligonucleotides. In transfection experiments for optimization of the formulations, it is useful to use reporter genes, such as luciferase or green fluorescence protein genes, for a sensitive and rapid in vitro evaluation of the results. Both MAALs and MMBs have shown a high efficiency of plasmid and siRNA delivery in vitro (20, 21, 25). Magnetic lipospheres loaded with siRNA down regulate reporter gene expression (eGFP) in GFP-expressing cells under flow conditions mimicking the bloodstream (25). Both lipospheres and microbubbles exhibit a significant response to the magnetic field and a high association with the cells after the transfection procedure. The magnetic lipospheres are efficient in vivo in animal models after either systemic or local administration, for example, as in a dorsal skinfold chamber mouse model, where magnetic targeting and sonoporation can be evaluated and where, additionally, intravital microscopy observation is also possible (20, 23).

The MAALs were used for the delivery of a VEGF plasmid and have shown a beneficial therapeutic effect in oversized skin flap survival in a rat model after application of an external, gradient magnetic field and ultrasound (26). A currently evolving aspect of our work is the combination of the therapeutic application of MMBs with the use of the medical ultrasound imaging in in vivo animal models. The protocols provided here may be used for further improvement of the MMB formulations and can be adapted to individual requirements.

2. Materials

The items listed below are used for both preparing acoustically active lipospheres and MMBs.

1. 2 mL Glass vials: Wheaton glass vials with aluminum cap with a rubber liner (product number 224222-01, Wheaton, Millville, NJ, USA) or Supelco clear glass vials with a screw top and polypropylene cap with a red rubber/PTFE septum (Sigma-Aldrich).
2. Perfluoropropane gas.
3. Mini BeadBeater (Biospec Products Inc., Bartlesville, OK, USA) or, alternatively, a CapMix™ capsule mixing device (3M ESPE AG, Seefeld, Germany).
4. Ultrasound water bath (Sonorex Digitec, Bandelin Electronics, Berlin, Germany).
5. 3-Way stopcock (Braun, Melsungen, Germany).

2.1. Preparation of Magnetic Acoustically Active Lipospheres

Unless stated otherwise, prepare all solutions using ultrapure water to obtain a resistivity of 18 MΩ·cm at 25°C. Unless otherwise indicated, prepare and store reagents at room temperature.

1. Metafectene (Biontex GmbH, Martinsried, Germany).
2. 1,2-Dioleoyl-*sn*-glycero-3-phosphoethanolamine (DOPE, Avanti Polar lipids, Alabaster, AL, USA).
3. 1,2-Dioleoyl-3-trimethylammonium-propane (DOTAP, Avanti Polar lipids, Alabaster, AL, USA).

4. Dilution buffer: 0.9% Sodium chloride, Glycerol, and Propylene glycol in an 8:1:1 v/v ratio, respectively.
5. Avanti® Mini-Extruder (Avanti Polar Lipids, Alabaster, AL, USA).
6. Polycarbonate membranes of 400, 200, and 100 nm (Avanti Polar Lipids, Alabaster, AL, USA).
7. Lipid stock solution 1 (DOPE:Metafectene at a molar ratio of 3:1): Weigh 13.2 mg (10.6 μmol) Metafectene and 26.8 mg (36.1 μmol) DOPE. Dissolve the lipids in a total of 2 mL of a 2:1 chloroform:methanol mixture. Remove the solvent by rotary evaporation to yield a thin lipid film on the sides of a round-bottomed flask. Dry the lipid film thoroughly to remove any residual organic solvent by placing the flask under high vacuum for 1 h. Hydrate the lipid film by adding 10 mL dilution buffer. Sonicate the lipid suspension in a bath sonicator (Bandelin) and filtrate it through the Avanti® Mini-Extruder using a 0.2 μm filter. The effluent should be slightly transparent.
8. Lipid stock solution 2 (DOPE:Metafectene at a 1:1 molar ratio): Prepare as described in Subheading 2.1, item 7, using 25.1 mg (20.1 μmol) Metafectene and 14.9 mg (20.1 μmol) DOPE.
9. Lipid stock solution 3 (DOPE:DOTAP at a molar ratio of 1:1): Prepare as described in Subheading 2.1, item 7, using 19.4 mg (27.74 μmol) DOTAP and 20.6 mg (27.74 μmol) DOPE.
10. FluidMAG-Tween 60 MNPs (chemicell GmbH, Berlin, Germany) with a nanomaterial concentration of 90 mg dry matter/mL and 31.5 mg Fe/mL. The particles have a mean maghemite crystallite size of 11 nm, a mean hydrated particle diameter of about 150 nm, and an iron content of 0.35 g Fe/g dry weight. Particles should have a saturation magnetization of the "core" M_s of 68.4 Am^2/kg Fe and an electrokinetic potential in water, ζ, of −8.9 ± 1.0 mV.
11. PEI-Mag2 MNP aqueous suspension (2.5 mg Fe/mL) coated with a mixture of 25 kDa branched polyethylenimine and fluorinated surfactant ZONYL® FSA synthesized and γ-sterilized as described in (30). PEI-Mag2 iron oxide nanoparticles had a mean magnetite crystallite size of 9 nm, a mean hydrated particle diameter of 28 ± 2, and an iron content of 0.56 g Fe/g dry weight. Particles should have a saturation magnetization of the "core" M_s of 62 Am^2/kg Fe and a ζ = +55.4 ± 1.6 mV. The concentration of PEI molecules assembled at the surface of PEI-Mag2 nanoparticles should be 0.14 g PEI/g Fe according to estimations based on X-ray photoelectron spectroscopy data (31).

12. PEI-Mag3 iron oxide nanoparticle aqueous suspension (4 mg Fe/mL) stabilized with fluorinated surfactant ZONYL® FSA (lithium 3-[2-(perfluoroalkyl) ethylthio] propionate) combined with a 25-kDa branched polyethylenimine (PEI-25_{Br}), similar to the PEI-Mag2 material (1). The particles have a mean magnetite crystallite size of 8 nm, a mean hydrated particle diameter of 29 ± 10, and an iron content of 0.21 g Fe/g dry weight. The saturation magnetization of the "core" M_s is 46 Am^2/kg Fe, and the electrokinetic (ζ) potential in water is $\zeta = +53.0 \pm 0.7$ mV. The concentration of PEI molecules assembled at the surface of PEI-Mag2 nanoparticles should be 0.3 g PEI/g Fe according to an estimation based on the X-ray photoelectron spectroscopy data.
13. Pal-D2-Mag1 MNP aqueous suspension (9 mg Fe/mL), synthesized as described (1): To modify the surface of the PalD2-Mag1 nanoparticles, the fluorinated surfactant ZONYL® FSE was combined with palmitoyl-dextran, PalD2. Palmitoyl-dextran (PalD2) with 32 palmitoyl groups per 100 dextran units was synthesized by means of esterification of Dextran-10 (Amersham Biosciences) with palmitoyl chloride as described elsewhere (32). The degree of esterification was determined as described (33). The particles had a mean magnetite crystallite size of 4 nm, a mean hydrated particle diameter of 55 ± 11, and an iron content of 0.51 g Fe/g dry weight. Saturation magnetization of the "core" M_s was 61 Am^2/kg Fe and the electrokinetic (ζ) potential in water was −53.7 ± 12.1 mV.
14. DNA stock solution: Luciferase plasmid p55pCMV-IVS-luc+ (pBLuc) encodes the firefly luciferase gene under the control of the cytomegalovirus (CMV) promoter (Plasmid Factory, Bielefeld, Germany) (5 mg/mL).
15. siRNAs: Green fluorescent protein (GFP) siRNA (Dharmacon, Lafayette, USA); Luciferase GL3-siRNA (Qiagen, Hilden, Germany); GFP22-siRNA-Rhodamine (Qiagen, Hilden, Germany); and negative control-siRNA, siGENOME Non-Targeting siRNA #1 (Dharmacon, Lafayette, USA).
16. siRNA solution: Reconstitute 5 nmol siRNA with 250 μL siRNA suspension buffer to obtain a 20 μM solution; heat the siRNA solution to 90°C for 1 min, incubate at 37°C for 60 min, and store in aliquots at −20°C.
17. Soybean oil (Hospital pharmacy, Klinikum rechts der Isar, Technical University of Munich).

2.2. Preparation of Magnetic Microbubbles

1. 1,2-Dipalmitoyl-*sn*-glycero-3-phosphocholine (DPPC, Avanti Polar Lipids, Alabaster, AL, USA).
2. 1,2-Distearoyl-*sn*-glycero-3-phosphoethanolamine-*N*-[biotinyl(polyethylene glycol)-2000] (ammonium salt) (DSPE-PEG(2000)-Biotin) (Avanti Polar Lipids, Alabaster, AL, USA).

3. Dilution buffer 0.9% Sodium chloride, Glycerol, and Propylene glycol in an 8:1:1 v/v ratio, respectively, as in Subheading 2.1, item 4.
4. 20 mM HEPES solution, pH 7.4.
5. Lipid stock solution 4: Prepare as described in Subheading 2.1, item 7, using 82.22 mg (112.01 μmol) DPPC and 17.78 mg (5.90 μmol) DSPE-PEG (2000)-Biotin.
6. FluidMAG-UC/C-11 and FluidMAG-UC/C-H iron oxide MNPs (chemicell GmbH, Berlin, Germany) with a concentration of 34 and 21 mg Fe/mL, a hydrodynamic diameter of 132 ± 57 nm and 153 ± 57 nm, and an electrokinetic potential of +20.8 ± 9.6 mV and +44.3 ± 6.2 mV, respectively.
7. PEI-Mag2 MNPs as described in Subheading 2.1, item 11.
8. DNA solution from Subheading 2.1, item 14.
9. A SonoVue® commercial microbubble set, including a syringe with 5 mL 0.9% NaCl reconstruction solution, and transfer system (BR1, Bracco, Italy) (34).
10. CombiMAG, 60 mg Fe/mL (chemicell GmbH, Berlin, Germany). The hydrodynamic diameter and electrokinetic potential of the particles when suspended in water were measured to be 96 ± 1 nm and +57.2 ± 1.7 mV, respectively. The iron content of the nanomaterial was 0.64 g/g dry matter (35).

2.3. Labeling of the Components of MAALs/MMBs

2.3.1. Fluorescent Labeling of the Components of MAALs/MMBs

1. DiOC18 (3,3′ dioctadecyloxacarbocyanine perchlorate, λ_{ex} 484/λ_{em} 501 nm) (Molecular Probes/Invitrogen, Karlsruhe, Germany): 2.5 mg/mL ethanol. Store 1 mL aliquots in the dark at −20°C.
2. YOYO-1 iodide stock: 1 mM solution in DMSO (Molecular Probes/Invitrogen). Store in aliquots in the dark at −20°C.
3. Fluorescence labeled plasmid DNA stock solution: Add 10 μL YOYO-1 iodide stock from step 2 to 40 μL (200 μg) of the DNA stock solution from Subheading 2.1, item 14 (corresponding to 1 dye molecule per 60 bp), and mix well. Incubate for 15 min in the dark. Store the solution at −20°C.
4. Rhodamine-labeled GFP-22 siRNA (λ_{ex} 550/λ_{em} 570 nm) labeled at the 3′-end of the sense strand (siRNA^{Rh}, Qiagen, Hilden, Germany). Store aliquots in the dark at −80°C.
5. Atto-550 fluorophore (λ_{ex} 554/λ_{em} 576 nm) (92835, Sigma) stock: 2 mg/mL DMSO. Store in the dark at −20°C.
6. 0.1 M Na-Borate buffer, pH 8.5.
7. Pierce cassette dialysis device with a cutoff of 3,500 MW.
8. 24-Well cell culture plates (Techno Plastic Products, Switzerland).
9. 24-Well magnetic plate (Chemicell, Berlin, Germany, and OZ Biosciences, Marseille, France).

2.3.2. Radioactive Labeling of siRNA and Plasmid

1. ^{125}I-labeled DNA solution prepared according to the protocol described in (1, 36): 90 μg DNA/mL water with a radioactivity of 2.4×10^4 CPM/μL.
2. ^{131}I-labeled siRNA solution prepared according to the protocol described in (1, 36): 118.3 μg siRNA/mL water with a radioactivity of 2.1×10^5 CPM/μL.
3. 3 M Sodium acetate buffer, pH 5.
4. Glycogen.
5. 70 and 95% ethanol.
6. DEPC-treated (RNase-free) water: It is suitable for use in the handling of RNA in the laboratory to reduce the risk of RNA degradation by RNases. It is prepared by incubating 500 mL distilled water with 0.5 mL 0.1%v/v diethylpyrocarbonate (DEPC) at room temperature for at least 1 h, autoclaved to inactivate traces of DEPC, and sterile filtered prior to use.

2.4. Determination of Loading of MAALs/MMBs with Magnetic Nanoparticles

1. 10% Hydroxylamine hydrochloride in water.
2. Ammonium acetate buffer: Dissolve 25 g ammonium acetate in 10 mL water, add 70 mL glacial acetic acid, and adjust volume to 100 mL with water.
3. 0.1% Phenanthroline solution: Dissolve 100 mg 1,10-phenanthroline monohydrate in 100 mL water and add 10 μL of concentrated hydrochloric acid.
4. 0.05 N $KMnO_4$ solution: Dissolve 0.790 g $KMnO_4$ in 100 mL water.
5. Iron stock solution: Dissolve 392.8 mg ammonium iron(II) sulfate hexahydrate in a mixture of 2 mL concentrated sulfuric acid and 10 mL water, add 0.05 N $KMnO_4$ dropwise until the pink color persists, and adjust the volume to 100 mL with water.
6. Standard iron solution: Dilute the iron stock solution 1:25 with water before carrying out the calibration measurements.

2.5. Determination of Size, Electrokinetic Potential, and Concentration Volume of MAALs/MMBs

1. Z2 Coulter counter (Beckman Coulter GmbH, Krefeld, Germany).
2. Isoton® II electrolyte solution (Beckman Coulter GmbH).
3. Dilution vials for Coulter® Counter, 20 mL-polystyrene square bottom style (Greiner Bio-One GmbH, Frickenhausen, Germany).
4. Coulter CC size Standards (Beckman Coulter GmbH).
5. Malvern Zetasizer 3000HS (Malvern, Worcestershire, UK) or Nano ZS (Malvern).
6. Folded capillary cell with 0.7 mL volume (Malvern, Worcestershire, UK).

2.6. Preparation of Gelatin-Clots with Embedded Bubbles for Transmission Electron Microscopy

1. Gelatin (catalog number G9382, Sigma GmbH; Taufkirchen, Germany), freshly prepared as a 10% solution in water.
2. 24-Well cell culture plate.
3. Electron microscopic grids (Plano GmbH, Wetzlar, Germany).
4. 0.5% Aqueous uranyl-acetate.
5. 0.3% Lead citrate.

2.7. Evaluation of Magnetic Responsiveness

1. Eight Ne-Fe-B permanent magnets, 18.0×16.0×4.0 mm (IBS Magnets).

2.8. Ultrasound Responsiveness

1. 1% Agarose solution in 45 mM Tris–borate, 1 mM EDTA pH 8.
2. A single-use Syringe (Omnifix®-F, B/Braun Melsungen AG, Melsungen, Germany) and a needle (B/Braun Melsungen AG, Melsungen, Germany).
3. Ultrasound device SIEMENS Acuson Antares equipped with a VFX 13-5 scanhead (center frequency: 10 MHz, imaging depth: 1 cm, MI = 0.8).
4. Tissue-mimicking phantom gel.
5. 0.9% Isotonic saline solution.
6. Ultrasonix Sonix RP research ultrasound system (Ultrasonix, Richmond, BC, Canada) with a C5-2/60 curved array (3.5 MHz center frequency).

2.9. Transfections with pDNA- or siRNA-Loaded MAALs

1. NCI-3T3 mouse fibroblasts (NIH 3T3) (ATCC).
2. NCI-H441 human pulmonary epithelial (H441) cells (ATCC).
3. HeLa cells stably expressing firefly luciferase (HeLa-Luci cells).
4. HeLa cells stably expressing eGFP (HeLa-GFP cells).
5. NIH 3T3 and HeLa culture medium: DMEM medium supplemented with 10% heat-inactivated FCS, 100 U/mL penicillin, and 100 mg/mL streptomycin.
6. H441 culture medium: Modified RPMI 1640 medium with 2 mM L-glutamine, 10 mM HEPES, 1 mM sodium pyruvate, 4.5 g/L glucose, and 1.5 g/L sodium bicarbonate supplemented with 10% heat-inactivated FCS, 100 U/mL penicillin, 100 mg/mL streptomycin, and 2 mM L-glutamine.
7. Trypsin/EDTA solution, 0.25%/0.02% (wt/vol).
8. Dulbecco's PBS w/o Ca^{2+} or Mg^{2+} solution.

2.10. Association of pDNA with Cells upon Transfection

1. Cells and reagents from Subheading 2.9, items 1–8.
2. ^{125}I-labeled plasmid DNA solution from Subheading 2.3.2, item 1.
3. Lysis buffer: 0.1% Triton X-100 in 250 mM Tris–HCl, pH 7.8.

2.11. Evaluation of Gene Expression In Vitro

1. Lysis buffer: 0.1% Triton X-100 in 250 mM Tris–HCl, pH 7.8.
2. Luciferin buffer: 35 mM D-luciferin, 60 mM DTT, 10 mM magnesium sulfate, and 1 mM ATP in 25 mM glycyl-glycine-NaOH buffer, pH 7.8.
3. Luciferase standard stock: 0.1 mg luciferase/mL (Roche Diagnostics) and 1 mg BSA/mL in 0.5 M Tris–acetate buffer, pH 7.5. Store in aliquots at −70°C.
4. BioRad protein assay reagent.
5. BSA stock solution: 1.5 mg/mL BSA in PBS. Store at 4°C.
6. Clear bottom, black-walled plate, 96-wells (Greiner Bio One).
7. Microplate reader Wallac 1420 Multilabel counter.
8. Microplate Scintillation & Luminescence Counter (Canberra Packard).

2.12. MTT-Based Cytotoxicity Assay

1. MTT solution: Dissolve 1 g thiazolyl blue tetrazolium bromide (MTT) and 5 g glucose in 1 L Dulbecco's PBS solution to obtain a final concentration of 1 mg MTT/mL and 5 mg/mL glucose. Make 30-mL aliquots in 50-mL centrifuge tubes and store the solution at −20°C.
2. MTT solubilization solution: 10% Triton X-100 in 0.1 N hydrochloric acid in anhydrous isopropanol. The solution can be stored at room temperature (15–25°C).
3. 96-Well format magnetic plate (OZ Biosciences, Marseille, France, and chemicell, Berlin, Germany).

3. Methods

3.1. Preparation of Magnetic Acoustically Active Lipospheres

MAALs, according to the procedures described below, are prepared from a mixture of zwitterionic lipid DOPE in combination with cationic Metafectene or DOTAP, soybean oil, core-shell MNPs possessing almost neutral, positive, or negative electrokinetic potential, and nucleic acids such as plasmid DNA and siRNA. With optimized ratios of the components described in this protocol, the formation of stable MAALs for most of the MNPs was possible only in combination with nucleic acids. Only palmitoyl dextran-stabilized MNPs allowed the formation of stable MAALs without nucleic acid that could be loaded with DNA or siRNA after preparation of the MAALs (in *trans*). Images and size distribution of plasmid- and siRNA-loaded MAALs are shown in Fig. 2. See also Note 1.

3.1.1. MAALs Loaded with FluidMAG-Tween 60 MNPs and Plasmid DNA (Tw-MAG-AAL/pDNA)

1. Put 11.1 μL FluidMAG-Tween 60 suspension into an Eppendorf tube (see Note 2) and add 90.5 μL (0.362 mg lipids) of lipid stock solution 1. Add 8 μL DNA stock solution (40 μg

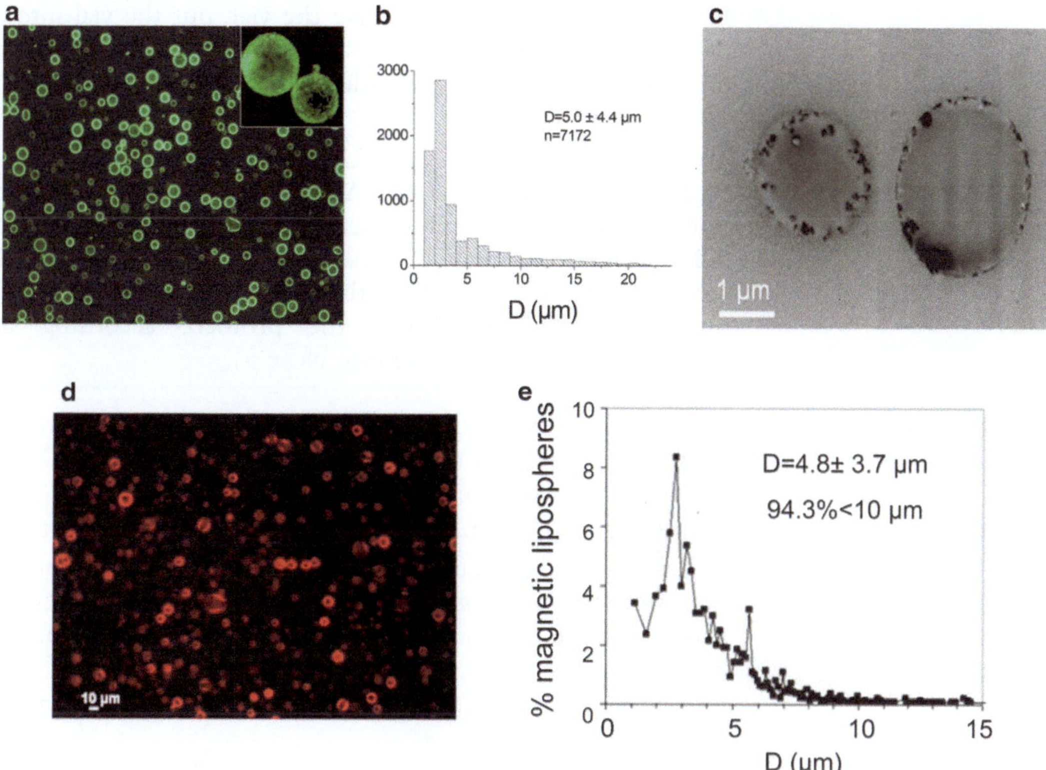

Fig. 2. Characteristics of the MAALs. (**a**) Fluorescence microscopy image (490 nm/509 nm) of the Tw-Mag-AAL-pBLuc/YOYO-1. (**b**) Size distribution of the Tw-Mag-AAL-pBLuc. (**c**) TEM image of the Tw-Mag-AAL-pBLuc embedded in 10% gelatin. Images (**a–c**) are reproduced with permission from WILEY-VCH Publisher's Adv. Funct. Mater. (20). (**d**) Fluorescence microscopy image (550 nm/570 nm) of the Tw-Mag-AAL-siRNA-Rho. (**e**) Size distribution of the Tw-Mag-AAL-siRNA based on quantitative analysis of the microscopy images. Images (**d** and **e**) are reproduced with permission from ACS Publications' NanoLetters (25).

plasmid DNA) and mix well. To form magnetic lipospheres associated with siRNA (Tw-MAG-AAL/siRNA), alternatively add 2.87 nmol siRNA (≈40 μg). Adjust the volume to 1 mL with dilution buffer (see Note 3).

2. Put 50 μL soybean oil into the Wheaton glass vial (see Note 4). Add the mixture from Subheading 3.1.1, step 1, to the soybean oil in the Wheaton glass vial.
3. Fill the headspace of the Wheaton glass vial from Subheading 3.1.1, step 2, with perfluoropropane gas using a 3-way stopcock, and agitate the vial for 60 s at 2,500 vpm using the Mini BeadBeater. Alternatively, a 2 mL Supelco vial can be used to prepare the mixture from Subheading 3.1.1, step 2, and agitated for 60 s at 4,650 vpm using the CapMix™ capsule-mixing device.
4. To remove the components that are not associated with MAALs, wash the lipospheres by multiple (up to three) centrifugation

flotations in 0.9% NaCl. Reverse the vial, put the vial into a 50 mL Falcon tube, centrifuge at 100×*g* for 5 min (980 rpm in a 5417R centrifuge, Eppendorf), and remove the underlying liquid with a syringe.

5. To prepare Tw-MAG-AAL/siRNA with a higher siRNA load of from 50 to 250 μg siRNA/mL, use between 113 and 566 μL lipid stock solution 1 containing between 0.452 and 2.26 mg lipids and from 6.25 to 31.25 μL of the siRNA stock solution, respectively. Adjust the volume to 1 mL with the dilution buffer and follow the protocol according to Subheading 3.1.1, steps 2–3 (see Note 5).

3.1.2. MAALs Loaded with PEI-Mag2 MNPs and Plasmid DNA or siRNA (PEI-Mag2-AAL/ pDNA(siRNA))

1. Put 46 μL lipid stock solution 2 (0.184 mg lipids) into an Eppendorf tube, add 40 μg nucleic acid (pDNA or siRNA), mix well with a pipette, and incubate for 10 min.
2. Transfer 200 μL PEI-Mag2 MNP suspension (0.5 mg iron) into the second Eppendorf tube, transfer the contents of the tube from Subheading 3.1.2, step 1, mix well with a pipette, and adjust the volume up to 1 mL with the dilution buffer (see Note 3).
3. Perform the steps as described in Subheading 3.1.1, steps 2–4.

3.1.3. MAALs Loaded with PEI-Mag3 MNPs and Plasmid DNA or siRNA (PEI-Mag3-AAL/ pDNA(siRNA))

1. Put 200 μL PEI-MAG3 particle suspension into an Eppendorf tube (see Note 2) and add 66 μL lipid stock solution 3. Add 8 μL DNA stock solution (40 μg plasmid DNA) and mix well. To form magnetic lipospheres coated with siRNA (Tw-MAG-AAL/siRNA), alternatively add 143.5 μL of the siRNA stock (20 μM) containing 2.87 nmol siRNA (≈ 40 μg for a siRNA with an MW of 14,000 Da). Adjust the volume to 1 mL with the dilution buffer (see Note 3).
2. Perform the steps as described in Subheading 3.1.1, steps 2–4.

3.1.4. MAALs Loaded with PalD2-Mag1 MNPs (PalD2-Mag1-AAL) with Plasmid DNA or siRNA Associated in Trans

1. Put 55.5 μL PalD2-Mag1 MNP suspension (see Note 2) containing 0.5 mg iron into an Eppendorf tube and add 66 μL (0.264 mg lipids) of lipid stock solution 3. Adjust the volume to 1 mL with the dilution buffer (see Note 3). Perform the steps as described in Subheading 3.1.1, steps 2–4.
2. To load nucleic acid in *trans*, take 100 μL of the liposphere suspension from the vial with a syringe, then take 8 μL DNA stock solution into the same syringe, and mix gently. Put the contents of the syringe back into the vial.

3.2. Preparation of Magnetic Microbubbles

MMBs, according to the procedures described below, are prepared from the mixture of DPPC and DSPE-PEG(2000)-Biotin lipids taken at a molar ratio of 95:5, different core-shell MNPs possessing rather high, positive electrokinetic potential, and plasmid DNA.

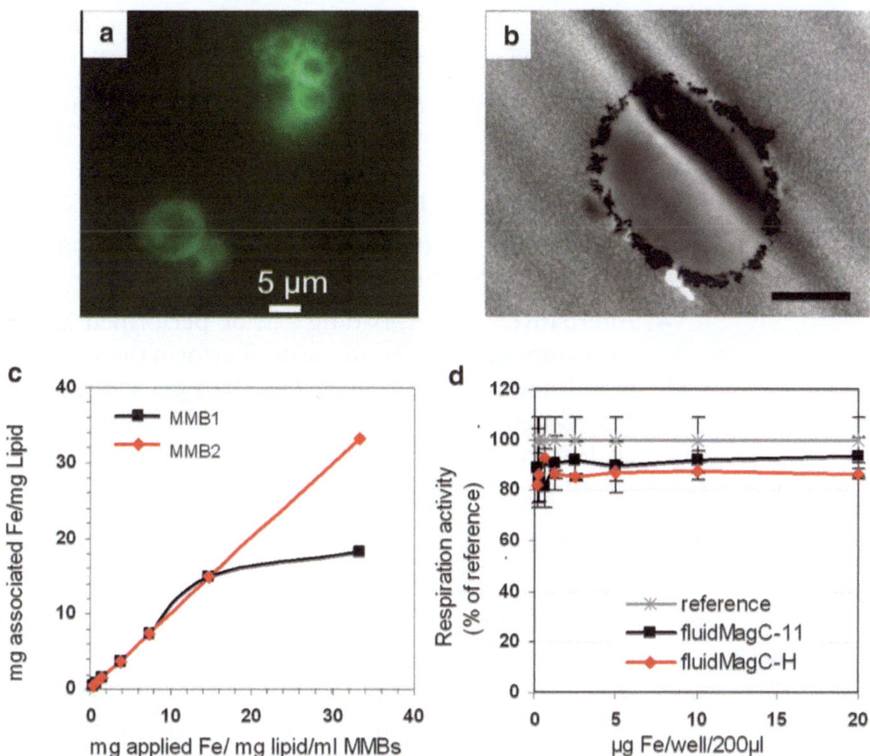

Fig. 3. Characteristics of MMBs. (**a**) Fluorescence microscopy image (490 nm/509 nm) of MMBs associated with a luciferase plasmid labeled with the intercalating fluorescence dye, YOYO-1. (**b**) TEM image of MMB1 embedded in 10% gelatin. (**c**) Association of MNPs with MMBs calculated from iron determination in the subnatant after magnetic sedimentation of the MMBs prepared with increasing concentration of MNPs. (**d**) Toxicity of fluidMagC-11 and fluidMagC-H nanoparticles in HeLa cells, as determined with the MTT-toxicity Assay. Reproduced with permission from the American Institute of Physics' AIP Conference Proceedings (21).

For PEI-Mag2-MBs, loading with plasmid DNA can be optimally performed upon MMB preparation or after preparation of MMBs in *trans*. FluidMAG-UC/C-11 and FluidMAG-UC/C-H allow loading of the MMBs with a high amount of MNPs (up to 15 mg Fe/mL MMB suspension, Fig. 3c) without sedimentation of the material. The loading concentrations of 10 mg Fe/mL and 5 mg Fe/mL for MMB1 and MMB2, respectively, show the best magnetic responsiveness without aggregation phenomena (Fig. 7). Images of MMBs are shown in Fig. 3a, b. With increasing loading of positively charged MNP, a gradual shift of the ζ-potential from negative to highly positive occurs, which enables high loading of MAALs with nucleic acids (Fig. 5) and an increase in magnetic responsiveness (Fig. 7).

3.2.1. MMBs Loaded with PEI-Mag2 MNPs and Plasmid DNA (PEI-Mag2-MB/DNA)

1. Put 200 μL (0.5 mg Fe) PEI-Mag2 nanoparticles into an Eppendorf tube and add 500 μL HEPES buffer.
2. Add 100 μL lipid stock 4 and 200 μL 1:10 diluted DNA stock solution (equivalent to 100 μg nucleic acid) to the suspension of MNPs in the Eppendorf tube from the previous step (Subheading 3.2.1, step 1).
3. Transfer the mixture into a 2 mL glass vial and proceed as described in Subheading 3.1.1, steps 2–4.
4. Alternatively, DNA loading can be performed after preparing the magnetic bubbles (in *trans*). Perform the steps as described in Subheading 3.2.1, steps 1–3. Using a syringe, add 100 μL 1:5 diluted DNA stock solution (equivalent to 100 μg nucleic acid) into the prepared MMB suspension (into the vial). Mix the suspension gently.

3.2.2. MMBs Loaded with FluidMAG-UC/C-11 (MMB1) or FluidMAG-UC/C-H (MMB2) MNPs and Plasmid DNA

1. Put 58.8 μL FluidMAG-UC/C-11 or 95.2 μL FluidMAG-UC/C-H MNPs (2 mg Fe) into an Eppendorf tube (see Note 2) and add 833.2 μL or 796.8 μL HEPES buffer, respectively.
2. Add 100 μL lipid stock solution 4 and 8 μL DNA stock solution (equivalent to 40 μg nucleic acid) to the suspension of MNPs in the Eppendorf tube from the previous step (Subheading 3.2.2, step 1).
3. Transfer the mixture into a 2 mL glass vial and proceed as described in Subheading 3.1.1, steps 2–4.
4. Alternatively, DNA loading can be performed after preparing the magnetic bubbles (in *trans*). Perform the steps as described in Subheading 3.2.2, steps 1–3, without adding the DNA. Using a syringe, add 8 μL DNA stock solution (equivalent to 40 μg nucleic acid) into the prepared MMB suspension (into the vial). Mix the suspension gently.

3.2.3. Loading of SonoVue® with Magnetic Nanoparticles and Plasmid DNA: Magnetic SonoVue®

The procedure can optionally be carried out in a hood under sterile conditions.

1. Sample 40 μL DNA stock solution (200 μg plasmid) into Eppendorf tube 1 and add 1 mL 0.9% NaCl included in the Sonovue® set provided by the manufacturer.
2. Sample 16.6 μL (1 mg Fe) CombiMAG MNPs into Eppendorf tube 2 and add 1 mL 0.9% NaCl included in the Sonovue® set provided by the manufacturer.
3. Put a needle on the syringe (provided by the manufacturer and included in the Sonovue® set); draw into the syringe the MNP suspension from the previous step and mix.
4. Use the transfer system (shown in Fig. 4a) to empty the syringe into the lyophilized SonoVue®. Shake the SonoVue® vial

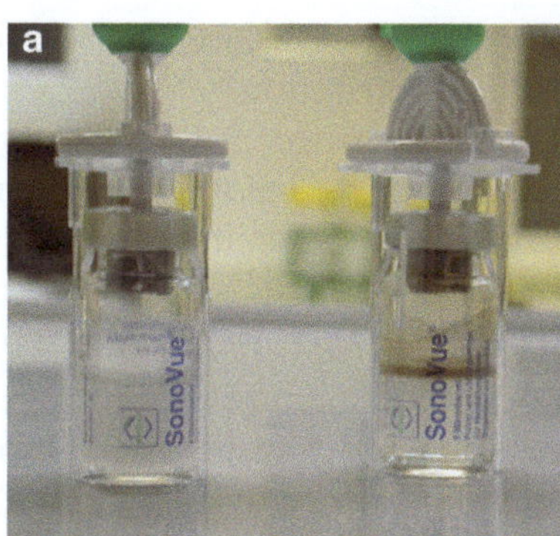

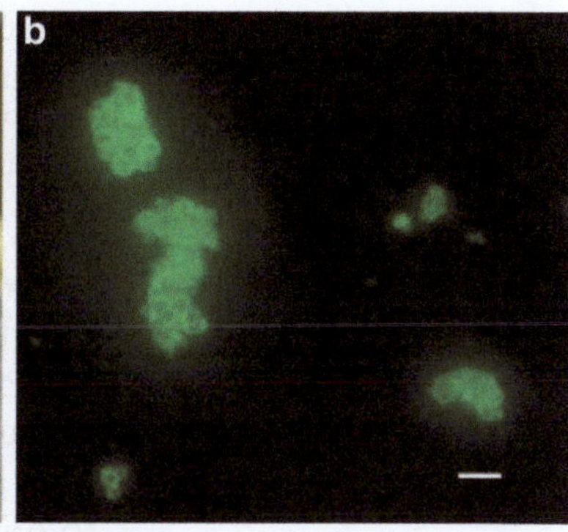

Fig. 4. Magnetic SonoVue®. (**a**) Macroscopic image of re-suspended nonmagnetic (*left*) and magnetic (*right*) commercial SonoVue® microbubbles. (**b**) Fluorescence microscopy image of magnetic SonoVue® microbubbles prepared after resuspension of the lyophilizate CombiMAG magnetic nanoparticle suspension and nucleic acid labeled with the intercalated dye, YOYO-1 (green fluorescence, excitation 490 nm/emission 509 nm). The magnetic SonoVue® microbubbles were magnetically sedimented for 10 min. The bar indicated in (**b**) is equal to 20 μm.

vigorously for 20 s to resuspend the bubbles and allow association with the MNPs. The resultant suspension contains MMBs (magnetic SonoVue®). The floating layer of the magnetic SonoVue® is shown in Fig. 4a (right).

5. To load magnetic SonoVue® with plasmid DNA, put a needle on said syringe and take the DNA solution from Eppendorf tube 1 (step 1). Use the transfer system to empty the syringe into the suspension of magnetic SonoVue® into the SonoVue vial (see Note 6). Association of plasmid DNA with magnetic SonoVue® is illustrated in the fluorescent microscopy image in Fig. 4b (see also Subheading 3.3, step 2).
6. To load SonoVue® with PEI-Mag2 nanoparticles, resuspend the lyophilized SonoVue according to the manufacturer's instructions in 5 mL 0.9% NaCl.
7. Take 20 μL of the PEI-Mag2 MNP suspension (equivalent to 50 μg Fe), add 1 mL reconstructed SonoVue® microbubbles, and then add 8 μL DNA stock solution.

3.3. Labeling the Components of MAALs/MMBs

3.3.1. Fluorescent Labeling of MAALs/MMBs

Fluorescence labeling of the components of the bubbles enables visualization of the bubbles, co-localization of the components, and evaluation of the loading capacity of the nucleic acids. Here, we describe a simple procedure to prepare lipids labeled with green fluorescent dye, or DiO for "DNA with intercalated YOYO-1 dye" (green fluorescence λ_{ex} 491/λ_{em} 509 nm), and labeling of the MNPs with a coating comprising PEI using Atto 550 NHS ester.

1. To generate a stock solution of the fluorescently labeled lipid mixture for preparation of MAALs loaded with PalD2-Mag1 MNPs, as described in Subheading 3.1.4, take 20 μL (0.053 μmol) of the DiOC18 solution from Subheading 2.3.1, item 1, and transfer it into a round-bottom flask for rotary evaporation.
2. Weigh 1.94 mg (2.8 μmol) DOTAP and 2.1 mg (2.8 μmol) DOPE; dissolve them in 0.5 mL of the chloroform:methanol mixture (2:1) and transfer them into the round-bottom flask with the fluorescence dye solution. Evaporate the chloroform to obtain a thin film of fluorescent lipid. Dry and hydrate the lipid film as described for lipid stock solution 1.
3. To prepare MAALs/MMBs with fluorescently labeled DNA, follow the protocol from Subheading 3.1 or 3.2 using 10 μL of the fluorescently labeled plasmid DNA solution from Subheading 2.3.1, item 3, instead of 8 μL of the unlabeled plasmid. The fluorescent microscopy image of the labeled MAALs is shown in Fig. 2a. To visualize nucleic acids in siRNA-loaded bubbles, commercially available, labeled siRNA products, such as Cy3-labeled siRNA or siRNA-Rho, can be used, which results in bright fluorescent images, such as the one shown in Fig. 2d.
4. To conjugate PEI-Mag2 MNPs with fluorescent dye Atto-550, mix 2 mL of the PEI-Mag2 nanoparticle suspension with 443 μL 0.1 M Na-borate buffer, pH 8.5, and add 57 μL Atto-550 solution. Incubate the resulting suspension at room temperature overnight, and dialyze against water using a Pierce cassette dialysis device with a 3,500 MW cutoff. The resulting suspension contains 2.7 mg Fe/mL and 47 μM dye (corresponding to 1.4%w based on iron weight or 15 dye molecules per insulated MNP). The bubbles prepared with the particles labeled according to the protocol described here show high fluorescence intensity and a high fluorescence stability without modifying the properties of the bubbles.
5. To visualize the fluorescently labeled MAALs/MMBs, dispense 5 μL homogenous MAALs/MMBs in 1 mL PBS and transfer into a well of a 24-well culture plate. To obtain sharp images, place the culture plate on a 24-well magnetic plate for 5 min to sediment the bubbles at the bottom of the well. For MAALs/MMBs, visualization with a 40× objective is recommended.

3.3.2. MAAL/MMBs with Radioactively Labeled siRNA and Plasmid

This step must be performed by authorized personnel and according to the rules and regulations for work with radioactive substances. See also Note 7.

In this section, we describe a protocol to prepare microbubbles loaded with radioactively labeled nucleic acids, which allows for high accuracy quantification of the MAAL/MMBs loading with DNA or siRNA. For this purpose, a total DNA-associated radioactivity of about 4×10^5 CPM/40 μg nucleic acid/mL is recommended.

In some cases, for in vivo application with imaging using gamma scintigraphy, which is used as a diagnostic procedure in nuclear medicine, it is necessary to achieve a high concentration and a specific radioactivity of about 4×10^8 CPM/250 μg nucleic acid/mL during preparation. This is why we also describe the procedure for concentrating the radioactively labeled nucleic acid by ethanol-precipitation.

Before loading the bubbles with radioactively labeled nucleic acid, measure the actual specific radioactivity of the ^{125}I-labeled DNA solution.

1. To prepare MAALs with radioactively labeled plasmid, follow any of the protocols described in Subheading 3.1 or 3.2, with the only difference being that radioactively labeled nucleic acid should be used. For example, with ^{125}I-labeled DNA (90 μg DNA/mL, with a radioactivity of 2.4×10^7 CPM/mL), take 16.6 μL (1.5 μg DNA) of radioactively labeled DNA solution and 7.7 μL DNA (38.5 μg DNA) stock solution for a total of 40 μg DNA with a radioactivity of 4×10^5.
2. To quantify the association of DNA with magnetic bubbles, allow the bubbles to float and take 500 μL of the underlying fluid. Alternatively, sediment the microbubbles magnetically and take a 500 μL sample from the supernatant with a needle.
3. Determine the radioactivity, $CPM_{measured}$, of the collected samples using a γ-counter (Wallac 1480 Wizard 3″, PerkinElmer Wallac, Freiburg, Germany). As a reference, simultaneously determine the specific radioactivity of the stock of the ^{125}I-labeled DNA solution (CPM/μL). Calculate the total radioactivity of the plasmid applied per mL preparation, CPM_{added}. Calculate the DNA association with MAALs as follows:

$$\text{Associated nucleic acid(\% of added)} = [(CPM_{added} - CPM_{measured} \times 2) / CPM_{added}] \times 100.$$

   Examples of the data for DNA association with MAALs are given in Table 1. Figure 5a shows the data on DNA association with MMBs as a function of the applied DNA doses.
4. Use 1 mCi of sodium ^{131}iodine to label 400 μg siRNA according to the previously described protocol (36). The procedure yields 3.5 mL of the preparation from ^{131}I-labeled siRNA solution with a concentration of 118.3 μg RNA/mL and a radioactivity of 2.1×10^5 CPM/μL.
5. To concentrate the labeled siRNA preparation (Note 7), transfer 300 μL of the ^{131}I-labeled siRNA into 11 Eppendorf tubes.

Table 1
Incorporation of plasmid DNA and magnetic nanoparticles to MAALs

Formulation	Preparation method	Fractionation method	Nucleic acid bound (% of input dose)	Magnetic nanoparticles bound[a] (% of input dose and μg Fe absolute)
Tw-Mag-MAALs	Plasmid integral part of preparation	Flotation (enforced by centrifugation)	97	92 (220 μg Fe)
	Plasmid integral part of preparation	Magnetic sedimentation	99	
	Plasmid added in *trans* after MAAL preparation	Flotation (enforced by centrifugation)	68 (aggregation)	n.d.
	Plasmid added in *trans* after MAAL preparation	Magnetic sedimentation	47 (aggregation)	
PEI-Mag2-MAALs	Plasmid integral part of preparation	Flotation (enforced by centrifugation)	38	88 (201 μg Fe)
	Plasmid integral part of preparation	Magnetic sedimentation	37	
	Plasmid added in *trans* after MAAL preparation	Flotation (enforced by centrifugation)	42 (aggregation)	n.d.
	Plasmid added in *trans* after MAAL preparation	Magnetic sedimentation	39 (aggregation)	

Reproduced with permission from WILEY-VCH Publisher: Adv. Funct. Mater. (20)
n.d. not determined
[a]The fractionation procedure for iron determination included both magnetic sedimentation and centrifugation as well as a washing step

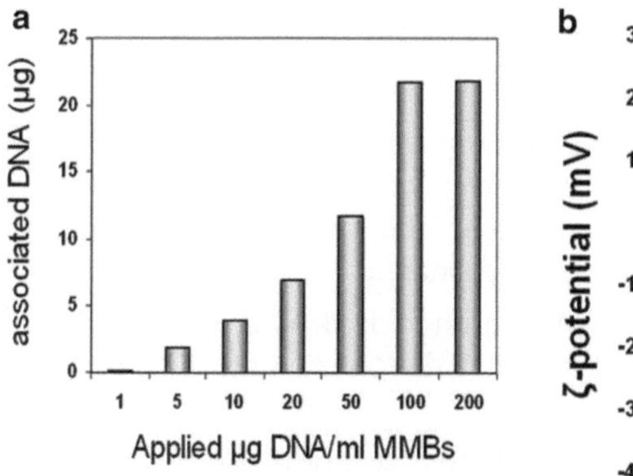

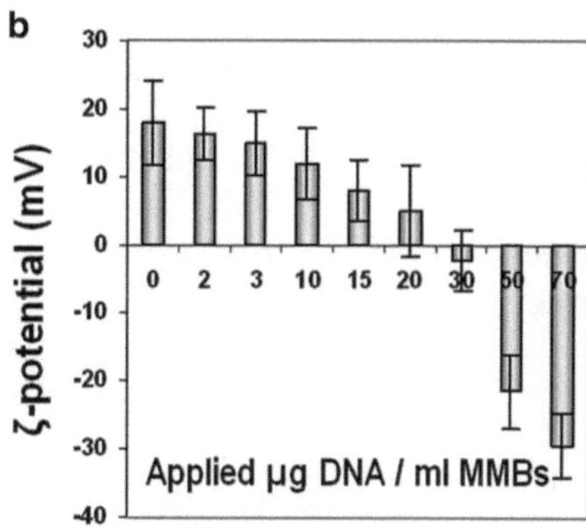

Fig. 5. Loading nucleic acids onto MMBs. (**a**) Association of nucleic acid with MMB1 is evaluated after preparation of magnetic microbubbles1 loaded with 10 mg Fe/mL by adding an increasing amount of ^{125}I-labeled nucleic acid to 10^7 MMBs (100 μL) up to a total volume of 200 μL and incubation for 15 min. After magnetic sedimentation with a permanent magnetic field ($\langle B \rangle = 130$–240 mT, $\langle \partial B/\partial Z \rangle = 70$–$120$ T/m) for 30 min, the nucleic acid not associated with MMBs is assayed by measuring the radioactivity in the supernatant using a γ-counter. (**b**) ζ-potential versus dosage of the applied plasmid DNA after preparation of magnetic microbubbles1 loaded with 10 mg Fe/mL by adding an increasing amount of nucleic acid per mL of the preparation and incubation for 15 min. Reproduced with permission from the American Institute of Physics' AIP Conference Proceedings (21).

Add 30 μL 3 M sodium acetate, pH 5, 2 μL glycogen, and 900 μL cold 95% ethanol, and incubate the tubes on ice for 15 min.

6. Centrifuge the tubes at 14,000 rpm at 4°C for 30 min, remove the supernatant, and measure the radioactivity in supernatant.
7. Add 600 μL cold 70% ethanol to each tube; shake gently, and centrifuge again with the same conditions for 5 min. Remove the supernatant and control the radioactivity of the pellet using a handheld radiation monitor.
8. Let the pellets dry for exactly 10 min, and carefully dissolve each pellet in 50 μL DEPC-H_2O per microcentrifuge tube (see Note 8). Pool the contents of all tubes.
9. Determine the radioactivity (CPM) of an aliquot of the concentrated radiolabeled nucleic acid using a gamma counter (Wallac 1480 Wizard 3″, PerkinElmer Wallac, Freiburg, Germany), and determine the nucleic acid concentration by measuring the absorbance at 260 nm. Calculate the specific activity of the concentrated labeled nucleic acid solution (CPM/μg DNA). In one of the experiments, we obtained an activity and siRNA concentration of 9.25×10^8 CPM/mL and 650 μg siRNA/mL, respectively.
10. To prepare MAALs with radioactively labeled siRNA, follow any of the protocols described in Subheading 3.1 or 3.2; the only difference is that radioactively labeled nucleic acid should

be used. In one of the experiments, we used 367 μL or 250 μg ^{131}I-labeled siRNA from Subheading 3.3.2, step 9.

3.4. Loading of MAALs/MMBs with Magnetic Nanoparticles

1. To quantify the binding MNPs to MAALs/MMBs, allow the bubbles to float and take 20 μL of the underlying fluid. Alternatively, sediment the microbubbles magnetically and take 20 μL of the supernatant.
2. Add 200 μL concentrated hydrochloric acid. Wait until the MNPs are completely dissolved, and adjust the volume to 5 mL with water.
3. Transfer 20 μL of the solution from the previous step (Subheading 3.4, step 2) to an Eppendorf tube. Add 20 μL concentrated hydrochloric acid, 20 μL hydroxylamine hydrochloride solution, 200 μL ammonium acetate buffer, 80 μL 1,10-phenanthroline solution, and 300 μL water. Mix well and allow to sit for 20 min.
4. To prepare a blank sample, proceed as described in step 3 using 20 μL water instead of the iron solution.
5. Measure the absorbance of the samples from step 3 at 510 nm against the blank using a spectrophotometer.
6. To construct a calibration curve for determining the iron concentration, add increasing amounts of the iron standard solution to microcentrifuge tubes (e.g., 50, 70, 90, up to 150 μL) and adjust the volume to 150 μL with water. Use 150 μL water instead of iron solution to prepare a blank sample. To each tube, add 20 μL concentrated hydrochloric acid, 20 μL 10% hydroxylamine hydrochloride solution, 200 μL ammonium acetate buffer, 80 μL 0.1% 1,10-phenanthroline solution, and 300 μL water. Mix well and allow to sit for 20 min. Measure the absorbance at 510 nm against the blank.
7. Plot the absorbance at 510 nm as a function of the iron concentration in the standard samples. Use linear regression as an approximation function to calculate the iron concentration in the MNP and magnetic liposphere/microbubble samples.

3.5. Determination of Size, Electrokinetic Potential, and Volume Concentration of MAALs/MMBs

Prior to use, gently rotate the MAAL or the MMB preparation to obtain a homogenous suspension (see Note 1).

1. Dispense 5 or 10 μL MAALs/MMBs into 20 mL Isoton®II diluent depending on the Beckmann Coulter aperture tube used, and determine the size and volume concentration of MAALs with the electrical sensing zone technique using a Z2 Coulter counter (Beckman Coulter GmbH) according to the manufacturer's instruction. For the MAAL/MMB concentration, the dilution factor is 1.6×10^5 and 4×10^5 for the 100 μm- and 50 μm-aperture tubes, respectively. Results for some of the preparations can be found in Table 2.

Table 2
Characteristics of selected MAALs fabricated with 40 μg/mL DNA or of selected MMBs without DNA loading

MAALs/MMBs	Hydrodynamic diameter (μm)	Electrokinetic potential in water (mV)	Electrokinetic potential in NaCl (mV)	Volume concentration (per mL preparation)
Tw-AAL-DNA	5.0 ± 4.4[a] 2.0 ± 0.5[b]	3.7 ± 0.8	4.9 ± 0.9	2.0×10^9
PEI-Mag2-AAL-DNA	3.0 ± 2.3[a] 2.0 ± 0.6[b]	37.4 ± 4.5 14.9 ± 5.2	28.9 ± 1.0	4.7×10^9
PEI-Mag3-AAL-DNA	2.1 ± 0.6[b]	7.6 ± 4.5	25 ± 1.1	2.0×10^{10}
Pal-D2-Mag-AAL (w/o DNA)	6.1 ± 1.5	-4.4 ± 6.2	-2.1 ± 3.5 (w/o DNA) -38.1 ± 5.3 (+DNA)	3.6×10^9
MMB1 (w/o DNA) (FluidMAG-UC/C-11 MNPs: 1 mg/mL)	1.7 ± 0.1	-13.9 ± 3.1	–	5.6×10^8
MMB2 (w/o DNA) (FluidMAG-UC/C-H MNPs: 1 mg/mL)	1.7 ± 0.3	-8.3 ± 3.4	–	4.2×10^8
PEI-Mag2-MMB (w/o DNA)	0.8 ± 0.3	28.8 ± 5.0	–	

[a]The MAAL size was determined by an image analysis program (S.CO LifeScience GmbH)
[b]The MAAL size was determined using a Z2 Coulter counter (Beckman Coulter GmbH)

2. To determine the size of MAALs, register images of the bubbles as described in Subheading 3.1 and perform the image analysis. The size distribution curve for MAALs is determined by an image analysis program, which was developed by S.CO LifeScience GmbH, Garching, Germany (http://www.sco-lifescience.com), shown in Fig. 2b.
3. To determine the ζ-potential and size distribution, dilute 5 μL MAALs in 1 mL water and perform measurements using a Nano ZS (Malvern) according to the manufacturer's instructions. Examples of the results are shown in Table 2.

3.6. Preparation of Gelatin-Clots with Embedded Bubbles for Transmission Electron Microscopy

1. Pipette 1 mL of the viscous gelatin solution from Subheading 2.6, item 1, into a well of a 24-well cell culture plate; take 20 μL MAALs/MMBs with a pipette (diluted 1:100 with water), and inject into the middle of the well filled with gelatin. Leave the gelatin clot to dry.

2. Thin sections (70–80 nm) from selected Epon blocs can be put onto electron microscopic grids (Plano GmbH, Wetzlar, Germany) and stained with 0.5% aqueous uranyl-acetate and 0.3% lead citrate for contrast. Samples can be observed with an EM 10 CR transmission electron microscope (Zeiss, Jena, Germany). Examples of the TEM images are given in Figs. 2c and 3b and show association of the MNPs with the surface of the vesicles.

3.7. Evaluation of Magnetic Responsiveness

To evaluate the sedimentation stability and the magnetically induced velocity (magnetic responsiveness) of MAALs or MMBs in gradient magnetic fields and to estimate MAAL/MMB loading with magnetic nanomaterial, a method can be employed based on the measurements of the time course of the turbidity of suspensions in a gravitational field and in applied inhomogeneous magnetic fields using the simple setup shown in Fig. 6a (36, 37).

1. Dilute the suspension of the bubbles with water to a final volume of 500 μL of the suspension that has an optical density between 0.4 and 1 at 650 nm.
2. To register the sedimentation stability (with no magnetic field applied), put 500 μL of the suspension into an optical cuvette, and measure the time course of the turbidity (optical density) at 650 nm in a "kinetic mode."
3. To register the magnetic responsiveness, position magnets on each side of the cuvette holder in spectrophotometer as shown

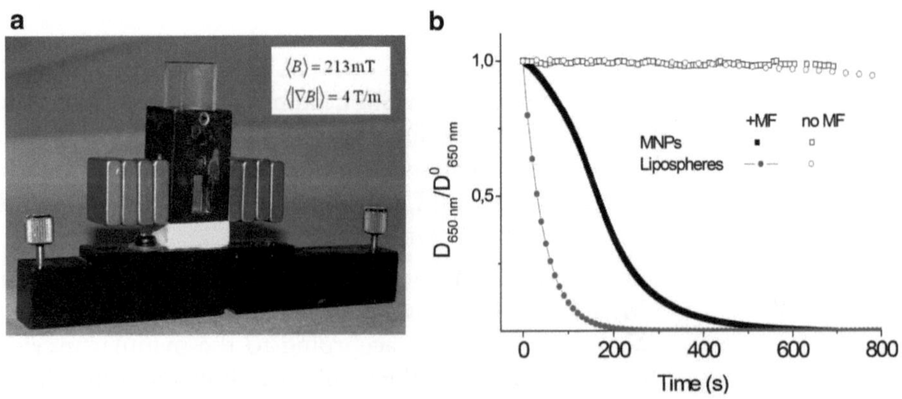

Fig. 6. Magnetic retention and magnetophoretic mobility of the MAALs. (**a**) Setup for the measurement of the magnetophoretic mobility (or magnetic responsiveness) of the MAALs. Two sets of four Nd-Fe-B permanent magnets (18.0 × 16.0 × 4.0 mm, IBS Magnets) are positioned symmetrically on either side of the cuvette holder and parallel to the light beam in a spectrophotometer. The average magnetic field and the gradient through the measuring window for the applied magnets' configuration were 213 mT and 4 T/m, respectively. (**b**) The normalized turbidity of the Tw-Mag-AAL/pBLuc lipospheres and the suspension of FluidMAG-Tween-60 MNPs upon application of a permanent inhomogeneous magnetic field and without the application of a magnetic field (no MF). Reproduced with permission from WILEY-VCH Publisher's Adv. Funct. Mater. (20).

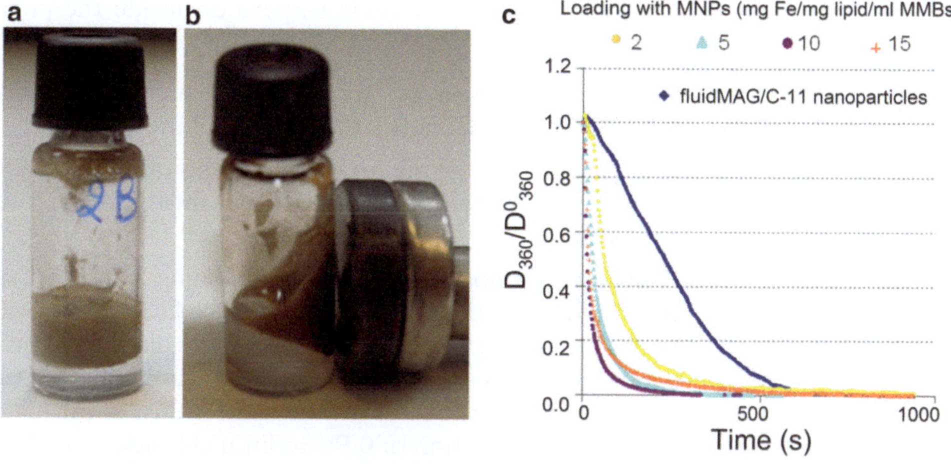

Magnetophoretic mobility of MMB1s and MNPs under magnetic field application			
	Loading with MNP (mg Fe/mg lipid/ml MMBs)	t_{90} (s)	Average magnetophoretic mobility (μm/s)
fluidMAG/C-11 nanoparticles	-	508	2
MMB1	2 5 10 15	197 98 53 94	5.1 10.2 18.7 10.6

Fig. 7. Magnetic properties of the MMBs. (**a** and **b**) Magnetic attraction of MMB1 preparation through a permanent magnetic field. (**c**) Time course of the normalized turbidity of MNPs (*dark blue* color) and microbubbles MMB1, as a representative sample, at 360 nm at different loadings of MNPs upon application of a permanent, inhomogeneous magnetic field of $\langle B \rangle = 213$ mT, $\partial B/\partial Z - 4$ T/m. (**c**) Demonstrates the magnetophoretic mobility of the MMBs, which depends on the MNP load compared to the magnetophoretic mobility of the MNPs. The time t_{90} required for the optical density to decrease by tenfold, or for sedimentation of 90% of the magnetic suspension, upon application of the defined gradient magnetic field is used to evaluate the average magnetophoretic mobility (given in the table). The highest mobility for MMB1 is observed at MNP loads of 10 mg Fe/mL. At higher loads of MNPs, an excess of unassociated MNPs was revealed (and is supported by the size distribution data) and aggregation of the particles and bubbles occurred. Reproduced with permission from the American Institute of Physics' AIP Conference Proceedings (21).

in Fig. 6a, put the cuvette with 500 μL of the suspension into the holder, and immediately start to measure the time course of the turbidity at 650 nm in a "kinetic mode."

4. To evaluate the efficient velocity, υ_z, of the MAALs or MMBs under a gradient magnetic field and to further calculate the average magnetic moment, *M*, of the magnetic lipospheres or microbubbles and estimate the number of MNPs, *N*, associated with the MAALs or MMBs, see Note 9 and (36, 37). Examples of the results are shown in Figs. 6b and 7.

3.8. Ultrasound Responsiveness

To quantify the ultrasound contrast of the MAALs or MMBs, simple and quick methods can be used using tissue-mimicking phantom gels with an injection of the bubble suspension into the gel phantom (Subheading 3.8, steps 1–3) or positioning a reservoir containing the bubble suspension inside the phantom (Subheading 3.8, steps 4–5).

1. Prepare a 1% agarose gel phantom with dimensions of $6.5 \times 10 \times 1.3$ cm from the 1% agarose solution.
2. Take 200–300 μL of the magnetic suspension into a single-use syringe and insert the needle deep into the agarose gel phantom.
3. Position an ultrasound probe from Subheading 2.8, item 3, on the gel and focus on the needle. Move the needle outwards without taking the needle out of the gel. When the ultrasound image is optimized, slowly inject the liposphere/microbubble suspension and use video to show the ultrasound contrast of the magnetic bubbles. Use injection of 0.9% sodium chloride as a reference. Ultrasound images are given in Fig. 8a, b to show the contrast effect of the MAALs.
4. Fill in the reservoir positioned inside a tissue-mimicking phantom gel with 35 mL 0.9% saline solution containing 2.5 μL of either an MAAL or an MMB suspension or the commercially available agent, SonoVue® (Bracco, Milano, Italy), as reference (shown in Fig. 8c).
5. An Ultrasonix Sonix RP research ultrasound system (Ultrasonix, Richmond, BC, Canada) with a C5-2/60 curved array (3.5 MHz center frequency) can be used to image the cross section of the reservoir as it is shown in Fig. 8d and described in previous study (20).

3.9. Transfections with pDNA- or siRNA-Loaded MAALs

1. Culture the cells in the appropriate growth medium at 37°C in a 5% CO_2 atmosphere.
2. For siRNA interference, the cells stably expressing the firefly luciferase gene (fLuc) and/or enhanced green fluorescent protein (eGFP) obtained upon retro- or lentiviral transduction, e.g., eGFP-H441, eGFP-HeLa, or fLuc-HeLa, can be used.
3. Prior to transfection, count the cells and seed them in the wells of 96-well flat-bottom culture plates at a density of 2×10^4 cells per well (in 150 μL modified RPMI 1640 medium) for NCI-H441 or eGFP-H441 cell lines. Adherent cells that divide more rapidly than H441 cells (such as NIH-3T3 or HeLa cells, eGFP-HeLa, or fLuc-HeLa cell lines) should be plated at a density of 0.8×10^4 per well (in 150 μL Dulbecco's modified Eagle medium, DMEM). Other cell lines could be also used in place of the aforementioned cells.
4. Place the plates in a cell culture incubator at 37°C in a 5% CO_2 atmosphere until transfection, which is usually 24 h later. The cells should be approximately 50% confluent at the time of transfection.

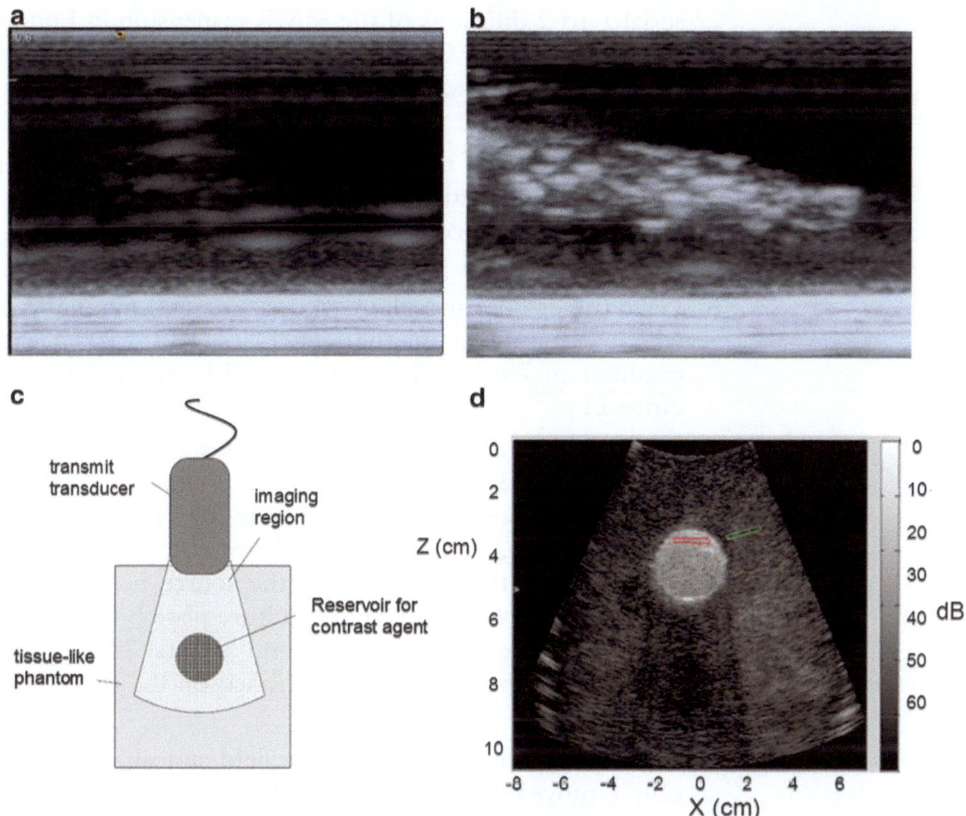

Fig. 8. Acoustic properties of MAALs and MMBs. Ultrasound images (**a**) of 0.9% NaCl as a reference and (**b**) Tw Mag-AAL (the *white* reflections represent the lipospheres) injected into 1% agarose gel phantoms; the images were taken using a SIEMENS Acuson Antares ultrasound device equipped with a VFX 13-5 scanhead (center frequency 10 MHz, imaging depth 1 cm, MI = 0.8). (**c**) Schematic presentation of the experimental apparatus for the ultrasound contrast quantification. (**d**) B-mode image obtained using an Ultrasonix Sonix RP research ultrasound system (Ultrasonix, Richmond, BC, Canada) with a C5-2/60 curved array (3.5 MHz center frequency) through pulse inversion in a region of interest, from which the contrast-to-tissue ratio (CTR) is measured (red frame: contrast agent region, green frame: tissue-like region). Reproduced with permission from the American Institute of Physics' AIP Conference Proceedings (21).

5. On the day of transfection, prepare the MAAL/MMB stock suspension according to the steps described in Subheadings 3.1 and 3.2 (see Note 10). Transfection studies can be performed using the plasmid p55pCMV-IVS-luc+ (pBLuc), coding for the firefly luciferase, or any other plasmid DNA, depending on the protein expression you want to evaluate.
6. Accounting for the loading of the bubbles in the stock suspension of 40 μg plasmid/mL, transfer 175 μL MAALs into the first tube and dilute to a final volume of 1,400 μL with complete cell culture medium. Additionally, prepare three

serial 1-to-2 dilutions of the MMB suspension in Eppendorf tubes containing 700 μL cell culture medium each by transferring 700 μL microbubble suspension from the first tube and so on.

7. Pipette 100 μL of the dilutions per well in the 96-well plate with the cells seeded. This results in 500, 250, 125, or 62.5 ng DNA per well and corresponds to 1.46, 0.73, 0.37, or 0.18 μg DNA/cm^2.
8. Place the cell culture plate on a 96-magnet magnetic plate, which generates a magnetic field of 130–240 mT and a gradient of 70–120 T/m at the cell layer for 15 min (see Note 11).
9. After 15 min of incubation, apply ultrasound at 2 W/cm^2 to selected wells for 30 s at 50% duty cycles using a Sonitron 2000D ultrasound device (Rich Mar Inc., Inola, OK, USA). Insert the ultrasound probe directly into the medium covering the cells with the lower end of the source being positioned approximately 4 mm above the culture plate bottom. As long as the cells are exposed to ultrasound (in total 30–40 min), the culture plate should remain positioned on the magnetic plate (see Note 12).
10. After application of the magnetic field and ultrasound to the cells, remove the magnetic plate and incubate the culture plate containing the transfected cells in a cell culture incubator at 37°C in a 5% CO_2 atmosphere. After 60–90 min, replace the medium in all wells with fresh complete medium with all additives. Keep the culture plate in the incubator until evaluation.
11. To evaluate the reporter gene expression or down regulation, culture plates must be incubated for 24–72 h after transfection. The present studies for detection of reporter gene expression or down regulation (as described in Subheading 3.11, steps 1–7) were performed 48 h post transfection.

    For siRNA interference experiments, prepare the MAAL suspension according to the steps described in Subheading 3.1 using Luci GL3 siRNA or GFP 22 siRNA duplex for luciferase or GFP knockdown, respectively, and use negative control siRNA.
12. Accounting for the loading of the bubbles in the stock suspension of 40 μg siRNA/mL, put 78.05 μL MAALs into the first tube and dilute up to a final volume of 1,400 μL with complete cell culture medium. Prepare an additional four serial, 1:2 dilutions of the MMB suspension in Eppendorf tubes containing 700 μL cell culture medium each by transferring 700 μL microbubble suspension from the first tube and so on.
13. Pipette 100 μL of the dilutions per well in the 96-well plate with the seeded cells. This should result in a siRNA concentration

per well of 64, 32, 16, 8, and 4 nM. Finally, follow Subheading 3.9, steps 8–11.

To test the transfection efficiency of equal doses of MMBs combined with different amounts of plasmid, prepare the MMB suspension according to Subheading 3.2.2, steps 1–3.

14. Perform a 1:100 dilution of the DNA stock solution, transfer 72 μL of this solution into well A1 of 96-well plate (preparation plate), and add 216 μL cell culture medium. Add 144 μL cell culture to wells C1–D1. Prepare 1:2 dilutions of the DNA solution in wells A1–D1 by transferring 144 μL DNA solution.
15. Add 36 μL MMBs to wells A1, B1, C1, and D1 and incubate for 10 min. Repeat steps 14–15 for each condition to be tested.
16. Pipette 50 μL from the A1–D1 wells from step 15 to the corresponding wells in a plate with seeded cells. Finally, follow Subheading 3.9, steps 8–11.

3.10. Association of pDNA with Cells Following Transfection

1. This step must be performed by authorized personnel and according to the rules and regulations for work with radioactive substances. See also Note 7.
2. Prepare MAALs loaded with radioactively labeled nucleic acid as described in Subheading 3.3.2.
3. Perform cell transfection according to Subheading 3.9, steps 1–10.
4. 1 h post transfection, carefully collect the supernatants from each well using a pipette, and transfer each sample, together with the pipette tip, into a scintillation vial.
5. Wash the cells with 0.2 mL PBS and again carefully collect the wash medium from each well into a scintillation vial.
6. Subsequently, lyse the cells with 200 μL Lysis buffer, and collect the cell lysate from each well into a scintillation vial.
7. Measure the radioactivity (CPM) in each vial (the cell lysates, the supernatants, and the wash fractions) using a γ-counter (Wallac 1480 Wizard 3″, PerkinElmer Wallac, Freiburg, Germany).
8. Calculate the amount of DNA (siRNA) associated with the cells as the amount of nucleic acid measured in the cell lysate as follows:

$$\text{Cell associated nucleic acid}(\%) = [\text{CPM}_{\text{cell lysate}} / (\text{CPM}_{\text{medium}} + \text{CPM}_{\text{PBS}} + \text{CPM}_{\text{cell lysate}})] \times 100.$$

Examples of the results for nucleic acid association with MAALs/DNA are shown in Fig. 9c.

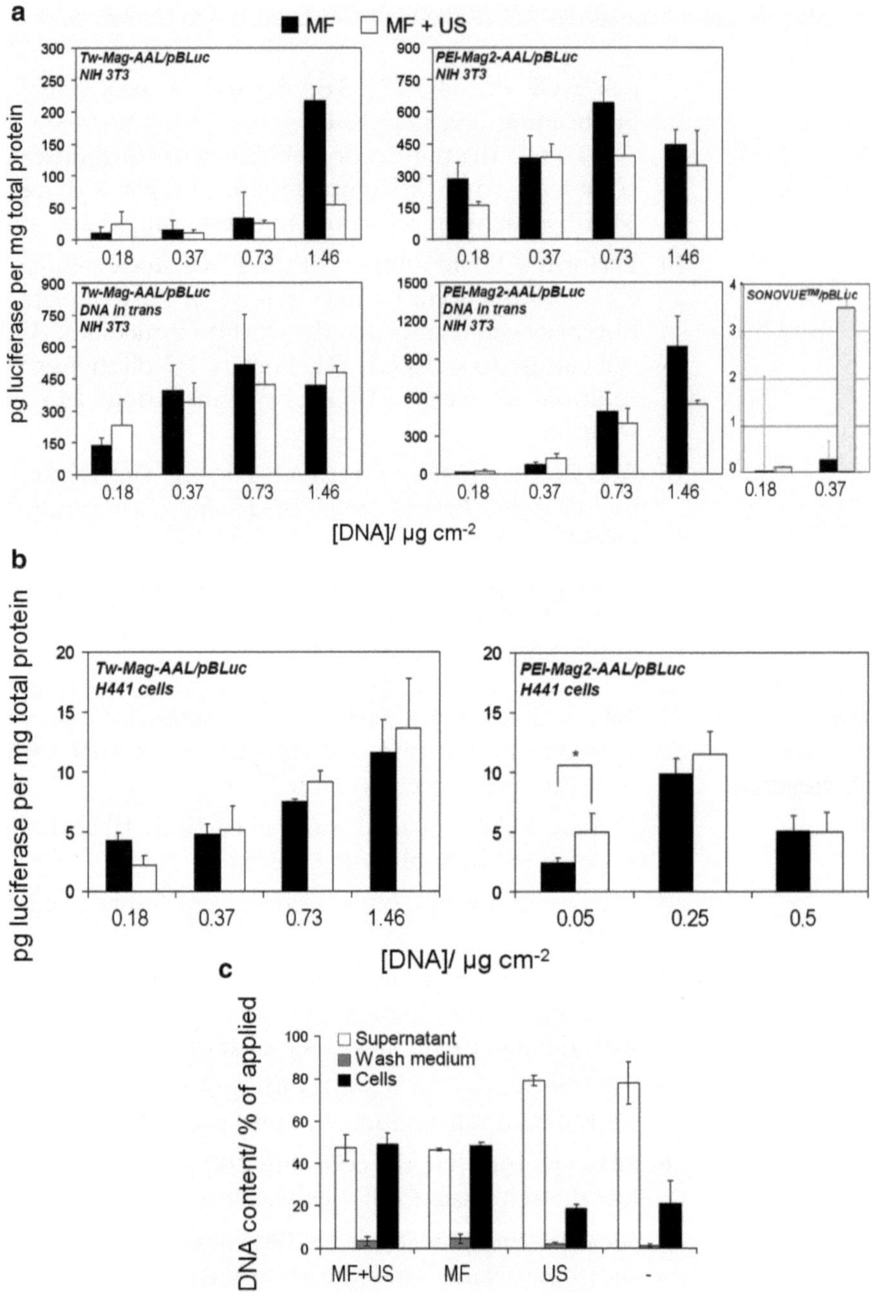

Fig. 9. The efficiency of DNA delivery with the MAALs in vitro. Luciferase expression in (**a**) NIH3T3 mouse fibroblasts 48 h post magnetofection (MF) or magnetofection combined with ultrasound treatment (MF + US). *Upper panel*: The cells were transfected with MAALs with plasmid DNA as an integral part of the preparation. *Lower panel*: The cells were treated with MAALs where the DNA was added after preparation of the magnetic lipospheres. In analogy, the plasmid was supplemented in *trans* to SonoVue®. The cells were exposed to Tw-Mag-AAL/pBLuc and PEI-Mag2-AAL/pBLuc magnetic lipospheres loaded with 40 μg plasmid/mL MAAL suspension or Tw-Mag-AAL/pBLuc and PEI-Mag2-AAL/pBLuc magnetic lipospheres mixed with 40 μg plasmid or 30 μL/cm² SonoVue® reconstituted with 0.9% NaCl and associated with plasmid DNA. Luciferase expression in (**b**) H441 lung epithelial cells exposed to Tw-Mag-AAL/pBLuc and PEI-Mag2-AAL/pBLuc magnetic lipospheres loaded with 40 g plasmid/mL MAAL suspension. (**c**) DNA association with 3T3 mouse fibroblasts 1 h post transfection with magnetic AALs (Tw-Mag-AAL/^{125}I-pBLuc) loaded with ^{125}I-labeled luciferase plasmid and the effect of the US, MF, or both factors (MF + US). Ultrasound was applied using a Sonitron 2000D device operating at 1 MHz and using a 3-mm ultrasound probe operating at 2 W/cm² and a 20% duty cycle for 30 s. The data are given as the mean ± SD. Reproduced with permission from WILEY-VCH Publisher's Adv. Funct. Mater. (20).

3.11. Evaluation of Gene Expression In Vitro, as Described in (38)

1. To prepare cell lysates from adherent cells, wash transfected adherent cells with 150 μL PBS per well using a multichannel pipette. Add 100 μL lysis buffer per well, incubate for 10 min at RT, and then place the culture plate on ice.
2. To quantify eGFP reporter gene expression in cell lysates, measure eGFP fluorescence at 485/535 nm in 100 μL cell lysate in a 96-well clear-bottom black-walled plate using a microplate reader.
3. To quantify luciferase reporter gene expression in cell lysates, transfer 50 μL cell lysate from each well into a clear-bottom black-walled plate. Add 100 μL luciferase buffer per well, and optionally mix with a pipette. Measure the chemiluminescence intensity (using a count time of 0.20 min with background correction) using the luminometer.
4. To construct a calibration curve to determine the amount of luciferase in transfected cell samples, add 50 μL lysis buffer per well to columns 1 and 3 of a black 96-well plate and 40 μL lysis buffer per well to columns 2 and 4. To well A1, add 30 μL lysis buffer and 20 μL luciferase standard stock (0.1 mg luciferase/mL and 1 mg BSA/mL in 0.5 M Tris–acetate buffer, pH 7.5). Pipette 50 μL from A1 to B1, mix well, and then from B1 to C1, etc. down to H1. From H1, continue the dilution series by transferring 50 μL to A3, and continue in column 3 down to G3, leaving H3 as blank. Pipette 10 μL each from columns 3–4 and from columns 1–2. Add 100 μL luciferase buffer to each of the wells of columns 2 and 4. Measure the chemiluminescence intensity as described above. Plot the logarithm of the luciferase content in the dilution series as a function of the logarithm of the measured luminescence intensity (light units). Use an approximation function (typically linear regression in this concentration range) to calculate the amount of luciferase in the transfected cell samples.
5. Determine the total protein content of the samples as follows. Add 150 μL water to each well in a flat-bottom 96-well plate. Using a multichannel pipette, transfer 10 μL of each of the cell lysates into the corresponding wells of the protein assay plate. Add 40 μL BioRad protein assay reagent to each well and mix carefully using a plate shaker or a multichannel pipette. Measure the absorbance at 590 nm using the Wallac 1420 Multilabel counter.
6. To construct a calibration curve to determine the total amount of protein in the transfected cell sample, add 25 μL lysis buffer per well in one row (e.g., row A) of a flat-bottom 96-well plate. Add 50 μL BSA stock solution to well 1 (e.g., A1). Mix well using a pipette. Transfer 50 μL from well 1 to well 2, mix, transfer 50 μL from well 2 to well 3, and so on to well 11, leaving well 12 as blank. Add 150 μL water per well in another row (e.g., row B). Transfer 10 μL from row A to row B. Add

40 μL BioRad reagent to each well, and mix carefully using a plate shaker or a multichannel pipette. Measure the absorbance at 590 nm (or 570 nm) using a microplate reader (e.g., a Wallac 1420 Multilabel counter with the measuring time set to 0.1 s). Plot the measured absorbance versus the protein content for each well. Use linear regression to derive a calibration function from which the protein content in the samples can be calculated. Calculate the total protein content per 10 μL cell lysate for every sample using the calibration curve.

7. Calculate the weight luciferase per weight total protein. The results can be plotted against the applied DNA concentration or dose per well in order to get a dose–response curve.
   Sample results of luciferase expression after transfection with MAALs or MMBs are given in Figs. 9, 10c, and 11.
   Luciferase or eGFP gene expression down-regulation is presented as a percentage of the micrograms of luciferase or green fluorescence protein per mg total protein normalized to the expression in untreated cells. Perform all experiments in triplicate.

3.12. MTT-Based Cytotoxicity Assay

1. Wash the transfected cells twice with 200 μL PBS per well, and discard the wash solutions. The washing procedure is important to remove any traces of magnetic nanomaterial from the cells because it can interfere with the results later.
2. Add 100 μL MTT solution per well, and incubate the culture plate in a cell culture incubator for 1.5–3 h (see Note 13). In the meantime, check the cells for the formation of insoluble violet formazan crystals.
3. Add 100 μL MTT solubilization solution per well to dissolve the formazan.
4. Cover the culture plate with Parafilm to avoid liquid evaporation, and incubate at RT overnight (with gentle shaking) until the formazan crystals completely dissolve.
5. Measure the optical density, *D*, of the MTT–formazan solution after solubilization at 590 nm using a microplate reader.
6. Non-transfected cells can be used as a 100% reference. Wells without cells where a mixture of 100 μL MTT reagent and 100 mL solubilization solution was pipetted can be used as a blank.

Fig. 10. (continued) The transfection was carried out in eGFP-HeLa cervical cancer cells (*upper left* diagram) and in fLuc-HeLa (*bottom left* diagram) in the absence and presence of ultrasound (+US). Ultrasound was applied using a Sonitron 2000D device operating at 1 MHz and using a 3 mm ultrasound probe operating at 2 W/cm^2 and a 50% duty cycle for 30 s. The transfection efficiency diagrams are accompanied with corresponding cell viability diagrams (MTT-toxicity assay) on the *right side*. In all experiments, MAALs/siRNA loaded with negative control siRNA, i.e., siRNA with random sequence, are included. The data are given as the mean ± SD. Reproduced with permission from ACS Publications' NanoLetters (25).

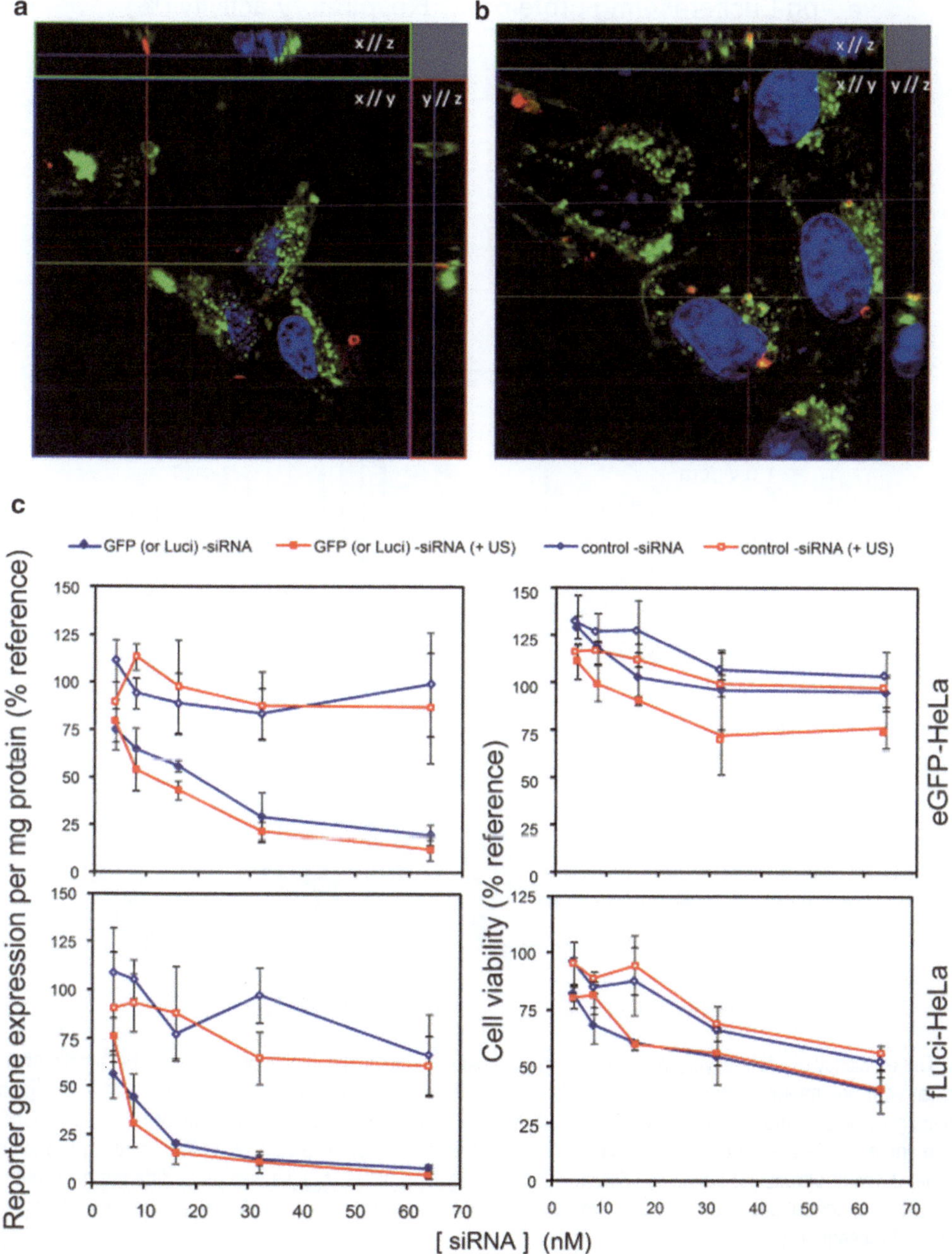

Fig. 10. The efficiency of siRNA delivery with the MAALs in vitro. Intracellular uptake of magnetic acoustically active lipospheres loaded with rhodamine-labeled siRNA (MAALs/siRNA-Rho). 3T3/NIH fibroblasts were transfected with MAALs/siRNA-Rho (the *red* color is from the rhodamine label) and images were taken of living cells over time with a confocal laser scanning microscope (**a**) immediately after addition and (**b**) after 60 h of incubation. Here, the nucleus and cellular membrane of the living cells have been stained in *blue* with DAPI and in *green* with WGA488 directly before imaging. An orthogonal view of the localization of MAALs/siRNA-Rho in the cells from different planes (*x*//*y*, *x*//*z*, and *y*//*z*) is shown. Immediately after addition of MAALs/siRNA-Rho, most of the lipospheres were located extracellularly. However, even at this early time point, some were found to be sitting on the cell membrane (see cross sections *y*//*z* and *x*//*z*). After 60 h, there was a notable accumulation of MAALs/siRNA-Rho inside the cells. (**c**) Down regulation of eGFP and fLuc gene expression compared to untreated cells with magnetic acoustically active lipospheres loaded with siRNA against eGFP (GFP-siRNA) or fLuc gene (Luci-siRNA) 48 h post magnetofection.

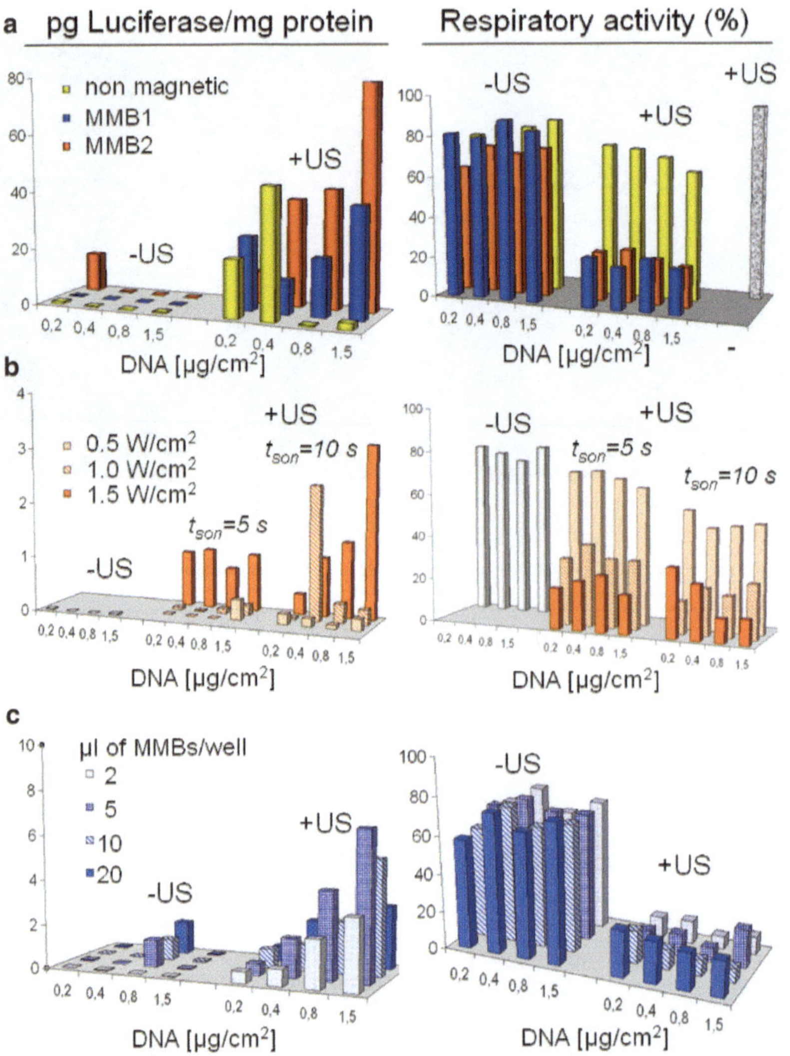

Fig. 11. The efficiency of DNA delivery with MMBs in vitro. Luciferase expression and cytotoxicity in HeLa cells after transfection (**a**) with nonmagnetic microbubbles, MMB1 and MMB2 upon magnetic field application with (+US) (optimized ultrasound parameters: intensity of 1.5 W/cm², 50% duty cycle, sonoporation of 10 s) and without (–US) ultrasound application. (**b** and **c**) Luciferase expression and cytotoxicity after transfection of HeLa cells with MMB1 and application of 50% duty cycle, different ultrasound intensities (W/cm²) and sonoporation times (t_{son}), as well as at different microbubble and plasmid DNA concentrations (μg DNA/cm²). Reproduced with permission from the American Institute of Physics' AIP Conference Proceedings (21).

7. Cell toxicity/viability, expressed in terms of cell respiration activity, is normalized to the reference data and is calculated using the following formula:

$$\text{Cell viability}(\%) = (D_{\text{sample}} - D_{\text{blank}}) / (D_{100\%\,\text{reference}} - D_{\text{blank}}) \times 100,$$

where D_{sample}, $D_{100\%\,\text{reference}}$, and D_{blank} are the optical densities at the maxima of the MTT–formazan absorption spectrum registered for a sample, untreated sample, and blank, respectively. Perform each experiment in triplicate.

Sample results of nucleic acid association with cells after transfection with MAALs or MMBs are given in Fig. 11.

4. Notes

1. The magnetic lipospheres and the MMBs float in a similar manner as the commercially available, non-MMBs. Keep the MMBs in the perfluoropropane atmosphere and take the required amount carefully with a needle. Do not inject air into the vial. It has been observed that exposure of the MMBs to air decreases the microbubbles' concentration to half in 2–3 h, on average.
2. Vortex MNPs and, optionally, sonicate them using an ultrasound water bath before their use for preparation of the MAALs or MMBs to avoid MNP aggregation and to obtain high-quality microbubbles.
3. There are no established rules on whether the mixing order of components (i.e., magnetic particles, nucleic acids, and enhancers) plays a major role in transfection efficiency. But the order of mixing, concentration of components, and medium composition can influence the characteristics of the bubbles and their loading with nucleic acids and/or MNPs.
4. Prior to use, warm the soybean oil to avoid pipetting mistakes. It has been observed and verified by measurements that the concentration of soybean oil in the MAAL preparations influences the MAALs' size.
5. The standard MAAL or MMB preparation in the vial has a volume of 1 mL. Preparations of 0.5 mL, or even of 0.25 mL, are also possible with a corresponding adjustment in the dose of the component. This is particularly useful when the loading drug is expensive.
6. Keep this order of mixing to avoid aggregation.
7. Use pipette tips with an aerosol filter to avoid radioactive contamination of the pipette.
8. The protocol described is a standard protocol for ethanol precipitation. Do not let the siRNA pellet air-dry for longer than 10 min to avoid difficulties in dissolving the siRNA in the diluent (RNAse-free H_2O).
9. The clearance velocity under the influence of a magnetic field gradient is related to the magnetophoretic mobilities of the magnetic particle assemblies or the magnetic lipospheres,

which, in turn, are proportional to their magnetic moments. The magnetic force that acts on magnetic particle assemblies or lipospheres comprising multiple magnetic particles in the presence of the magnetic field gradient B is given by $F_m = (M\cdot\ )B$. The total magnetic moment, M, is the product of the effective magnetic moment, m_{eff}, of the MNP under the magnetic field, B, and the total number, N, of magnetic particles associated with the liposphere ($M = N \times m_{eff}$). Above the saturation magnetization of the MNPs, which is achieved with fields exceeding 200 mT for magnetite, the magnetic force experienced by the magnetic dipole is a linear function of the field gradient. Magnetic lipospheres must move in the direction of the maximum magnetic field and are subject to a hydrodynamic drag force that can be described by Stokes law as $F_d = -3\pi\eta D_h \upsilon$, where η is the viscosity of the carrier liquid, D_h is an average hydrodynamic diameter of the objects, and υ_z is the velocity. In a stationary regime, the hydrodynamic drag force counterbalances the magnetic force. The average magnetic moment, M, can then be calculated from the magnetically induced velocity in a magnetic field gradient as described by Wilhelm et al. (39). At a magnetic field $\langle B \rangle$ of 213 mT, the magnetization of the nanoparticles, according to the experimentally measured magnetization curve, corresponds to 97% of the saturation value. Therefore, the effective magnetic moment of each particle is $m_{eff} = (0.97 M_s) P_{part}^{Fe}$, where M_s is the specific saturation magnetization per unit iron weight and P_{part}^{Fe} is the content of iron in one particle with a known core size. Therefore, the number of particles associated with a liposphere can be calculated from the magnetophoretic mobility, υ_z, which is estimated from the clearance curves as $N = \frac{3\pi D_h \upsilon_z}{\nabla B m_{eff}}$. An average velocity, υ_z, under a magnetic field gradient was evaluated from the magnetic responsiveness curves as $\upsilon_z = \langle L \rangle / t_{0,1}$. Here, $\langle L \rangle = 1$ mm is the average path of the complexes' movement, perpendicular to the measuring light beam symmetrical in both directions to the surface of the magnets arranged from both sides of the 4-mm-wide optical cuvette, and $t_{0,1}$ is the time required for a tenfold decrease in optical density.

10. Preferably, use freshly prepared magnetic lipospheres or MMBs for transfection experiments.
11. Keep the exposure time to the magnetic field the same for all experiments with the same cell line. Avoid overexposure to the magnetic field, which might negatively influence the transfection results.
12. For the in vitro experiments, use a vertically displaceable flat surface, and put the culture plate on it to ensure accurate and convenient positioning of the plate for ultrasound treatment.

13. The incubation time of the MTT reagent can vary from one cell line to the next and should therefore be optimized. The optical density at 550–590 nm for the untreated cells (used as a 100% reference) after the formazan crystal solubilization should be between 0.3 and 1.0.

Acknowledgements

The authors would like to thank Dr. Christian Bergemann (chemicell GmbH, Berlin) for providing a large selection of MNPs for the preparation of the MAALs. The authors greatly appreciate the contribution of Prof. Alexander L. Klibanov from the Department of Biomedical Engineering, Division of Cardiovascular Medicine, University of Virginia, as well as the contributions of Prof. Georg Schmitz and Mrs Karin Hensel from the Institute of Medical Engineering, Department for Electrical Engineering and Information Sciences, Ruhr-University Bochum, for their input on the evaluation of ultrasound responsiveness of MAALs/MMBs. The ultrasound images in phantom gel were obtained with the kind contribution of Priv. Doz. Dr. Klaus Woertler from the Institute of Radiology, Klinikum rechts der Isar, TU Munich. The authors would also like to thank Priv. Doz. Dr. Axel Walch, Dr. Shingi Takenaka, and Mrs. Luise Jennen of the Helmholtz Zentrum München, German Research Center for Environmental Health, for the TEM images of the MAALs and MMBs.

This work was supported by the Nanobiotechnology program (www. nanobio.de) of the German Federal Ministry of Education and Research, Nanobiotechnology grants 13N8186 and 13N8538. Financial support of the German Research Foundation DFG Research Unit FOR917 (Project PL 281/3-1), BMBF project ELA 10/002, and German Excellence Initiative via the "Nanosystems Initiative Munich (NIM)" is gratefully acknowledged.

References

1. Mykhaylyk O, Antequera YS, Vlaskou D, Plank C (2007) Generation of magnetic nonviral gene transfer agents and magnetofection in vitro. Nat Protoc 2:2391–2411
2. Plank C, Anton M, Rudolph C, Rosenecker J, Krotz F (2003) Enhancing and targeting nucleic acid delivery by magnetic force. Expert Opin Biol Ther 3:745–758
3. Fechheimer M, Boylan JF, Parker S, Sisken JE, Patel GL, Zimmer SG (1987) Transfection of mammalian cells with plasmid DNA by scrape loading and sonication loading. Proc Natl Acad Sci USA 84:8463–8467
4. Kim HJ, Greenleaf JF, Kinnick RR, Bronk JT, Bolander ME (1996) Ultrasound-mediated transfection of mammalian cells. Hum Gene Ther 7:1339–1346
5. Bao SP, Thrall BD, Miller DL (1997) Transfection of a reporter plasmid into cultured cells by sonoporation in vitro. Ultrasound Med Biol 23:953–959

6. Newman CM, Lawrie A, Brisken AF, Cumberland DC (2001) Ultrasound gene therapy: on the road from concept to reality. Echocardiography 18:339–347
7. Tata DB, Dunn F, Tindall DJ (1997) Selective clinical ultrasound signals mediate differential gene transfer and expression in two human prostate cancer cell lines: LnCap and PC-3. Biochem Biophys Res Commun 234:64–67
8. Lawrie A, Brisken AF, Francis SE, Tayler DI, Chamberlain J, Crossman DC, Cumberland DC, Newman CM (1999) Ultrasound enhances reporter gene expression after transfection of vascular cells in vitro. Circulation 99:2617–2620
9. Hernot S, Klibanov AL (2008) Microbubbles in ultrasound-triggered drug and gene delivery. Adv Drug Deliv Rev 60:1153–1166
10. Ward M, Wu J, Chiu JF (1999) Ultrasound-induced cell lysis and sonoporation enhanced by contrast agents. J Acoust Soc Am 105:2951–2957
11. Wu J, Ross JP, Chiu JF (2002) Reparable sonoporation generated by microstreaming. J Acoust Soc Am 111:1460–1464
12. Tachibana K, Uchida T, Ogawa K, Yamashita N, Tamura K (1999) Induction of cell-membrane porosity by ultrasound. Lancet 353:1409
13. Basta G, Venneri L, Lazzerini G, Pasanisi E, Pianelli M, Vesentini N, Del Turco S, Kusmic C, Picano E (2003) In vitro modulation of intracellular oxidative stress of endothelial cells by diagnostic cardiac ultrasound. Cardiovasc Res 58:156–161
14. Dijkmans PA, Juffermans LJ, Musters RJ, van Wamel A, ten Cate FJ, van Gilst W, Visser CA, de Jong N, Kamp O (2004) Microbubbles and ultrasound: from diagnosis to therapy. Eur J Echocardiogr 5:245–256
15. Pitt WG, Husseini GA, Staples BJ (2004) Ultrasonic drug delivery–a general review. Expert Opin Drug Deliv 1:37–56
16. Taniyama Y, Tachibana K, Hiraoka K, Aoki M, Yamamoto S, Matsumoto K, Nakamura T, Ogihara T, Kaneda Y, Morishita R (2002) Development of safe and efficient novel nonviral gene transfer using ultrasound: enhancement of transfection efficiency of naked plasmid DNA in skeletal muscle. Gene Ther 9:372–380
17. Bekeredjian R, Chen SY, Frenkel PA, Grayburn PA, Shohet RV (2003) Ultrasound-targeted microbubble destruction can repeatedly direct highly specific plasmid expression to the heart. Circulation 108:1022–1026
18. Unger EC, Porter T, Culp W, Labell R, Matsunaga T, Zutshi R (2004) Therapeutic applications of lipid-coated microbubbles. Adv Drug Deliv Rev 56:1291–1314
19. Tsutsui JM, Xie F, Porter RT (2004) The use of microbubbles to target drug delivery. Cardiovasc Ultrasound 2:23
20. Vlaskou D, Mykhaylyk O, Krötz F, Hellwig N, Renner R, Schillinger U, Gleich B, Heidsieck A, Schmitz G, Hensel K, Plank C (2010) Magnetic and acoustically active lipospheres for magnetically targeted nucleic acid delivery. Adv Funct Mater 20:3881–3894
21. Vlaskou D, Pradhan P, Bergemann C, Klibanov AL, Hensel K, Schmitz G, Plank C, Mykhaylyk O (2010) Magnetic microbubbles: magnetically targeted and ultrasound-triggered vectors for gene delivery in vitro. AIP Conference Proceedings 1311:485–494
22. Plank C, Vlaskou D, Schillinger U, Mykhaylyk O, Brill T, Rudolph C, Huth S, Krötz F, Hirschberger J, Bergemann C (2005) Localized nucleic acid delivery using magnetic nanoparticles, *Eur.* Cells Mater 10(Suppl 5):8
23. Hellwig N, Plank C, Vlaskou D, Bridell H, Sohn HY, Pohl U, Krotz F (2005) Ultrasound-enhanced microbubble-magnetofection: a new approach for targeted delivery of nucleotides in vivo. J Vasc Res 42:86–87
24. Vlaskou D, Mykhaylyk O, Giunta R, Neshkova I, Hellwig N, Kroetz F, Bergemann C, Plank C (2006) Magnetic microbubbles: new carriers for localized gene and drug delivery. Mol Ther 13:S290–S290
25. del Pino P, Munoz-Javier A, Vlaskou D, Rivera Gil P, Plank C, Parak WJ (2010) Gene silencing mediated by magnetic lipospheres tagged with small interfering RNA. Nano Lett 10:3914–3921
26. Holzbach T, Vlaskou D, Neshkova I, Konerding MA, Wortler K, Mykhaylyk O, Gansbacher B, Machens HG, Plank C, Giunta RE (2010) Non-viral VEGF(165) gene therapy - magnetofection of acoustically active magnetic lipospheres ('magnetobubbles') increases tissue survival in an oversized skin flap model. J Cell Mol Med 14:587–599
27. Unger EC, McCreery TP, Sweitzer RH, Caldwell VE, Wu Y (1998) Acoustically active lipospheres containing paclitaxel: a new therapeutic ultrasound contrast agent. Invest Radiol 33:886–892
28. Stride E, Porter C, Prieto AG, Pankhurst Q (2009) Enhancement of microbubble mediated gene delivery by simultaneous exposure to ultrasonic and magnetic fields. Ultrasound Med Biol 35:861–868
29. Lentacker I, De Smedt S, Demeester J, Van Marck V, Bracke M, Sanders N (2007) Lipoplex-loaded microbubbles for gene

delivery: a Trojan horse controlled by ultrasound. Hum Gene Ther 18:1046–1046

30. Mykhaylyk O, Vlaskou D, Tresilwised N, Pithayanukul P, Möller W, Plank C (2007) Magnetic nanoparticle formulations for DNA and siRNA delivery. J Magn Magn Mat 311:275–281
31. Sanchez-Antequera Y, Mykhaylyk O, Thalhammer S, Plank C (2010) Gene delivery to Jurkat T cells using non-viral vectors associated with magnetic nanoparticles. Int J Biomedical Nanoscience and Nanotechnology 1:202–229
32. Suzuki M, Mikami T, Matsumoto T, Suzuki S (1977) Preparation and antitumor activity of o-palmitoyldextran phosphates, o-palmitoyldextrans, and dextran phosphate. Carbohydr Res 53:223–229
33. Snyder F, Stephens N (1959) A simplified spectrophotometric determination of ester groups in lipids. Biochim Biophys Acta 34:244–245
34. Schneider M (1999) Characteristics of SonoVuetrade mark. Echocardiography 16: 743–746
35. Sanchez-Antequera Y, Mykhaylyk O, van Til NP, Cengizeroglu A, de Jong JH, Huston MW, Anton M, Johnston IC, Pojda Z, Wagemaker G, Plank C (2011) Magselectofection: an integrated method of nanomagnetic separation and genetic modification of target cells. Blood 117:e171–e181
36. Mykhaylyk O, Zelphati O, Hammerschmid E, Anton M, Rosenecker J, Plank C (2009) Recent advances in magnetofection and its potential to deliver siRNAs in vitro. Methods Mol Biol 487:111–146
37. Mykhaylyk O, Zelphati O, Rosenecker J, Plank C (2008) siRNA delivery by magnetofection. Curr Opin Mol Ther 10:493–505
38. Mykhaylyk O, Sanchez-Antequera Y, Vlaskou D, Hammerschmid E, Anton M, Zelphati O, Plank C (2010) Liposomal magnetofection. Methods Mol Biol 605:487–525
39. Wilhelm C, Gazeau F, Bacri JC (2002) Magnetophoresis and ferromagnetic resonance of magnetically labeled cells. Eur Biophys J 31:118–125

Chapter 16

Synthesis of Lipidic Magnetic Nanoparticles for Nucleic Acid Delivery

Yoshihisa Namiki

Abstract

Lipidic magnetic nanoparticles are a rapid and reliable magnet-guided nucleic acid delivery system, which we have named as "LipoMag" (Namiki et al., Nat. Nanotechnol 4:598–606, 2009). LipoMag is composed of a cationic lipid shell, an oleic acid binder, and an iron oxide nanocrystal core. Through the electrostatic force, positively charged LipoMag and negatively charged nucleic acid form complexes, termed "mag-lipoplex." Under a magnetic field, LipoMag bearing nucleic acid can be controllably accumulated in target cells. Here we describe the procedure for the preparation of LipoMag which can be used for the transfection of nucleic acids, such as plasmid DNA and small interfering RNA.

Key words: Lipidic magnetic nanoparticle (LipoMag), Gene therapy, Transfection, Plasmid DNA, Small interfering RNA

1. Introduction

External energy sources may be used to confer a variety of functions on nanoparticles. Among them, the magnetic field is one of the most useful, because magnetic nanoparticles can be attracted by a permanent magnetic field (1–5), heated by an AC magnetic field (6) and detected by a magnet sensor (7). On the other hand, therapeutic nucleic acids can be used as a disease-specific and non-invasive treatment option for various kinds of diseases. However, an efficient nucleic acid delivery system has not yet been developed. We have recently developed self-assembling type lipidic magnetic nanoparticles, called "LipoMag," for the delivery of nucleic acids to target lesions. LipoMag nanoparticles are constructed via a hydrophobic interaction with a cationic lipid shell and oleic acid-coated magnetic nanocrystal core (1, 2, 8). In this chapter, we

Manfred Ogris and David Oupicky (eds.), *Nanotechnology for Nucleic Acid Delivery: Methods and Protocols*,
Methods in Molecular Biology, vol. 948, DOI 10.1007/978-1-62703-140-0_16, © Springer Science+Business Media, LLC 2013

describe the procedure for preparing LipoMag as a delivery for plasmid DNA and small interfering RNA (siRNA).

2. Materials

Use analytical grade reagents and nuclease-free distilled water (DW) for the synthesis of the magnetic fluid and the lipidic magnetic nanoparticles.

2.1. Magnetic Fluid Component

1. Ammonium hydroxide (25 %, w/v).
2. Iron (II) chloride solution: Dissolve 7.95 g Iron (II) chloride tetrahydrate by adding 8 mL DW in a 50 mL conical tube (Becton, Dickinson and Company (BD), NJ, USA).
3. Iron (III) chloride solution: Dissolve 21.62 g Iron (III) chloride hexahydrate by adding 21 mL DW in a 50 mL conical tube.
4. Iron chloride solution: Mix the iron (II) chloride solution and the iron (III) chloride solution and fill up to 50 mL with DW.
5. 10 % Oleic acid/dodecane solution: Mix 5 mL oleic acid with 45 mL dodecane.
6. An electronic overhead stirrer (RW16 basic) and stirrer blade (25 mm) (IKA-Works, Inc., NC, USA).
7. An oil bath (for example BO600, Yamato Scientific Co., Ltd., Tokyo, Japan).

2.2. Lipidic Magnetic Nanoparticle Components

1. Chloroform.
2. *O,O'*-Ditetradecanoyl-*N*-(α-trimethylammonioacetyl) diethanol-amine chloride (DC-6-14) (Sogo Pharmacochemical (Tokyo, Japan)).
3. Dioleoylphosphatidylethanolamine (DOPE) (Sigma, St. Louis, MO).
4. DC-6-14/chloroform solution: Transfer 20 mg DC-6-14 to a 15 mL glass test tube with screw cap. Add 10 mL chloroform, seal argon gas in this tube, tighten the screw cap, and completely dissolve DC-6-14. Store at −80 °C.
5. DOPE/chloroform solution: Transfer 20 mg DOPE to a 15 mL glass test tube with screw cap. Add 10 mL chloroform, seal argon gas in this tube, tighten the screw cap, and completely dissolve DOPE. Store at −80 °C.
6. A pear-shaped flask (10 mL).
7. Argon gas.

8. A vortex mixer (Vortex-Genie 2, Scientific Industries Inc., NY, USA).
9. A cup-horn type sonicator (Sonifier 250D, Branson, CT, USA).
10. A rotary evaporator (RE301, Yamato Scientific Co., Ltd.).
11. A water bath (BM400, Yamato Scientific Co., Ltd.).
12. A vacuum pump (PX-51, Yamato Scientific Co., Ltd.).
13. A French press (Thermo Fisher Scientific Inc., MA, USA).
14. A sintered magnetic metal filter (9, 10) (Giken parts, Nara, Japan).

3. Methods

Using a modified peptization technique (11, 12), we initially prepare a magnetic fluid composed of iron oxide nanocrystals coated with oleic acid, dispersed in chloroform. The self-assembling type lipidic magnetic nanoparitcles are synthesized through the hydrophobic interaction between cationic lipid and oleic acid using both sonication and vacuum procedures.

3.1. Preparation of Chloroform-Based Magnetic Fluid

1. Using a stirrer blade-attached electronic overhead stirrer, mix 50 mL iron chloride solution with 50 mL ammonium hydroxide under vigorous agitation (200 rpm), and black magnetic slurry is obtained.
2. This obtained mixture is heated to 95 °C by an oil bath, and 10 % oleic acid/dodecan (50 mL) is added under vigorous agitation (200 rpm) using a stirrer (see Note 1).
3. A distinct phase separation between the upper organic portion will appear in which the oleic acid-coated iron oxide nanocrystals are stably dispersed, and there will be a lower aqueous portion containing ammonium chloride.
4. Remove most of the lower aqueous portion using a glass pipette.
5. Continue heating until the remaining water is completely evaporated and then collect the approximately 70 % upper organic portion.
6. Add 2 mL of the collected upper organic portion to 100 mL acetone in order to flocculate the oleic acid-coated iron oxide nanocrystals.
7. Collect this flocculate using a permanent magnet (1 T) while the supernatant liquid is poured off.

8. To eliminate excess oleic acid, wash this flocculate with 50 mL acetone and repeat step 7.
9. Completely remove the acetone from this washed flocculate under reduced pressure (50 mmHg) using a rotary evaporator.
10. Weigh out 2 mg of the resulting dried oleic acid-coated iron oxide nanocrystals and mix this amount with 1 mL chloroform.
11. Finally, a chloroform-based magnetic fluid (2 mg/mL) will be obtained.

3.2. Preparation of Lipidic Magnetic Nanoparticles

1. Mix 344.8 μL DC-6-14/chloroform solution (2 mg/mL), 155.2 μL DOPE/chloroform solution (2 mg/mL), 200 μL chloroform-based magnetic fluid (2 mg/mL), and 300 μL chloroform in the pear-shaped flask.
2. Add 1 mL DW and charge argon gas into the pear-shaped flask and seal it with the glass stopper.
3. Agitate the flask on a vortex mixer for 1 min (maximum speed) and sonicate the flask with a cup-horn type sonicator at maximum output for 2 min.
4. Immediately attach the flask to a rotary evaporator and rotate at maximum speed.
5. Evaporate the contents at 50 mmHg with strong sonication at 37 °C using a water bath and a clear brown-colored suspension will be obtained (Fig. 1, see Note 2).
6. Remove the remaining chloroform from this suspension at 10 mmHg for 10 min (see Note 3).
7. Treat the suspension with a French Press at 40,000 lb/in.^{2}.
8. Attach the magnet on the sintered magnetic metal filter (Fig. 2).
9. Introduce the French Press-treated suspension to this filter (see Note 4).
10. Add 3 mL DW to this filter twice.
11. Remove the magnet from the sintered magnetic metal filter.
12. Add 1 mL DW, insert and quickly push the plunger, and collect the eluted liquid.
13. Add the total volume of collected liquid to the magnetic filter and similarly collect the re-eluted liquid.
14. Repeat step 13 twice (i.e., the washing step).
15. Finally, the purified lipidic magnetic nanoparticles will be obtained.

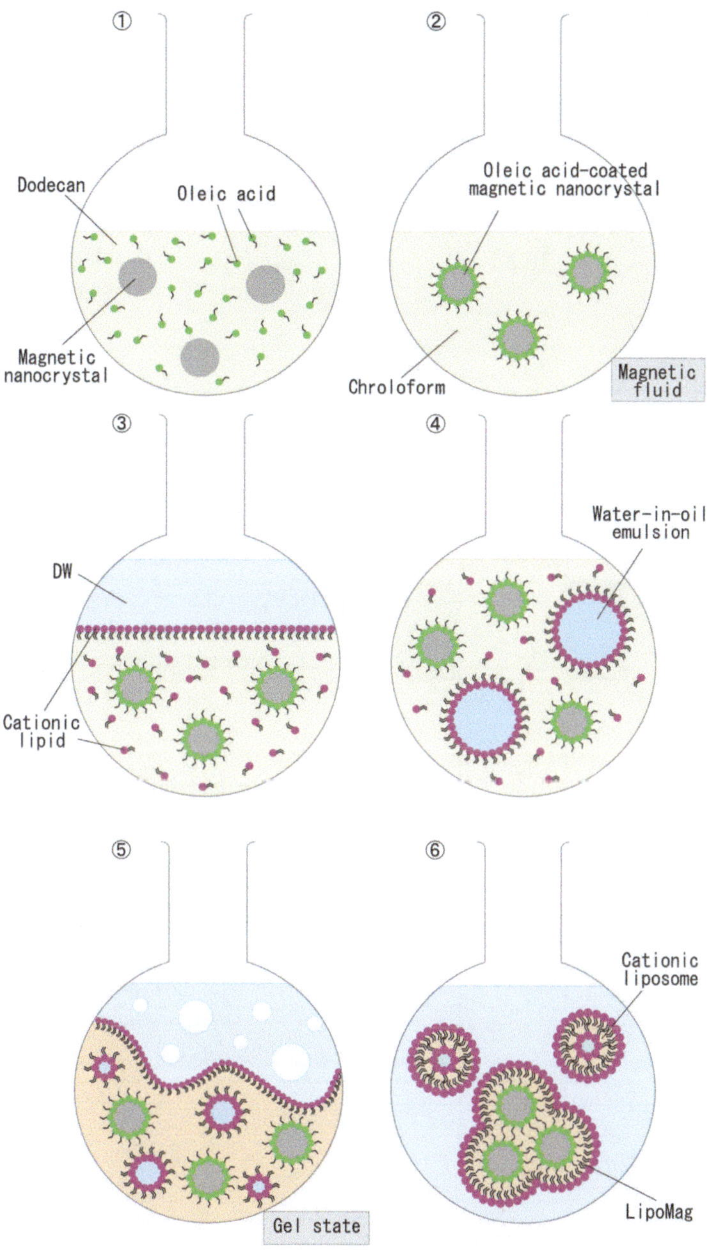

Fig. 1. Schematic representation of the preparation of the lipidic magnetic nanoparticles. (1–2) The magnetic fluid is composed of the oleic acid (*green*)-coated magnetite nanocrystals (*gray*) dispersed in chloroform (*yellow*). The magnetic fluid and phospholipids (*pink*) dissolved in chloroform were mixed. (3) When this mixture and DW (*blue*) were poured into a pear-shaped flask, an upper layer of aqueous phase and a lower layer of the organic phase containing the oleic acid-coated magnetic nanocrystals appeared. (4) Microdroplets were stabilized by a phospholipid single layer, and "inverted micelles" formed after vigorous agitation and sonication. (5) A transformation of these inverted micelles into a viscous gel state was induced by removal of chloroform by strong vacuuming. (6) The hydrophobic groups of the amphiphilic phospholipids and oleic acid joined together to form the hydrophilic, lipidic magnetic nanoparticles after the induction of gel-collapse by a continuous vacuum.

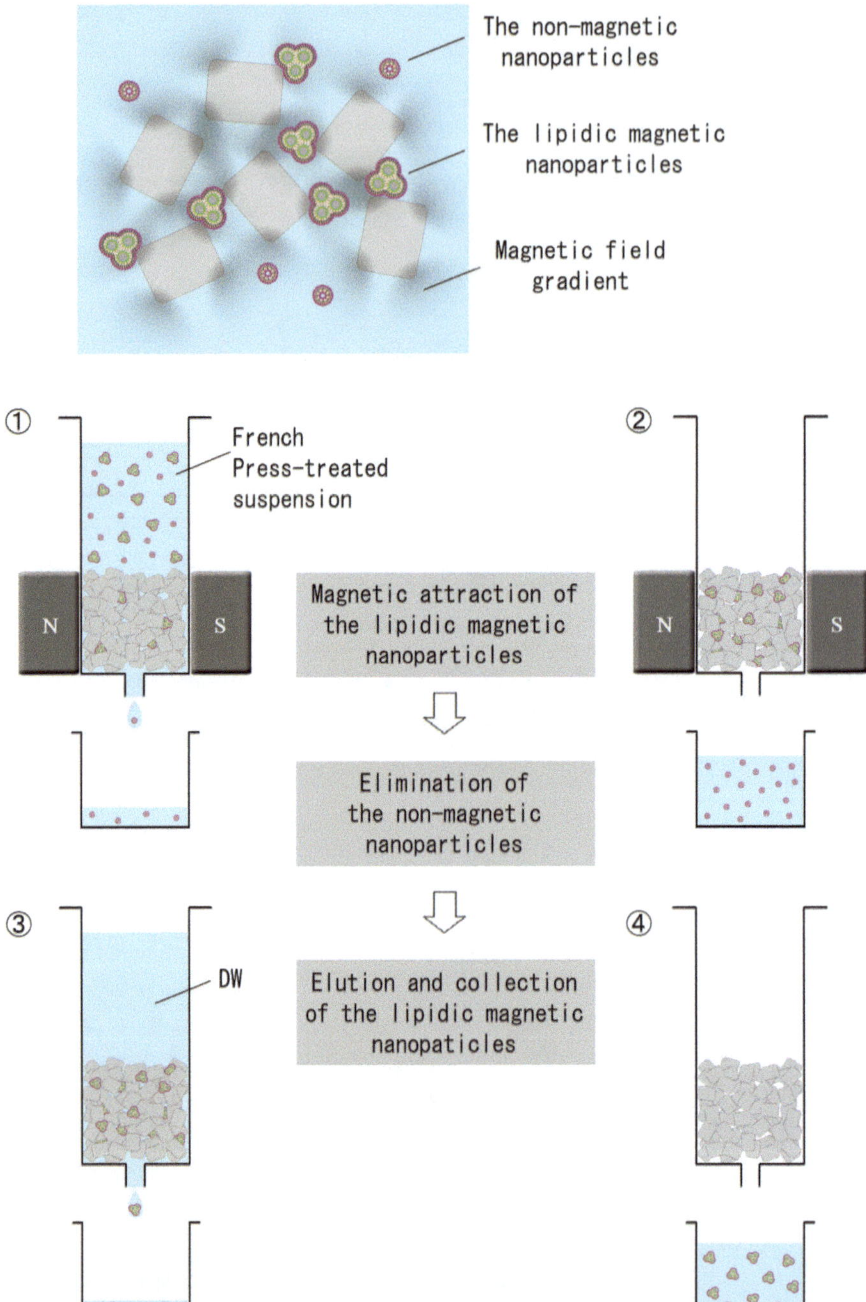

Fig. 2. Schematic representation of the preparation of the magnetic filter used for the purification of the lipidic magnetic nanoparticles. The sintered magnetic filter is composed of magnetically attractive stainless steel (SUS 430) granules (0.4 mm). This magnetic filter is inserted in the cylinder (2.5 mL). N-pole and S-pole of twin neodymium magnets, which are placed horizontally, are oppositely arranged in the closed magnetic circuit.

3.3. Quantitative Analysis of Lipidic Magnetic Nanoparticle Components

1. The amount of oleic acid in the oleic acid-coated iron oxide can be measured using a gas chromatograph.
2. The amount of iron oxide in the LipoMag can be measured using an atomic absorption photometer after the treatment with hydrochloric acid.
3. The phosphate in the LipoMag can be quantified using the modified Malachite green reagent, BIOMOL GREEN Reagent (BIOMOL, Research laboratories, PA, USA), according to the manufacturer's protocol.

4. Notes

1. In this process, the iron oxide nanocrystals become coated with hydrophilic ammonium oleate, and this ammonium oleate starts to decompose at 78 °C. Upon continuous heating, the iron oxide nanocrystals will change so as to be coated with hydrophobic oleic acid with the evaporation of the ammonia gas.
2. In this process, chloroform will be evaporated off to yield a gel, and the gel will collapse to give a clear brown-colored suspension.
3. In this process, several hundred microliters of DW should be added under atmospheric pressure before the contents become dry.
4. In this process, the iron oxide-free lipid vesicles (cationic liposomes) can be removed from the crude French Press-treated suspension.

Acknowledgment

I We dedicate this work to the late T. Terada and the late K. Nariai. I We would like to acknowledge Yukiko Ishii, Yoko Yumoto, Yuki Nagase, and Shuichi Nakagawa for their assistance. This work was supported by a Funding Program for Next Generation World-Leading Researchers (LS114) from JSPS, an Industrial Technology Research Grant Program (08C46049a) from the NEDO of Japan, Grant-in-Aid for Scientific Research (#22300180) from JSPS, and the Iketani Science and Technology Foundation (2012).

References

1. Namiki Y, Namiki T, Yoshida H, Ishii Y, Tsubota A, Koido S, et al. A novel magnetic crystal-lipid nanostructure for magnetically guided *in vivo* gene delivery. Nat Nanotechnol. 2009;4:598–606.
2. Namiki Y, Fuchigami T, Tada N, Kawamura R, Matsunuma S, Kitamoto Y, et al. Nanomedicine for cancer: lipid-based nanostructures for drug delivery and monitoring. Acc Chem Res. 2011;44:1080-93.
3. Fuchigami T, Kawamura R, Kitamoto Y, Nakagawa M, Namiki Y. Ferromagnetic FePt-nanoparticles/polycation hybrid capsules designed for a magnetically guided drug delivery system. Langmuir. 2011;27:2923–8.
4. Fuchigami T, Kawamura R, Kitamoto Y, Nakagawa M, Namiki Y. A magnetically guided anti-cancer drug delivery system using porous FePt capsules. Biomaterials. 2012;33:1682–7.
5. Namiki Y, Namiki T, Ishii Y, Koido S, Nagase Y, Tsubota A, et al. Inorganic-Organic Magnetic Nanocomposites for use in Preventive Medicine: A Rapid and Reliable Elimination System for Cesium. Pharm Res. 2012;29:1404–18.
6. Kitamoto Y, He J. Chemical synthesis of FePt nanoparticles with high alternate current magnetic susceptibility for biomedicalapplications. Electrochim Acta. 2009;54:5969–72.
7. Koh I, Josephson L. Magnetic nanoparticle sensors. Sensors. 2009;9:8130–45.
8. Namiki Y, Namiki T. Synthesis of lipidic magnetic nanoparticles as nucleic acid and drug delivery system. Japanese Patent. 2008; 4,183,047.
9. Namiki Y. Sintered magnetic filter for the purification of magnetic nanoparticles. Japanese Patent Application. 2011;083367.
10. Namiki Y, Matsunuma S, Inoue T, Koido S, Tsubota A, Namiki T, et al. Magnetic nanostructures for biomedical application. In: Masuda Y, editor. Nanocrystal. Rijeka, Croatia: Sciyo. 2011;349–372.
11. Rosenweig RE. Magnetic fluids. Int Sci Tech. 1966;55:48–56.
12. Reimers GW, Khalafalla SE. Production of magnetic fluids by peptization techniques. US Patent. 1974;3,843,540.

Chapter 17

Lipopeptide Delivery of siRNA to the Central Nervous System

Mark D. Zabel

Abstract

RNA interference is a relatively new tool used to silence specific genes in diverse biological systems. The development of this promising new technique for research and therapeutic use in studying and treating neurological diseases has been hampered by the lack of an efficient way to deliver siRNA transvascularly across the blood–brain barrier (BBB) to the central nervous system (CNS). Here we describe a method for delivering siRNA to the CNS by complexing it to a peptide that acts as a neuronal address by binding to acetylcholine receptors (AchRs). Adding cationic liposomes to the complex protects it from serum nucleases and proteases en route. When injected intravenously, these liposome–siRNA–peptide complexes resist serum degradation, effectively cross the BBB, and deliver siRNA to AchR-expressing cells to suppress protein expression in the CNS.

Key words: RNAi, siRNA, Delivery, Central nervous system, Blood–brain barrier, Liposomes, Peptides, Complexes, Prions

1. Introduction

Ribonucleic acid interference (RNAi) is a relatively new discovery in cell biology that diverse organisms use to help regulate the amount of messenger RNA (mRNA) available for translation. Dicer, an endogenous mammalian protein complex that mediates RNAi, cleaves double-stranded or short hairpin RNA (shRNA) into functional, small interfering RNA (siRNA). The RNA-induced silencing complex (RISC) guides siRNA to its complementary mRNA target, which is then digested by argonaut, an endonuclease present in the RISC. A concomitant reduction in protein expression ensues. Researchers have begun exploiting RNAi as a tool for silencing specific genes in biological systems (1–4). RNAi

Manfred Ogris and David Oupicky (eds.), *Nanotechnology for Nucleic Acid Delivery: Methods and Protocols*, Methods in Molecular Biology, vol. 948, DOI 10.1007/978-1-62703-140-0_17,

has shown potential efficacy in treating several diseases including hepatitis (5, 6), cancer (7), ocular disorders (8), chronic pain (9), and neuropathologies induced by viruses (10) and prions (11). Initial attempts to use RNAi in the central nervous system (CNS) primarily used intracranial (ic) injections of lentiviral vectors encoding shRNA. Lentivirus encoding shRNA against mRNA transcripts has demonstrated decreased expression of the cognate protein (12, 13), but several aspects of this method limit its efficacy. Inadequate neuroinvasion by lentiviral vectors across the blood–brain barrier (BBB) necessitates their direct delivery to the CNS via stereotactic intracranial injection of shRNA lentivirus directly into the brain. Even then, lentiviral infection and shRNA expression were limited to a small area around the injection site, resulting in spatially limited knockdown of protein expression. This method also lacks desired temporal control of protein expression, since once brain cells are infected with lentivirus, it will likely irreversibly suppress protein expression in these cells. Moreover, despite improvements in lentiviral vector technology and construction (14–17), concerns over the oncogenic capacity of lentiviral delivery systems remain (18–20).

Nonviral strategies to deliver siRNA molecules to cells include using cationic peptides capable of crossing plasma membranes (21–23). Conjugating siRNA to these cell-penetrating peptides via thiol linkages allows for efficient dissociation of siRNA–peptide complexes by reduction of the disulfide bond in the cytoplasm (24, 25). While this delivery method holds therapeutic promise, it still does not solve two important problems of drug delivery to the brain: cell specificity and transport of its cargo across the BBB. Kumar et al. developed a transvascular method to deliver siRNA across the BBB specifically to neuronal cells in the brain via intravenous (iv) injection (10). This method involves complexing siRNA to a short peptide derived from the rabies virus glycoprotein that binds specifically to acetylcholine receptors (AchRs) on neuronal cells (26, 27). Adding nine D-Arginines to the carboxyl terminus of this peptide (RVG-9r) enabled it to electrostatically interact with siRNA and specifically deliver siRNA to neurons in mouse brains to suppress protein expression and protect against fatal viral encephalitis. This significant advance in siRNA delivery to the CNS was tempered by the lack of direct detection of siRNA in the brains of a majority of treated mice, suggesting that these naked complexes may have degraded significantly in the blood during transport. Efficient peptide-mediated delivery of siRNA to the brain after iv injection relies on protecting the complex from serum nuclease and protease degradation en route.

Complexing or encapsulating siRNA with liposomes has been shown to protect siRNA from degradation and improve delivery through the vasculature (28–33). Adding polyethylene glycol (PEG) to lipids increases bioavailability and circulation time in blood (34–36) by decreasing immune clearance (37). PEGylated liposomes increased delivery of doxorubicin and cisplatin in experimental tumor models (38, 39) and have been used in clinical trials to treat a variety of cancers (40), including AIDS-related Kaposi's sarcoma (41, 42) and advanced malignant solid tumors (43), including those found in ovarian (44) and breast (45) cancer.

While protecting therapeutic agents from serum degradation represents a crucial step in transvascular delivery to target sites in the brain, liposomes alone do not provide cell-specific delivery, nor do they ensure transport across the BBB. Coupling a monoclonal antibody against glial fibrillary acidic protein (GFAP) to PEGylated liposomes successfully targeted them to astrocytes in cell culture, but still failed to deliver them across the BBB to astrocytes in mouse brains (46). An effective, efficient delivery system must do all three: (1) provide a molecular address to the agent for delivery to neuronal cells, (2) gain access to the brain by crossing the BBB, and (3) protect the agent from serum degradation en route. Using siRNA as a therapeutic agent and prion diseases as models, we have developed a therapeutic delivery system that addresses all of these issues. By complexing siRNA targeting mRNA transcripts encoding PrP^{C} to cationic liposomes, and then adding the RVG-9r peptide, we have created liposome–siRNA–peptide complexes (LSPCs) that effectively cross the BBB and deliver PrP siRNA to AchR-expressing cells to suppress PrP^{C} expression in vivo.

2. Materials

2.1. LSPC Reagents

1. siRNA: Designed using siRNA scales (47) and synthesized, pre-annealed, and lyophilized by Qiagen (see Note 1). Resuspend in sterile RNAse-free water (supplied by Qiagen) to 20 μM according to the manufacturer's directions. Store in single-use aliquots at –70°C (see Note 2).
2. Peptides: RVG-9r and control RVM-9r peptides (Table 1) were synthesized and purified by high-performance liquid chromatography by Global Peptide (www.globalpeptide.com).

Table 1
Delivery peptide sequences

RVG-9r[a]	YTIWMPENPRPGTPCDIFTNSRGKRASNGGGGrrrrrrrrr
RVM-9r[b]	MNLLRKIVKNRRDEDTQKSSPASAPLDGGGGrrrrrrrrr

[a]Lower case r denotes D-arginine racemer
[b]Control

Resuspend in sterile phosphate-buffered saline (PBS), pH 7.4 to 400 μM, or approximately 2 mg/mL. For example, resuspend 100 mg of peptide in 50 mL PBS. Store at –70°C in single-use aliquots (typically 400 μL, see Note 2).

3. 10% Sterile sucrose: Weigh 1 g of sucrose and mix into 9 mL of sterile deionized water.

4. Liposomes: Prepare cationic liposomes by mixing *N*-[1-(2,3-Dioleoyloxy)propyl]-*N*,*N*,*N*-trimethylammonium chloride (DOTAP, Sigma-Aldrich, St. Louis, MO) and cholesterol (NOF America, Irvine, CA) in chloroform at 1:1 molar ratio to a final concentration of 2 mM. Weigh and mix 14 mg of DOTAP and 7.7 mg of cholesterol into 10 mL of chloroform (see Note 3) in a round-bottom, 15-mL glass tube. Place the tube in a vacuum desiccator and dry overnight to a thin film (see Note 4).

Rehydrate lipids to a final concentration of 10 mM in sterile 10% sucrose. Add 2 mL of sucrose solution to the glass tube containing the lyophilized lipids and mix by pipetting. Incubate for 1 h at 50°C, and then at room temperature for 2 h. Mix and serially filter through 1, 0.45, and 0.2 μm pore filters (Pall Life Sciences). Store in sterile tubes at 4°C for up to several months until ready to use.

Dilute liposomes to 200 μM working concentration 1:100 in sterile PBS immediately prior to use. For example, briefly mix stock liposomes by pipetting and add 2–198 μL sterile PBS. Mix by pipetting and put 100 μL into two 200 μL thin-walled PCR tubes (VWR). Seal tubes shut with parafilm and place into the tube holder of MP4000 horn sonicator (Qsonica). Sonicate liposomes for 40 s at 75% maximum power (see Note 5). The liposomes can be used immediately for complexation, or stored for several weeks at 4°C (see Note 6).

2.2. LSPC Delivery Supplies

1. Mice: Choose mouse lines and strains appropriate for the experimental design. Various strains of wild-type mice can be purchased from Charles River or Jackson labs (see Note 7).
2. 26–30 Gauge insulin syringes.
3. Heat lamp.
4. Rodent anesthesia machine dispensing Isoflurane/Oxygen (Parkland Scientific, see Note 8).

3. Methods

3.1. Preparing LSPCs

Complex siRNA, liposomes, and peptides using filter barrier tips and sterile technique in a cell culture hood. The following preparation will produce enough LSPCs to treat five mice.

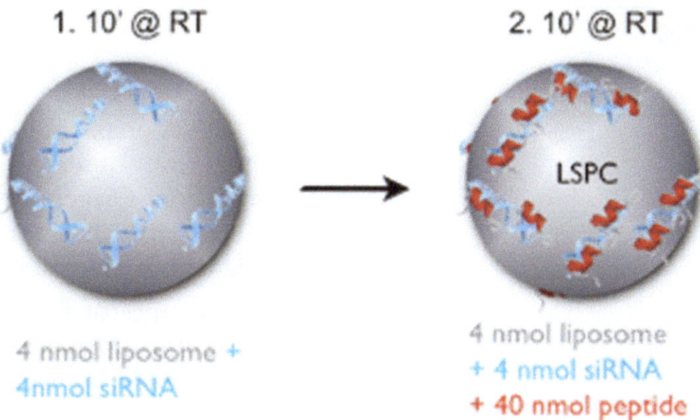

Fig. 1. LSPC formulation.

1. Thaw 1 mL of stock siRNA and 400 μL stock peptides on ice.
2. Add 100 μL of 200 μM stock liposomes (sonicate first if necessary, see Note 6) to 1 mL of 20 μM siRNA and mix gently by pipetting.
3. Incubate at room temperature for 10 min (Fig. 1).
4. Add 500 μL of peptide, mix gently by pipetting, and incubate at room temperature for 10 min (see Note 9).

3.2. Injecting LSPCs

1. Warm mice in their cage by placing it 4–6 in. below a heat lamp for 5–10 min (see Note 10).
2. While warming the mice, turn on the anesthesia machine such that oxygen flows at a rate of 3–4 L/min. Turn on the isoflurane at 3 L/min.
3. When the tail vasculature is clearly visible, place a single mouse in the anesthesia chamber until its respiration visibly slows to 1–2 breaths per second (see Note 11).
4. Meanwhile, load 300 μL of LSPCs into a sterile insulin syringe (see Note 12).
5. Place the mouse on a flat surface with its face in the nose cone attached to the anesthesia machine (see Note 13).
6. Starting with the mouse laying on its ventral side (dorsal side up), locate one of the two lateral veins. Starting as far posterior as possible, rotate the tail slightly and insert the needle with the bevel up at a shallow 10–15° angle 1–2 mm into the vein. Slowly press the plunger to inject the LSPC solution (see Note 14).

4. Notes

1. We have used siRNA prepared for us by several biotechnology companies, but have observed optimal knockdown using siRNA prepared from Qiagen. Additionally, Qiagen can label siRNA with Alexa fluorochromes to monitor LSPC trafficking for determining delivery kinetics (see Note 15).
2. We typically treat cages of five mice at a time and therefore aliquot 1 mL of 200 μM stock siRNA, enough to treat five mice. Aliquots should be adjusted to meet researchers' experimental requirements.
3. Accurately weighing these amounts requires a microbalance. Alternatively, prepare a 20 mM stock by mixing 140 mg of DOTAP and 77 mg of cholesterol into chloroform to a final volume of 10 mL, and then dilute 1 mL of this solution into 9 mL of chloroform to a final concentration of 2 mM. Store the 20 mM stock at room temperature protected from air and light for up to several months.
4. Desiccation time may vary. Most importantly, the lipids must be completely dry. For faster desiccation, the solution can be split into several tubes.
5. Do not sonicate more than 100 μL per tube using this sonicator. Other sonicators may be used, but volumes may have to be increased significantly if a probe sonicator is used. Horn sonicators produce the best, most consistent results with the small volumes used here.
6. If used later, sonicate liposomes again prior to complexation.
7. White mice (FVB, C3H, CD-1, for example) tend to be much easier to inject intravenously in the tail than brown or black mice (129sv, C57/Bl6).
8. The machine should have an incubating chamber and a nose cone. We recommend isoflurane, but alternative anesthesia, such as ketamine, can be used. While tail vein injections can be done using physical restraint, we strongly recommend using chemical restraint to avoid sudden movements from the mouse that can compromise the injection and waste valuable time and reagents.
9. LSPCs are ready to use in 10 min, but are stable for several hours at room temperature.
10. This step is crucial for successful tail vein injections, especially in dark-colored mice. The cage should reach an ambient temperature between 32 and 35°C. The lateral tail veins and/or the ventral artery should be clearly visible before proceeding. This usually coincides with salivation, so look for moisture

around the mouth and nose. Monitor mice under the heat lamp every 2 min for signs of distress or hyperthermia. Do not leave a mouse cage under the heat lamp for more than 15 consecutive minutes.

11. This should take 1–3 min. Isoflurane is very safe and accidental overdoses are very rare. Distressed mice can be given pure oxygen to help revive them if necessary.
12. If bubbles appear, hold the syringe vertically with the needle pointing up and flick the syringe until the bubbles dislodge and float to the top of the syringe, toward the needle. Carefully keeping the needle tip above the syringe barrel, place it on the inside of the tube containing the LSPCs and gently push the syringe plunger until the bubbles are evacuated (i.e., when a small amount of liquid emerges from the needle). Load additional LSPC mixture into the syringe to 300 μL if necessary.
13. Mice can be placed in physical restraints at this time if desired, but we recommend simultaneous chemical restraint.
14. Curling the tail over a finger can help orient the tail in a position that facilitates inserting the needle at the shallow 10–15° angle required for entering the vein. You should, but not always, feel the needle entering the vein lumen. You should feel little or no resistance on the plunger if you are in the lumen, and you will often see the vein blanche as the solution displaces the blood. If you feel resistance or see the tail swell just anterior to the injection site, stop immediately. You are most likely not in the vein. Apply direct pressure to the injection site for a few seconds and then attempt the injection again further anterior to the previous site. One typically has three opportunities for a successful injection per vein. If the injection is unsuccessful in both veins, turn the mouse over and try injecting into the ventral artery. If this fails, or if the vasculature is no longer visible, stop. Allow the mouse to recover, proceed to the next mouse, and try injecting the former mouse after attempting the others.
15. To monitor LSPC trafficking and determine delivery kinetics, label peptides with DyLight 649 or a similar fluorochrome (Alexa dyes, for example) according to the manufacturer's directions (Pierce). Use a 3 kDa molecular weight cutoff Microcon centrifugal filter (Millipore) or a similar device to exchange labeling buffer for storage buffer (10 mg/mL bovine serum albumin in sterile PBS).
16. LSPCs can be monitored in live mice by injecting them with fluorochrome-labeled LSPCs, which can be visualized using a Xenogen (Ivis) or a similar live animal imager. We observed LSPC delivery in as little as 2 min using this technique (Fig. 2) and detected LSPCs up to 10 days after injection (11).

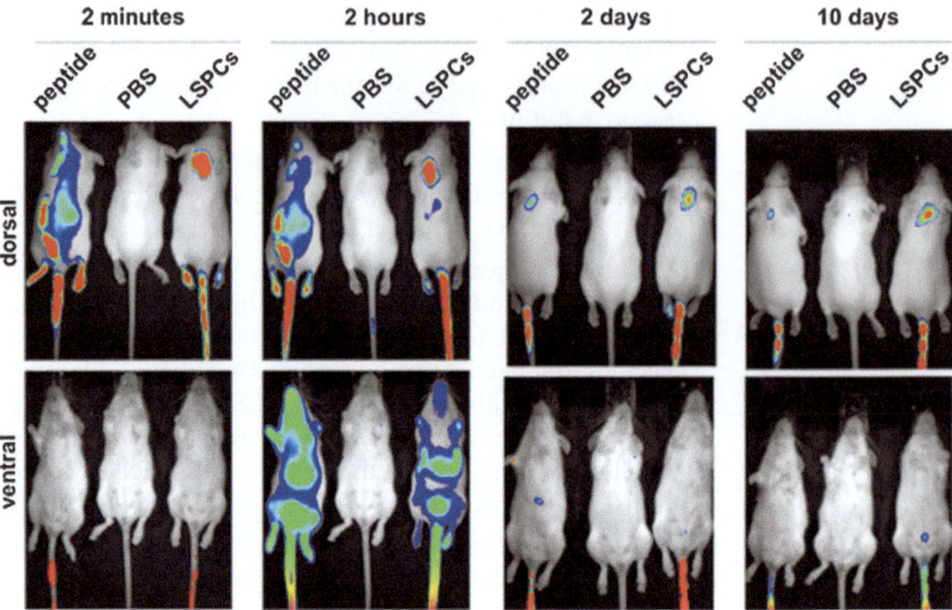

Fig. 2. LSPC delivery to the CNS in live animals. Mice were injected iv into the tail vein with PBS alone (*middle mouse* in each panel), PrP siRNA complexed to the RVG-9r peptide (*first mouse* in each panel), or PrP LSPCs (*third mouse* in each panel). siRNA was detected in the brains of mice in as little as 2 min after injection. LSPCs targeted siRNA more specifically to the brain, where after at least 10 days post injection, we detected more LSPCs than siRNA complexed with peptide alone.

17. LSPC localization can be observed in the CNS by fluorescent microscopy (11). Euthanize injected mice and remove and snap-freeze their brains in OCT (Tissue Tek) in liquid nitrogen. Using a cryostat cooled to −20°C (Leica), cut 8-μm-thick brain sections and mount them on glass slides. Fix in ice-cold acetone for 10 min and air-dry overnight. Slides can be stained with additional markers (the protein knockdown target, for example) or DAPI before mounting in ProLong Gold (Invitrogen) or a similar anti-fade mounting media. Visualize LSPCs in brain sections using a microscope capable of detecting the fluorochromes used (Fig. 3a–t).
18. Quantitative analysis of LSPC delivery can be performed on brain cells isolated from injected mice by flow cytometry (11). Sacrifice mice and remove their brains. Strain an approximately 1-mm-thick brain section through a 45 μm nylon mesh (Fisher Scientific) to produce a single cell suspension. Cells can be stained with additional cell marker of interest (the protein knockdown target) and analyzed using a Cyan (BD Pharmingen) or a similar flow cytometer (Fig. 3u–w).

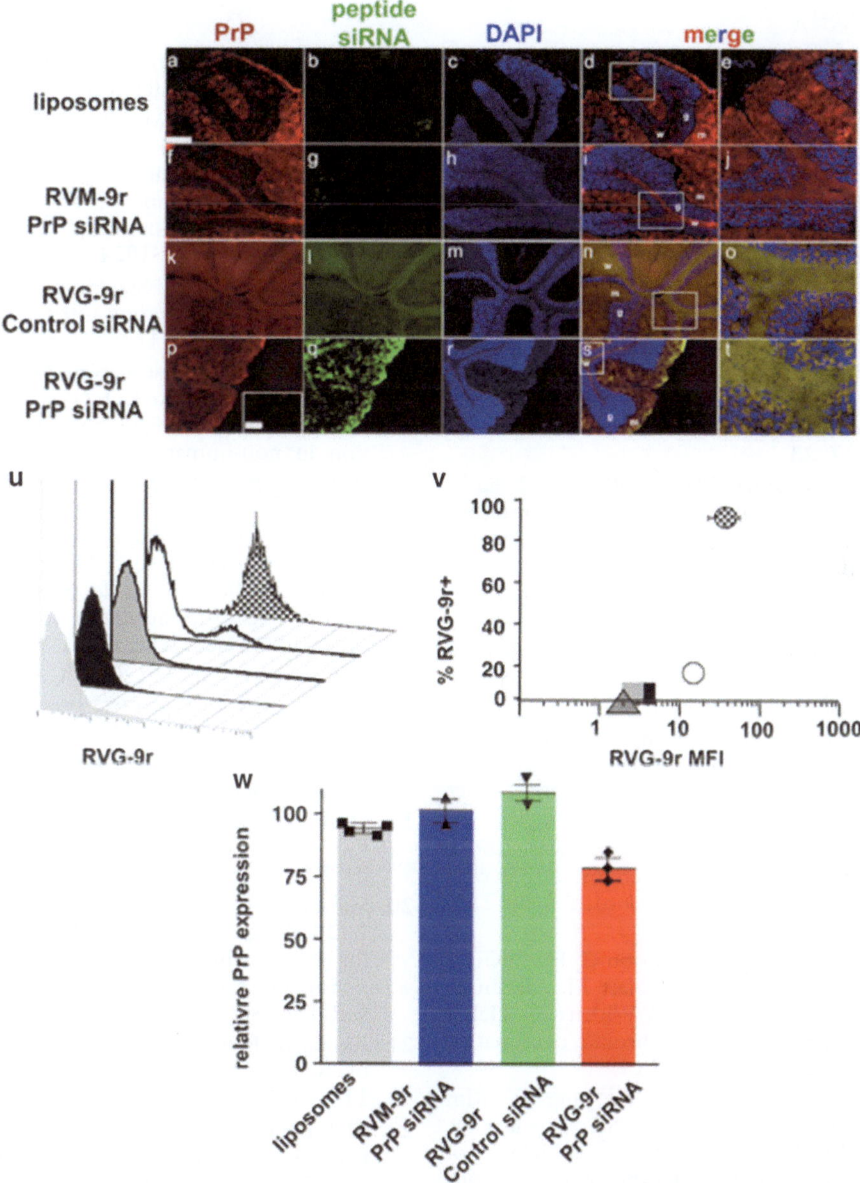

Fig. 3. LSPC delivery to PrP^C-expressing cells in vivo (11). FVB mice were intravenously injected with liposomes alone (panels **a–e**) or LSPCs containing DyLight 488-labeled peptides and Alexa 488-labeled siRNAs as follows: PrP siRNA-control RVM-9r (**f–j**), Control siRNA-RVG-9r (**k–o**), or PrP siRNA-RVG-9r (**p–t**). Mice were sacrificed 24 h later, dissected, and their brains snap frozen and 8 μm sections thereof stained with DyLight 649-labeled anti-PrP antibody (**a**, **f**, **k**, and **p**) and DAPI (**c**, **h**, **m**, and **r**) and visualized by fluorescence microscopy. Panels (**e**), (**j**), (**o**), and (**t**) depict magnifications of the *boxed areas* in panels (**d**), (**i**), (**n**), and (**s**). *Inset* in panel (**p**) depicts a serial section stained with an irrelevant isotype control antibody. Scale bars, 50 μm. (**u** and **v**) FACS analysis of PrP-RVG-9r LSPC delivery to brain cells (*checkered peak* (**u**) and *circle* (**v**)), kidney cells (*white peak* and *circle*), splenocytes (*dark grey peak* and *triangle*), and hepatocytes (*black peak* and *square*). (**w**) Cumulative data showing relative PrP expression on brain cells treated with liposomes alone or with LSPCs containing PrP-RVM-9r, control-RVG-9r, or PrP-RVG-9r. *Symbols* associated with each bar indicate values for individual mice.

References

1. Fire A, Xu S, Montgomery MK, Kostas SA, Driver SE, Mello CC (1998) Potent and specific genetic interference by double-stranded RNA in Caenorhabditis elegans. Nature 391:806–811
2. McCaffrey AP, Meuse L, Pham TT, Conklin DS, Hannon GJ, Kay MA (2002) RNA interference in adult mice. Nature 418:38–39
3. Paddison PJ, Caudy AA, Hannon GJ (2002) Stable suppression of gene expression by RNAi in mammalian cells. Proc Natl Acad Sci USA 99:1443–1448
4. Paddison PJ, Hannon GJ (2002) RNA interference: the new somatic cell genetics? Cancer Cell 2:17–23
5. Giladi H, Ketzinel-Gilad M, Rivkin L, Felig Y, Nussbaum O, Galun E (2003) Small interfering RNA inhibits hepatitis B virus replication in mice. Mol Ther 8:769–776
6. McCaffrey AP, Nakai H, Pandey K, Huang Z, Salazar FH, Xu H, Wieland SF, Marion PL, Kay MA (2003) Inhibition of hepatitis B virus in mice by RNA interference. Nat Biotechnol 21:639–644
7. Pai SI, Lin YY, Macaes B, Meneshian A, Hung CF, Wu TC (2006) Prospects of RNA interference therapy for cancer. Gene Ther 13:464–477
8. Campochiaro PA (2006) Potential applications for RNAi to probe pathogenesis and develop new treatments for ocular disorders. Gene Ther 13:559–562
9. Dorn G, Patel S, Wotherspoon G, Hemmings-Mieszczak M, Barclay J, Natt FJ, Martin P, Bevan S, Fox A, Ganju P, Wishart W, Hall J (2004) siRNA relieves chronic neuropathic pain. Nucleic Acids Res 32:e49
10. Kumar P, Wu H, McBride JL, Jung KE, Kim MH, Davidson BL, Lee SK, Shankar P, Manjunath N (2007) Transvascular delivery of small interfering RNA to the central nervous system. Nature 448:39–43
11. Pulford B, Reim N, Bell A, Veatch J, Forster G, Bender H, Meyerett C, Hafeman S, Michel B, Johnson T, Wyckoff AC, Miele G, Julius C, Kranich J, Schenkel A, Dow S, Zabel MD (2010) Liposome-siRNA-peptide complexes cross the blood-brain barrier and significantly decrease PrP on neuronal cells and PrP in infected cell cultures. PLoS One 5:e11085
12. Pfeifer A, Eigenbrod S, Al-Khadra S, Hofmann A, Mitteregger G, Moser M, Bertsch U, Kretzschmar H (2006) Lentivector-mediated RNAi efficiently suppresses prion protein and prolongs survival of scrapie-infected mice. J Clin Invest 116:3204–3210
13. White MD, Farmer M, Mirabile I, Brandner S, Collinge J, Mallucci GR (2008) Single treatment with RNAi against prion protein rescues early neuronal dysfunction and prolongs survival in mice with prion disease. Proc Natl Acad Sci USA 105:10238–10243
14. D'Costa J, Mansfield SG, Humeau LM (2009) Lentiviral vectors in clinical trials: current status. Curr Opin Mol Ther 11:554–564
15. Dissen GA, Lomniczi A, Neff TL, Hobbs TR, Kohama SG, Kroenke CD, Galimi F, Ojeda SR (2009) In vivo manipulation of gene expression in non-human primates using lentiviral vectors as delivery vehicles. Methods 49:70–77
16. Lundberg C, Bjorklund T, Carlsson T, Jakobsson J, Hantraye P, Deglon N, Kirik D (2008) Applications of lentiviral vectors for biology and gene therapy of neurological disorders. Curr Gene Ther 8:461–473
17. Nanou A, Azzouz M (2009) Gene therapy for neurodegenerative diseases based on lentiviral vectors. Prog Brain Res 175:187–200
18. Mulligan RC (1993) The basic science of gene therapy. Science 260:926–932
19. Quinonez R, Sutton RE (2002) Lentiviral vectors for gene delivery into cells. DNA Cell Biol 21:937–951
20. Subramanya S, Kim S-S, Manjunath N, Shankar P (2010) RNA interference-based therapeutics for human immunodeficiency virus HIV-1 treatment: synthetic siRNA or vector-based shRNA? Expert Opin Biol Ther 10:201–213
21. Akhtar S, Benter IF (2007) Nonviral delivery of synthetic siRNAs in vivo. J Clin Invest 117:3623–3632
22. Mathupala SP (2009) Delivery of small-interfering RNA (siRNA) to the brain. Expert Opin Ther Pat 19:137–140
23. Meade BR, Dowdy SF (2008) Enhancing the cellular uptake of siRNA duplexes following noncovalent packaging with protein transduction domain peptides. Adv Drug Deliv Rev 60:530–536
24. Chiu YL, Ali A, Chu CY, Cao H, Rana TM (2004) Visualizing a correlation between siRNA localization, cellular uptake, and RNAi in living cells. Chem Biol 11:1165–1175
25. Muratovska A, Eccles MR (2004) Conjugate for efficient delivery of short interfering RNA (siRNA) into mammalian cells. FEBS Lett 558:63–68

26. Lafon M (2005) Rabies virus receptors. J Neurovirol 11:82–87
27. Lentz TL, Burrage TG, Smith AL, Crick J, Tignor GH (1982) Is the acetylcholine receptor a rabies virus receptor? Science 215:182–184
28. Barichello JM, Ishida T, Kiwada H (2010) Complexation of siRNA and pDNA with cationic liposomes: the important aspects in lipoplex preparation. Methods Mol Biol 605:461–472
29. Leng Q, Scaria P, Lu P, Woodle MC, Mixson AJ (2008) Systemic delivery of HK Raf-1 siRNA polyplexes inhibits MDA-MB-435 xenografts. Cancer Gene Ther 15:485–495
30. Leng Q, Woodle MC, Lu PY, Mixson AJ (2009) Advances in systemic siRNA Delivery. Drugs Future 34:721
31. Morrissey DV, Lockridge JA, Shaw L, Blanchard K, Jensen K, Breen W, Hartsough K, Machemer L, Radka S, Jadhav V, Vaish N, Zinnen S, Vargeese C, Bowman K, Shaffer CS, Jeffs LB, Judge A, MacLachlan I, Polisky B (2005) Potent and persistent in vivo anti-HBV activity of chemically modified siRNAs. Nat Biotechnol 23:1002–1007
32. Rothdiener M, Muller D, Castro PG, Scholz A, Schwemmlein M, Fey G, Heidenreich O, Kontermann RE (2010) Targeted delivery of siRNA to CD33-positive tumor cells with liposomal carrier systems. J Control Release 144:251–258
33. Zimmermann TS, Lee AC, Akinc A, Bramlage B, Bumcrot D, Fedoruk MN, Harborth J, Heyes JA, Jeffs LB, John M, Judge AD, Lam K, McClintock K, Nechev LV, Palmer LR, Racie T, Rohl I, Seiffert S, Shanmugam S, Sood V, Soutschek J, Toudjarska I, Wheat AJ, Yaworski E, Zedalis W, Koteliansky V, Manoharan M, Vornlocher HP, MacLachlan I (2006) RNAi-mediated gene silencing in nonhuman primates. Nature 441:111–114
34. Allen TM, Hansen C, Martin F, Redemann C, Yau-Young A (1991) Liposomes containing synthetic lipid derivatives of poly(ethylene glycol) show prolonged circulation half-lives in vivo. Biochim Biophys Acta 1066:29–36
35. Harrington KJ, Rowlinson-Busza G, Syrigos KN, Uster PS, Abra RM, Stewart JS (2000) Biodistribution and pharmacokinetics of 111In-DTPA-labelled pegylated liposomes in a human tumour xenograft model: implications for novel targeting strategies. Br J Cancer 83:232–238
36. Baru M, Carmel-Goren L, Barenholz Y, Dayan I, Ostropolets S, Slepoy I, Gvirtzer N, Fukson V, Spira J (2005) Factor VIII efficient and specific non-covalent binding to PEGylated liposomes enables prolongation of its circulation time and haemostatic efficacy. Thromb Haemost 93:1061–1068
37. Papahadjopoulos D, Allen TM, Gabizon A, Mayhew E, Matthay K, Huang SK, Lee KD, Woodle MC, Lasic DD, Redemann C et al (1991) Sterically stabilized liposomes: improvements in pharmacokinetics and antitumor therapeutic efficacy. Proc Natl Acad Sci USA 88:11460–11464
38. Harrington KJ, Rowlinson-Busza G, Uster PS, Stewart JS (2000) Pegylated liposome-encapsulated doxorubicin and cisplatin in the treatment of head and neck xenograft tumours. Cancer Chemother Pharmacol 46:10–18
39. Harrington KJ, Rowlinson-Busza G, Syrigos KN, Vile RG, Uster PS, Peters AM, Stewart JS (2000) Pegylated liposome-encapsulated doxorubicin and cisplatin enhance the effect of radiotherapy in a tumor xenograft model. Clin Cancer Res 6:4939–4949
40. Harrington KJ, Lewanski CR, Stewart JS (2000) Liposomes as vehicles for targeted therapy of cancer. Part 2: clinical development. Clin Oncol (R Coll Radiol) 12:16–24
41. Goebel FD, Goldstein D, Goos M, Jablonowski H, Stewart JS (1996) Efficacy and safety of Stealth liposomal doxorubicin in AIDS-related Kaposi's sarcoma. The International SL-DOX Study Group. Br J Cancer 73:989–994
42. Northfelt DW, Dezube BJ, Thommes JA, Miller BJ, Fischl MA, Friedman-Kien A, Kaplan LD, Du Mond C, Mamelok RD, Henry DH (1998) Pegylated-liposomal doxorubicin versus doxorubicin, bleomycin, and vincristine in the treatment of AIDS-related Kaposi's sarcoma: results of a randomized phase III clinical trial. J Clin Oncol 16:2445–2451
43. Zamboni WC, Ramalingam S, Friedland DM, Edwards RP, Stoller RG, Strychor S, Maruca L, Zamboni BA, Belani CP, Ramanathan RK (2009) Phase I and pharmacokinetic study of pegylated liposomal CKD-602 in patients with advanced malignancies. Clin Cancer Res 15:1466–1472
44. Muggia FM, Hainsworth JD, Jeffers S, Miller P, Groshen S, Tan M, Roman L, Uziely B, Muderspach L, Garcia A, Burnett A, Greco FA, Morrow CP, Paradiso LJ, Liang LJ (1997) Phase II study of liposomal doxorubicin in refractory ovarian cancer: antitumor activity and toxicity modification by liposomal encapsulation. J Clin Oncol 15:987–993
45. Ranson MR, Carmichael J, O'Byrne K, Stewart S, Smith D, Howell A (1997) Treatment of advanced breast cancer with sterically stabilized liposomal doxorubicin: results of a multicenter phase II trial. J Clin Oncol 15: 3185–3191

46. Chekhonin VP, Zhirkov YA, Gurina OI, Ryabukhin IA, Lebedev SV, Kashparov IA, Dmitriyeva TB (2005) PEGylated immunoliposomes directed against brain astrocytes. Drug Deliv 12:1–6
47. Matveeva O, Nechipurenko Y, Rossi L, Moore B, Saetrom P, Ogurtsov AY, Atkins JF, Shabalina SA (2007) Comparison of approaches for rational siRNA design leading to a new efficient and transparent method. Nucleic Acids Res 35:e63

Chapter 18

Flow Cytometry-Based Cell Type-Specific Assessment of Target Regulation by Pulmonary siRNA Delivery

Olivia M. Merkel, Leigh M. Marsh, Holger Garn, and Thomas Kissel

Abstract

Pulmonary siRNA delivery has attracted strong interest and has been reported to successfully mediate target gene knockdown in a number of disease models. However, the nature of the epithelial cells that eventually take up siRNA and the question if other lung cell types may also be transfected have so far been neglected. Therefore, we describe here a flow cytometry-based method using transgenic enhanced green fluorescence protein-expressing mice (EGFP mice) for the differentiation of transfected lung cell populations based on their antigen expression.

Key words: siRNA delivery, Lung, EGFP mice, Flow cytometry, Knockdown, Epithelial cells, Type II pneumocytes, Differentiation

1. Introduction

Since its discovery and translation into functional genomics in mammalian cells (1), RNA interference (RNAi), a Nobel-Prize-winning technology, has attracted strong interest to evaluate its therapeutic potential (2). However, rapid degradation by nucleases and fast excretion following systemic injection of short interfering RNA (siRNA) (3) impede the development of RNAi-based therapies. In fact, the majority of clinically investigated siRNA-based drugs exploit local delivery routes such as intraocular, intradermal, and pulmonary administration (2). Considering the large number of reports in the literature addressing successful pulmonary siRNA delivery (4) and the variety of lung disorders ranging amongst the ten leading causes of death worldwide (5), this specific local administration route is highly promising from a translational point of view.

Manfred Ogris and David Oupicky (eds.), *Nanotechnology for Nucleic Acid Delivery: Methods and Protocols*, Methods in Molecular Biology, vol. 948, DOI 10.1007/978-1-62703-140-0_18, © Springer Science+Business Media, LLC 2013

The advantages of local noninvasive targeting are self-evident: immediate availability, decreased dose and systemic side effects, and the fact that several obstacles such as interactions with serum proteins can be avoided by pulmonary delivery (6). Despite the favorably vast surface area and strong perfusion of the lung (7), several barriers must be overcome for efficient pulmonary siRNA delivery. These hurdles are of an anatomic nature as reflected in the pulmonary architecture, and also physiologic represented by active clearance processes and effective immune responses (8). While mucociliary clearance, cough clearance, and the clearance by macrophages constantly eliminate foreign material from the lung (9), respiratory mucus (10) and airway surface liquid (11) additionally hinder efficient siRNA delivery into lung tissue.

To achieve specific therapeutic effects, siRNA not only needs to be directed to its target region (12), but it also has to be delivered into its target cell populations (13).

Despite the availability of several animal models of pulmonary diseases, little attention has been paid to the elucidation of the nature of the epithelial cells that are transfected, and whether other lung cell types also take up siRNA after pulmonary delivery (14). Transgenic Actin-enhanced green fluorescence protein (EGFP)-expressing mice, often used in the literature (15–18), are an animal model almost predestined for such investigations. While in the first reports EGFP knockdown after intranasal delivery of siRNA and short interfering locked nucleic acids (siLNA) was only determined by counting EGFP-positive cells in paraffin sections (17) or by means of a physical fractionator probe (16), there are currently two reports where the success of EGFP knockdown after intratracheal administration of siRNA was measured by flow cytometry in lung homogenates (13, 16) obtained as previously described (19). A recent study employed cell sorting of CD45-positive cells from lung homogenates of EGFP-expressing mice to differentiate the knockdown in leukocytes and lung macrophages (20). Accordingly, a flow cytometry-based analysis can be utilized to differentiate myeloid from endothelial cells and type II alveolar epithelial cells (AEC-II) (21) to determine which cell types in the lung take up siRNA for functional RNAi. In parallel, the knockdown efficiency can be measured by quantifying the EGFP fluorescence in the FITC-channel for each population.

2. Materials

2.1. siRNA Formulation

Due to increased stability and high activity accompanied by decreased immunogenicity, 2′-*O*-methylated 25/27mer DsiRNA targeting EGFP (22) and negative control DsiRNA from Integrated

DNA Technologies (Leuven, Belgium) are recommended for in vivo use.

1. siEGFP: Sense: 5′-pACCCUGAAGUUCAUCUGCACCA-CdCdG, antisense: 3′-mAmCmUGmGGmACmUUmCAmA GmUAmGAmCGUGGUGGC.
2. siNegCon: Sense: 5′-pCGUUAAUCGCGUAUAAUACGCG UdAdT, antisense: 3′-mCmAmGCmAAmUUmAGmCGmC AmUAmUUmAUGCGCAUAp.
3. A cationic lipid or a polymer for nanoformulations of siRNA, for example 25 kDa branched polyethylenimine (PEI; BASF, Ludwigshafen, Germany).

2.2. In Vivo Transfection

1. C57BL/6-Tg(CAG-EGFP)1Osb/J Actin-EGFP-expressing mice, 6 weeks old, both genders.
2. EGFP-negative mice (e.g., C57BL/6) as control.
3. One tubus for each formulation made of the flexible part of a BD Insyte 24-gauge catheter (BD Biosciences, Heidelberg, Germany) attached to a 24-gauge needle (BD Biosciences, Heidelberg, Germany). Alternatively, the Mouse Intubation Pack (Hallowell Engineering & Manufacturing Corporation, Pittsfield, MA, USA) can be used.
4. Tweezers wrapped with Parafilm™.
5. Intubation stand, e.g., Rodent Tilting Work Stand (Hallowell Engineering & Manufacturing Corporation, Pittsfield, MA, USA).
6. Syringes, 1 ml (BD Biosciences, Heidelberg, Germany).
7. Microliter pipet adjustable or fixed at 50 μl, corresponding pipet tips.
8. Cold Light Illumination Source (e.g., Leica L2).

2.3. Bronchoalveolar Lavage

1. Surgery table and surgical instruments.
2. Venflon catheters, 18-gauge (BD Biosciences, Heidelberg, Germany).
3. Centrifuge cups, 1.5 ml, and tubes, 15 ml.
4. Syringes, 1 and 20 ml.
5. PBS buffer containing 2 mM EDTA.
6. Complete Protease Inhibitor Cocktail (Roche Diagnostics, Mannheim, Germany): Dissolve one tablet in 50 ml PBS buffer.

2.4. Lung Homogenates

1. Low melting point agarose (Promega, Mannheim, Germany): 1% in PBS, boiled in a microwave oven.
2. Dispase (Roche, Mannheim, Germany), working concentration: 2 U/ml in HEPES-buffered saline (50 mM HEPES/KOH, pH 7.4, 150 mM NaCl).
3. DNase (Roche, Mannheim, Germany), working concentration: 100 U/ml in RPMI (PAA, Cölbe, Germany) containing 25 mM HEPES.
4. Wash buffer: PBS supplemented with 2 mM EDTA and 0.5%v/v fetal calf serum.
5. Cell culture plates, 6-well.
6. Petri dishes, 6 cm.
7. Nylon Cell Strainers, 100 and 40 μm (BD Falcon, Heidelberg, Germany).
8. Tubes, 50 ml.
9. Venofix™ 0.8×20 mm (B. Braun, Melsungen, Germany) butterfly tube attached to a 20 ml syringe.
10. PBS buffer.
11. CasyTon™ isotonic salt solution (Schaerfe Systems, Reutlingen, Germany).
12. Casy TT™ cell counter (Schaerfe Systems, Reutlingen, Germany) (see Note 1).

2.5. Cell Fixation and Permeabilization

1. Paraformaldehyde: 1%v/v in PBS buffer, freshly prepared.
2. Wash buffer: PBS supplemented with 2 mM EDTA and 0.5%v/v fetal calf serum.
3. Saponin buffer: 0.3%w/v saponin (Carl Roth, Karlsruhe, Germany) in PBS buffer.
4. FACS tubes.

2.6. Antibodies

1. Rat anti-mouse CD16/CD32 (Fc-Block; BD Biosciences, Heidelberg, Germany), 1:100 dilution in 0.3% saponin buffer.
2. Rabbit polyclonal anti-mouse pro-surfactant protein-C (proSP-C) antiserum, reactive with human and mouse proSP-C (Abcam, Cambridge, UK), 1:500 dilution in 0.3% saponin buffer.
3. Rat anti-mouse CD45-PerCP/Cy5.5, clone 30-F11 (BioLegend, Uithoorn, The Netherlands), 1:50 dilution in 0.3% saponin buffer.
4. Rat anti-mouse CD31-PE, clone MEC13.3 (BioLegend, Uithoorn, The Netherlands), 1:100 dilution in 0.3% saponin buffer.
5. Secondary antibody: Alexa Fluor® 647 F(ab′)2 fragment of goat anti-rabbit IgG (H+L)×2 (Invitrogen, Karlsruhe, Germany) 1:500 dilution in 0.3% saponin buffer.

6. Isotype controls: Rabbit immunoglobulin, rat IgG2a κ-PE (Biolegend), rat IgG2b κ PerCP/Cy5.5 (Biolegend).
7. FACS Canto II (BD Biosciences, Heidelberg, Germany) or a similar flow cytometer equipped with the following lasers and filter settings: GFP: excitation: 488 nm, emission filter: 530 nm/30; PerCP/Cy5.5: excitation: 488 and 635 nm, emission filter: 670 nm/LP; PE: excitation: 488 nm, emission filter: 585 nm/42; Alexa Fluor®: excitation: 635 nm, emission filter: 670 nm/20.

3. Methods

3.1. In Vivo Transfection

1. Prepare siRNA containing nanoparticles or complexes according to your standard protocol. For example, mix 50 μg siRNA with 40 μg PEI for a charge ratio of 6 PEI amines per phosphate. Pipet vigorously and then incubate for 20 min at room temperature. The maximum volume that can be instilled per mouse is 50 μl. The nanoparticle formulation should therefore contain the desired dose of siRNA in a total volume of 50 μl (see Note 2).
2. Deeply anesthetize EGFP-mice intraperitoneally with Xylazine/Ketamine and fix one mouse after another in a supine position. Place the gooseneck light on the lower part of the thorax, gently pull the tongue out with Parafilm™-wrapped tweezers, and clear the throat so that the trachea can be recognized as a light spot. Use magnification if necessary and move the light source until the trachea clearly appears as light spot. Gently intubate the mouse (see Note 3).
3. Pipet up to 50 μl into the tubus, then attach a syringe filled with 100 μl air, and push the air into the tubus, thereby instilling the material into the mouse lung.
4. Check for vital signs, return the mouse into the cage, and instill the next mouse as described.
5. The treatment can be performed repeatedly, e.g., on day 1 and day 3.
6. Treat at least 5–8 animals per group (siEGFP, scrambled control sequence, vehicle only, and EGFP-negative mice).

3.2. Bronchoalveolar Lavage

1. Sacrifice animals 48 h after the last treatment by cervical dislocation and exsanguination. Fix the mouse in a supine position and cut open the thorax without affecting the heart or the lung. Remove the thymus and all tissue surrounding the trachea. Gently cut a 2 mm hole in the trachea (see Note 4) and cannulate the trachea using an 18-gauge Venflon catheter.

2. Rinse the lung with 1 ml fresh PBS/cOmplete via the tracheal catheter and store the bronchoalveolar lavage (BAL) fluid (BALF) of each animal on ice. Spin the BALF down for 5 min at 350 g (see Note 5).
3. To perfuse the lung, first cut the left atrium, then insert the butterfly into the right ventricle (only insert ~3 mm), and gently inject 20 ml PBS until lung appears white.

3.3. Flow Cytometric Quantification of EGFP Knockdown

1. To obtain lung homogenates, inflate the lungs with 1 ml dispase (2 U/ml) followed by 0.5 ml 1% low-melting-point agarose (alternatively lungs may be prepared for histological analysis, see Note 6).
2. Secure the trachea and excise the lungs and incubate them in an additional 3 ml dispase per lung for 30 min at 37°C in a 6-well plate.
3. Place the lungs into a 6 cm petri dish filled with 6 ml RPMI containing 100 U/ml DNase and 25 mM HEPES. Remove the lung tissue from the trachea by gentle tapping with the back of a pair of tweezers (see Note 7).
4. Incubate for 10 min at room temperature with slight agitation.
5. Add 4 ml medium with a 10 ml serological pipet.
6. To further disarticulate the tissue, homogenize by pipetting up and down with a 1 ml pipet (see Note 8).
7. Pipet the lung homogenate up and down again with a 10 ml serological pipet and transfer it to a 100 μm nylon cell strainer placed on a 50 ml tube. Add 5 ml medium without DNase and process the tissue through the cell strainer.
8. Pass the lung homogenate through the 40 μm cell strainer into a fresh 50 ml tube.
9. Centrifuge the single cell suspension for 10 min at 350 g.
10. Discard the supernatant and resuspend the pellet in 5 ml fresh medium.
11. To determine the amount of cells, dilute the suspension 1:2,000 in 10 ml CasyTon™ and count the cells using a Casy TT cell counter (see Note 1).
12. Centrifuge the single cell suspension for 10 min at 350 g.
13. Discard the supernatant and resuspend the pellet in 4 ml fresh 1% paraformaldehyde solution and incubate for 15 min on ice.
14. Centrifuge the fixed cells for 10 min at 350 g.
15. Discard the supernatant and resuspend in 5 ml wash buffer.
16. Centrifuge cells for 10 min at 350 g.

17. Discard the supernatant and resuspend the pellet in 4 ml wash buffer.
18. Transfer about 200,000 cells per lung sample into FACS tubes.
19. Prepare additional samples as blank and isotype controls.
20. Centrifuge the aliquots of 200,000 cells per FACS tube for 10 min at 350 g.
21. Discard the supernatant, add 100 μl 0.3% saponin buffer per sample for permeabilization, mix, and incubate for 15 min at 4°C.
22. Centrifuge FACS tubes for 5 min at 350 g.
23. Discard the supernatant, add 10 μl diluted Fc-block, add 20 μl of the appropriate dilutions of the antibodies against CD45, CD31, and pro-SPC (or isotype controls), vortex, and incubate for 25 min at 4°C in the dark.
24. Centrifuge the stained cells for 5 min at 350 g.
25. Discard the supernatant, wash with 100 μl 0.3% saponin buffer twice, and centrifuge the cells a third time for 5 min at 350 g.
26. Discard the supernatant, add 20 μl of the appropriate dilution of the secondary antibody (including the isotype control tube), mix, and incubate for 25 min at 4°C in the dark.
27. Centrifuge the stained cells for 5 min at 350 g.
28. Discard the supernatant, wash with 100 μl 0.3% saponin buffer twice, and centrifuge the cells a third time for 5 min at 350 g.
29. Discard the supernatant, wash with 100 μl once with wash buffer, centrifuge the cells again for 5 min at 350 g, and resuspend them in 200 μl wash buffer.
30. Adjust the laser power at the flow cytometer using the isotype controls and EGFP-negative samples.
31. Gate the cells to exclude debris and cell clumps. Count at least 30,000 events.
32. Compensate the fluorophores to remove the spectral overlap so that three different populations of myeloid cells (CD45+), endothelial cells (CD31+), and type II pneumocytes (pro-SPC+) can be distinguished (see Note 9).
33. Gate single populations as shown in Fig. 1. Use density plots for better differentiation, if necessary.
34. Determine EGFP expression in every single population in the FITC channel (see Note 10) and record mean fluorescence for each population as a major readout parameter as shown in Figs. 2 and 3.

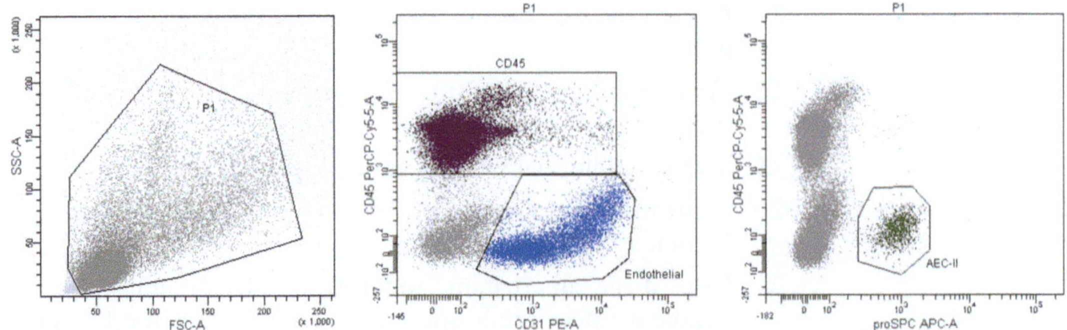

Fig. 1. Dot plots obtained after fluorescence compensation. Myeloid CD45-positive cells are gated in the *middle panel*, shown in *purple*. Endothelial, CD31-positive cells are gated in the *middle panel*, shown in *blue*. Type II alveolar epithelial cells (AEC-II) which are positive for pro-surfactant protein-C (pro-SPC) are gated in the *right panel*, shown in *green*.

4. Notes

1. Alternatively use a hemocytometer counting chamber.
2. Instillation of volumes greater than 50 μl can drown the mice whereas it has to be noted that a certain volume is always lost in the tubus. Therefore volumes smaller than 30 μl are not recommended either.
3. Deep intubation leads to treatment of the left lung lobes only. Therefore the tubus should only be introduced a few millimeters deep.
4. If the trachea is cut through, it easily disappears and can hardly be cannulated. Therefore, only cut a tiny hole of 2 mm in the anterior part of the trachea.
5. To determine the total number of leukocytes captured by lavage, resuspend the pellet in PBS containing 1% BSA, dilute the suspension 1:2,000 in 10 ml CasyTon™, and count the cells using a Casy TT™ cell counter. Cytospins can be prepared with aliquots of 100,000 cells spun on glass slides (100 g, 5 min) and stained for differentiation with Diff-Quick (DADE Diagnostics, Unterschleissheim, Germany). For evaluation, count 100 cells and classify concerning morphology and staining. Cytokine and chemokine secretion can be measured in the supernatant by ELISA or multiplex ELISA (e.g., Luminex System, Bio-Rad Laboratories, Germany) as previously described (15).
6. For microscopic assessment of histology and EGFP knockdown, some lungs (e.g., 2 out of 8 per group) can be prepared for histological analysis, leaving another six for flow cytometric

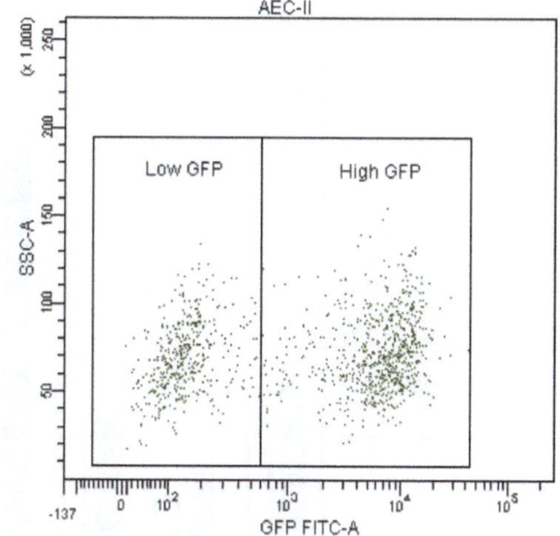

		% Parent	EGFP Mean Fluorescence
	P1	88.6	414
	AEC-II	3.8	4842
	Low EGFP	37.6	160
	High EGFP	62.6	7645
	CD45	56.2	194
	Endothelial	28.9	327

b: siEGFP

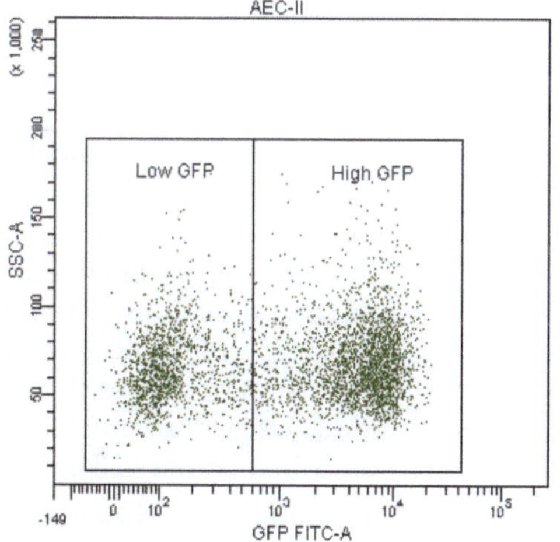

		% Parent	EGFP Mean Fluorescence
	P1	89.2	309
	AEC-II	4.0	3999
	Low EGFP	35.4	154
	High EGFP	64.8	6084
	CD45	53.4	110
	Endothelial	29.9	262

Fig. 2. The treatment with nanoparticles containing the scrambled negative control sequence (**a**: siNegCon) or the specific siRNA sequence (**b**: siEGFP) affects the EGFP fluorescence which can be measured for every population as shown in the statistic tables below the dot plots. Amongst AEC-II, highly EGFP-expressing cells (High GFP: shown in the *right part* of the dot plot) and only slightly highly EGFP-expressing cells (Low GFP: shown in the *left part* of the dot plot) can be found and have to be evaluated separately.

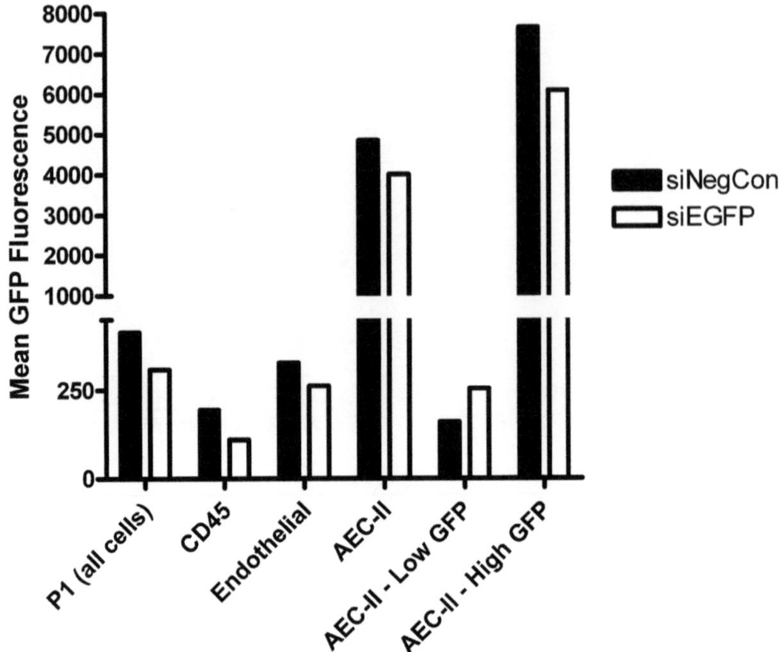

Fig. 3 Mean fluorescence values of EGFP expression in every single cell population can be summarized in bar graphs. The knockdown efficiency in every single population can then be easily recognized.

assessment of the knockdown. For microscopy, inflate the lungs with 6% paraformaldehyde and store them in 6% paraformaldehyde in the dark for 24 h, before embedding in paraffin. Deparaffinize and slice sections of 3 μm. Counterstain the tissue sections with 1 μg/ml 4′,6-diamidino-2-phenylindole (DAPI) (Molecular Probes, Invitrogen, Karlsruhe, Germany) or stain with hematoxylin and eosin (H&E) for confocal laser scanning or light microscopy, respectively, as described before (13).

7. The lung tissue should not be cut or the very sensitive type II pneumocytes will be lost.
8. Slightly cutting off the tip of 1 ml pipet tip helps the homogenization step.
9. Older cytometers require compensation to be performed prior to data acquisition; in this case spectral overlap can be determined using samples stained with a single antibody.
10. As can be seen in Figs. 1 and 2, it is possible that the EGFP expression within one population varies strongly. In this case, subpopulations need to be gated to account for these differences in EGFP expression and to quantify the fluorescence in every single population correctly.

Acknowledgment

This work was supported by "Meditrans," an Integrated Project funded by the European Commission under the Sixth Framework (NMP4-CT-2006-026668). The authors are grateful to Dr. Tanja Dicke and Matthias Schiller for expert support.

References

1. Elbashir SM et al (2001) Duplexes of 21-nucleotide RNAs mediate RNA interference in cultured mammalian cells. Nature 411:494–498
2. Watts JK, Corey DR (2010) Clinical status of duplex RNA. Bioorg Med Chem Lett 20:3203–3207
3. Dykxhoorn DM, Palliser D, Lieberman J (2006) The silent treatment: siRNAs as small molecule drugs. Gene Ther 13:541–552
4. Lam JK, Liang W, Chan HK (2012) Pulmonary delivery of therapeutic siRNA. Adv Drug Deliv Rev 64(1):1–15
5. WHO (2008) The top 10 causes of death (No. 310), WHO, New York.
6. Durcan N, Murphy C, Cryan SA (2008) Inhalable siRNA: potential as a therapeutic agent in the lungs. Mol Pharm 5:559–566
7. Cryan SA, Sivadas N, Garcia-Contreras L (2007) In vivo animal models for drug delivery across the lung mucosal barrier. Adv Drug Deliv Rev 59:1133–1151
8. Oberdorster G, Oberdorster E, Oberdorster J (2005) Nanotoxicology: an emerging discipline evolving from studies of ultrafine particles. Environ Health Perspect 113:823–839
9. Birchall J (2007) Pulmonary delivery of nucleic acids. Expert Opin Drug Deliv 4:575–578
10. Ferrari S et al (2001) Mucus altering agents as adjuncts for nonviral gene transfer to airway epithelium. Gene Ther 8:1380–1386
11. Rosenecker J et al (2003) Interaction of bronchoalveolar lavage fluid with polyplexes and lipoplexes: analysing the role of proteins and glycoproteins. J Gene Med 5:49–60
12. Dames P et al (2007) Targeted delivery of magnetic aerosol droplets to the lung. Nat Nanotechnol 2:495–499
13. Jiang HL et al (2009) The suppression of lung tumorigenesis by aerosol-delivered folate-chitosan-graft-polyethylenimine/Akt1 shRNA complexes through the Akt signaling pathway. Biomaterials 30:5844–5852
14. Davies LA et al (2007) Identification of transfected cell types following non-viral gene transfer to the murine lung. J Gene Med 9:184–196
15. Merkel OM et al (2009) Nonviral siRNA delivery to the lung: investigation of PEG-PEI polyplexes and their in vivo performance. Mol Pharm 6:1246–1260
16. Glud SZ et al (2009) Naked siLNA-mediated gene silencing of lung bronchoepithelium EGFP expression after intravenous administration. Oligonucleotides 19:163–168
17. Howard KA et al (2006) RNA interference in vitro and in vivo using a novel chitosan/siRNA nanoparticle system. Mol Ther 14:476–484
18. Nielsen EJ et al (2010) Pulmonary gene silencing in transgenic EGFP mice using aerosolised chitosan/siRNA nanoparticles. Pharm Res 27:2520–2527
19. Corti M, Brody AR, Harrison JH (1996) Isolation and primary culture of murine alveolar type II cells. Am J Respir Cell Mol Biol 14:309–315
20. Beyerle A et al (2011) Comparative in vivo study of poly(ethylene imine)/siRNA complexes for pulmonary delivery in mice. J Control Release 151(1):51–56
21. Marsh LM et al (2009) Surface expression of CD74 by type II alveolar epithelial cells: a potential mechanism for macrophage migration inhibitory factor-induced epithelial repair. Am J Physiol Lung Cell Mol Physiol 296:L442–L452
22. Rose SD et al (2005) Functional polarity is introduced by Dicer processing of short substrate RNAs. Nucleic Acids Res 33: 4140–4156

Chapter 19

Liver-Targeted Gene Delivery Through Retrograde Intrabiliary Infusion

Xuan Jiang, Yong Ren, John-Michael Williford, Zhiping Li, and Hai-Quan Mao

Abstract

Retrograde intrabiliary infusion (RII) has recently been characterized as a safe and effective administration route for liver-targeted gene delivery. Efficient transgene expression in the liver has been achieved by infusing a variety of gene vectors including adenovirus, retrovirus, lipoplexes, polyplexes, and naked DNA through the common bile duct. Here, we describe the RII technique and key infusion parameters for delivering plasmid DNA and DNA nanoparticles to the rat liver. After RII of plasmid DNA, the level of transgene expression in rat liver is comparable to that achieved by hydrodynamic injection of plasmid DNA, which is considered to be "gold standard" for liver-targeted gene delivery. RII has also been shown to significantly enhance the gene delivery efficiency by polymer/DNA nanoparticles in comparison with intravenous and intraportal infusions. This method induces minimal level of cytotoxicity and damage to the liver and bile duct. Due to these advantages, RII has the potential to be used for delivering various gene vectors in clinical setting through the endoscopic retrograde cholangiopancreatography procedure.

Key words: Retrograde intrabiliary infusion, Plasmid DNA, Liver-targeted gene delivery

1. Introduction

The liver is one of the most important targets for gene delivery as it is associated with many metabolic genetic disorders and it is susceptible to viral infection and malignancies. Moreover, transgene-encoding proteins expressed in the liver have direct access to systemic circulation, making the liver attractive for *in situ* production of recombinant therapeutics in treating "off-site" diseases (1–3). A number of gene carriers have been studied for gene delivery to the liver, including viral and nonviral gene carriers. Viral vectors, particularly adenovirus (4, 5) and lentivirus (6), have

Manfred Ogris and David Oupicky (eds.), *Nanotechnology for Nucleic Acid Delivery: Methods and Protocols*, Methods in Molecular Biology, vol. 948, DOI 10.1007/978-1-62703-140-0_19, © Springer Science+Business Media, LLC 2013

mediated high level of transgene expression in the liver. However, viral vectors are associated with severe problems including host immune response (7), intrinsic toxicity of the viral proteins (8), and limited packaging size (9). Nonviral gene carriers have been proposed as safer alternatives to viruses because of ease of synthesis, flexibility in the size of the transgene to be delivered, and, most importantly, low toxicity and minimal host immune response (10, 11). Liposomes and cationic polymers are the two mostly studied nonviral carriers for liver gene delivery. To achieve active targeting to the liver, liposomes and polymeric gene carriers have been modified with hepatocyte-specific ligands to enhance receptor-mediated cellular uptake of gene carriers by hepatocytes (12, 13). Despite the improvement of the targeting effect of gene carriers, the efficiency of liver gene delivery mediated by nonviral carriers is still relatively low, and remains to be further optimized.

In addition to developing safe and highly efficient gene delivery carriers, an effective and clinically applicable delivery route is also critical to the success of liver-targeted gene delivery. To date, three administration routes have been developed for liver-targeted gene delivery. Among them, intravenous injection represents the easiest administration route. Following intravenous injection, gene carriers travel through systemic circulation and encounter macrophages residing in the lung or the liver. Therefore, significant level of uptake by these macrophages can be expected. Compared with intravenous injection, portal vein infusion offers a more direct administration of gene carriers to the liver. However, gene carriers delivered through both of these routes will need to cross sinusoidal fenestrae (average 170 nm in diameter and occupy about 6–8% of the sinusoidal surface area (14)) to reach hepatocytes. Concurrently, macrophages lining sinusoidal endothelium of the liver (Kupffer cells) may scavenge gene carriers and lead to the reduction in gene delivery efficiency.

Recently, retrograde intrabiliary infusion (RII) has been developed as a safe and efficient alternative to intravenous injection and portal vein injection. RII is achieved by infusion of gene carriers into the biliary tree under mild pressure, which leads to direct delivery of complexes to hepatocytes and transport of complexes to vascular compartment via sinusoidal drainage (15). The advantage of this delivery route is the direct access to hepatocytes, bypassing Kupffer cells lining the endothelium. The biliary system consists of 7–10 orders of cholangiographically visible bile ducts spanning the expanse of the hepatic parenchyma (16) with a surprisingly large and distensible volume (~29 mL for human liver) (17). The exposure of hepatocytes via RII occurs through bile canaliculi, which measure 1–1.25 μm in diameter in the region of the portal triads (Rappaport's zone 1) and 0.5–1 μm in the region of central veins (zone 3) (18). The ultrastructural surface of an entire normal biliary tree for a human liver is around 3,000 cm^2.

This large surface area and wide distribution of the biliary system throughout the liver provide direct access to nearly all hepatocytes in liver parenchyma via bile canaliculi, before the delivered drug or gene is exposed to KCs (19). More importantly, RII can be easily adopted in clinical settings due to the small infusion volume (<2% of body weight) and feasibility of using endoscopic retrograde cholangiopancreatography (ERCP), a routine bile duct canulation procedure (20, 21).

Tominaga et al. (22) reported that the efficiency of RII in delivering adenovirus (Ad5) encoding LacZ gene in rats was similar to that by intraportal infusion. Surprisingly, re-administration of Ad by RII on days 35 and 70 successfully induced similar levels of transgene re-expression in the liver, contrasting results observed following intraportal infusion, which failed to show gene expression upon re-administration. Otsuka et al. (15) showed that multilamellar liposomes delivered by RII were 100 times more efficient than those delivered by intraportal infusion. We demonstrated that RII is an efficient administration route for the delivery of plasmid DNA or polymer/DNA nanoparticles to the liver. Here, we describe the optimized infusion parameters for delivering plasmid DNA and polymer/DNA nanoparticles through RII leading to efficient transgene gene expression in rat liver.

2. Materials

Plasmid DNA: VR1255C plasmid DNA (6.4 Kb) encoding firefly luciferase driven by CMV promoter (Vical, San Diego, CA).

Infusion Needle: 32 Gauge, plastic hub (Hamilton, Reno, NV).

Extension line: 20-in., luered (Strategic Applications Inc., Libertyville, IL).

Anesthetics: Ketamine (100 mg/mL), Xylazine (100 mg/mL).

Analgesics: Ketoprofen, freshly diluted from 100 to 5 mg/mL with 1× phosphate-buffered saline (PBS) before use.

Heating pad: Deltaphase® isothermal pad (Braintree Scientific, Braintree, MA).

Surgical instruments: Rodent surgery board, Mayo scissors, microsurgical scissors, scalpels, blunt-tipped forceps, Ryder needle holders, microsurgical needle holders, gauze pads.

Antiseptics: 70% ethylene glycol, 10% Povidone-iodine aqueous solution.

Wound closure: 2-0, 4-0, and 10-0 resorbable sutures, wound clips (9 mm), wound clip applicator, and remover.

Lysis buffer: Dilute 10 mL of report lysis buffer (5×, Promega, Madison, WI) with 40 mL of DI water prior to use.

Luciferase assay solution: Reconstitute a vial of lyophilized luciferase assay reagent by dissolving in 10 mL of Luciferase Assay Buffer (Promega, Madison, WI).

Luciferin solution: Dissolve 1 g of VivoGlo™ D-luciferin (Promega, Madison, WI) in 33 mL of 1× PBS.

3. Methods

3.1. Preparation of Rats

1. The rat is anesthetized by intraperitoneal injection of Ketamine (100 mg/kg) and Xylazine (10 mg/kg).
2. The anesthetized rat is placed on a rodent surgery board. The body of the rat is restrained by tying four limbs to the limb holders on the surgical board with rubber bands.
3. The hair on the abdominal wall of rats is carefully removed with electric clippers. The exposed skin is then disinfected with 70% ethanol and 10% povidone-iodine consecutively.

3.2. Intrabiliary Infusion of Plasmid DNA

1. The abdomen of the rat is opened through a midline incision from about half length of the abdominal posterior to the xiphoid cartilage with a scalpel.
2. The duodenum and a small part of intestine are pulled out to the right and placed on a gauze pad moistened with PBS, on the abdomen. The major lobes of the liver are pushed back out of the way against the diaphragm, and a moistened gauze pad is placed against the major liver lobes.
3. An ideal infusion site is usually on the part of bile duct that is above pancreatic tissue (see Note 1) and 0.5–1 cm below where the hepatic ducts join to form common bile duct (Fig. 1). Once the infusion site is determined, blunt-tipped forceps are carefully pushed through the connective tissue under this part of bile duct. A small length of a 2-0 suture is then pulled through under the bile duct and a loose ligature is made.
4. Before the insertion of the infusion needle, the connective tissue around the infusion site should be carefully stripped off with fine-tipped forceps (see Note 2).
5. After the part of the bile duct for infusion is fully exposed, the infusion needle is carefully pushed into the bile duct (see Note 3) past the loose ligature but below where the hepatic ducts join to form common bile duct (see Note 4). This process is facilitated by the use of a binocular dissecting microscope. The ligature is then tightened and a square knot is made to secure the needle in the bile duct.

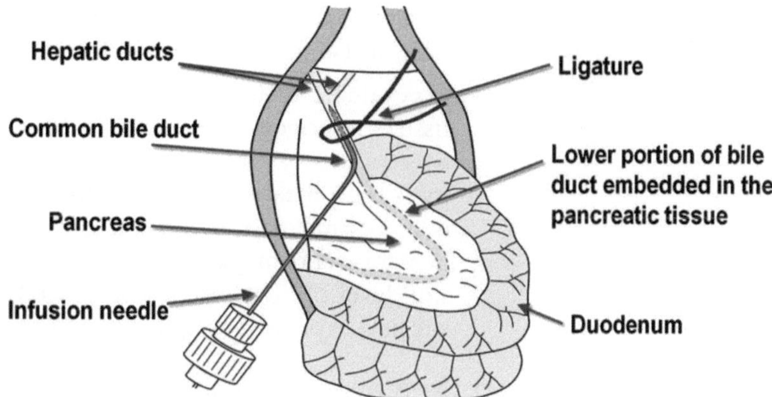

Fig. 1. Insertion of infusion needle into bile duct.

6. Using a syringe pump, 4 mL of plasmid DNA solution containing 20 μg of DNA is infused through bile duct to the rat liver at a rate of 12 mL/h (see Note 5). After infusion, the square knot around the bile duct is removed and the needle is carefully withdrawn from the bile duct.
7. After the needle is removed from the bile duct, the infusion site should be carefully examined under microscope to ensure that there is no leakage of bile from the needle hole (see Note 6).
8. The abdominal wall of the rat is closed in two layers. The peritoneum and muscle are closed using a continuous running suture technique with a 4-0 resorbable suture. The skin layer is then closed using wound clips.

3.3. Postoperative Care

1. It usually takes 1–1.5 h for the rat to recover from anesthesia. To prevent body heat loss, the rat should be placed on a preheated isothermal pad to help maintain the body temperature of the rat at around 37°C during the period of recovery.
2. To reduce postoperative pain, Ketoprofen (5 mg/kg) is administered subcutaneously to the rat every 24 h for the first 3 days after infusion.

3.4. Evaluation of Transgene Expression After Intrabiliary Infusion of Plasmid DNA

3.4.1. Determination of Luciferase Expression Level with a Luminometer

1. At different time points after intrabiliary infusion, rats are euthanized by CO_2. Major organs (liver, heart, lung, spleen, and kidney) are harvested and washed with PBS pH 7.4. Each individual organ is then weighed and transferred to a 50-mL conical centrifuge tube.
2. To each 50-mL conical centrifuge tube, 5 mL of 1× lysis buffer is added. The organs are then homogenized with a tissue homogenizer.
3. The tissue homogenate is frozen at –80°C for 30 min and thawed in a room-temperature water bath. An aliquot of tissue

homogenate (~1 mL) is transferred to a 1.5-mL microcentrifuge tube and centrifuged at 16,000 × g for 5 min at 4°C.

4. To evaluate luciferase expression, 20 μL of supernatant from tissue homogenate is mixed with 100 μL of luciferase assay reagent in a 1.5-mL microcentrifuge tube. The luciferase activity is then measured by a luminometer. The luciferase activity is converted to the amount of luciferase using recombinant luciferase (Promega, Madison, WI) as the standard. The expression level of luciferase is normalized against the weight of tissue and expressed as nanogram luciferase/gram tissue.

3.4.2. Determination of Luciferase Expression Level with In Vivo Bioluminescence Imaging

1. At 4 h after intrabiliary infusion, the rat is placed in a clear plexiglass anesthesia box (2.5–3.5% isoflurane). After the rat is fully anesthetized, 1 mL of D-luciferin solution (30 mg/mL) is injected intraperitoneally.
2. After the injection of D-luciferin, the rat is transferred from the box to the nose cones attached to the manifold in the imaging chamber of an IVIS Spectrum Imaging System. The rat is positioned with abdomen side up.
3. After 5 min, the bioluminescence of the rat is imaged from 15 s to 1 min (see Note 7) on the IVIS Spectrum Imaging System. The level of luciferease expression is calculated by dividing the total photon counts in the region of interest (ROI) by exposure time, and expressed as photon counts/ROI/s. As shown in Fig. 2, plasmid DNA delivered through intrabiliary infusion mediates high level of luciferase expression in the liver. More

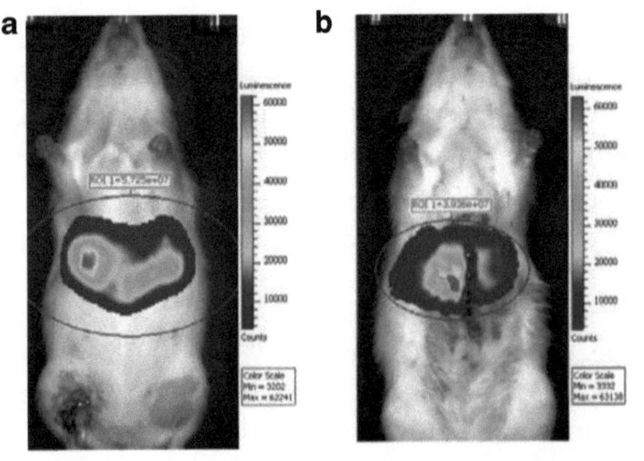

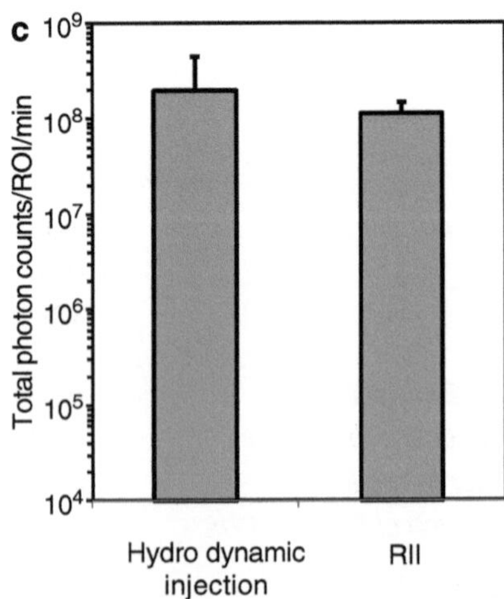

Fig. 2. Distribution of luciferase expression at 4 h after (**a**) hydrodynamic injection and (**b**) intrabiliary infusion of 20 μg VR1255 plasmid DNA (driven by CMV promoter) encoding firefly luciferase characterized with in vivo BLI as described in Subheading 3.4.2. The exposure time was set as 30 s. (**c**) Quantitative comparison of luciferase expression in rat liver at 4 h determined by in vivo BLI. Each bar represents the average ± standard deviation ($n = 3$).

importantly, the level of transgene expression is comparable to that mediated by the same dose of plasmid DNA delivered through hydrodynamic injection (see Note 8).

4. After the imaging is complete, the rat is returned to its cage. It will normally regain consciousness within 5 min.

4. Notes

1. The infusion site should be above where the bile duct enters duodenum because pancreatic tissue is not associated with bile duct above this point. If infusion is carried out below this point, the introduction of ligature with blunt-tip forceps may damage pancreatic tissue and cause leakage of bile from pancreatic ducts after infusion.
2. The surface of this part of common bile duct is usually covered by a thin layer of connective tissues, which makes it difficult to see the bile duct under microscope. Besides, the tip of infusion needle tends to slip on this layer of connective tissue during the insertion of the needle. When the connective tissue surrounding the common bile duct is being removed, great care should be taken not to damage the underlying vascular network on the common bile duct (also known as extrahepatic peribiliary plexus). Any damage to extrahepatic peribiliary plexus around infusion site will significantly slow down the healing process of needle hole caused by infusion, and thus potentially lead to leakage of bile post infusion.
3. To secure the infusion needle during infusion, the front part of the infusion needle (around 1/3 of the needle length) can be bent around 90° to the rest part of the needle so that the needle hub can be taped to the surgery board.
4. To secure the needle in the bile duct and prevent backflow of infused plasmid DNA solution into duodenum, the infusion needle should be pushed past the ligature so that a tie can be made around the needle. However, one should avoid pushing the needle into either branch of the hepatic ducts because this will lead to the infusion of plasmid DNA solution preferentially into specific lobes of the rat liver.
5. The infusion rate can be varied from 12 to 48 mL/h without causing obvious damage to the rat liver. However, higher infusion rate did not lead to significantly improved transgene expression in rat liver (Fig. 3).
6. The needle hole can be sealed without suture if the damage to bile duct is minimal. On rare occasions, the needle hole may

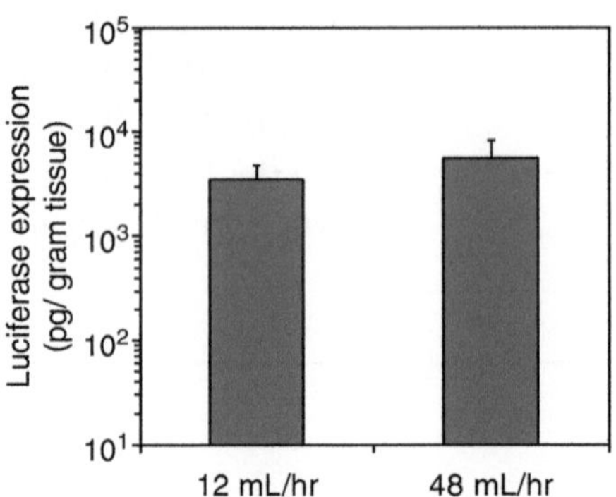

Fig. 3. Effect of infusion rate on the transgene expression mediated by plasmid DNA in the rat liver. VR1255 plasmid DNA at a dose of 20 μg was infused at different rates through the bile duct of rats. The rat livers were harvested at 24 h after infusion and homogenized in 1× lysis buffer. The expression level of luciferease in the rat liver was determined as described in Subheading 3.4.1 with a luminometer. Bars represent the average ± standard deviation ($n = 3$).

not close by itself due to the damage caused by the insertion of the needle. As the flow of bile is established, yellowish bile may slowly leak out from the needle hole shortly after the needle is withdrawn (usually within 2 min). Leakage of bile into abdominal cavity may cause serious inflammation or even death. Therefore, if bile leakage is observed after infusion, the needle hole should be carefully closed by one or two stitches with 10-0 sutures under a dissecting microscope.

7. Due to high luciferase expression level after intrabiliary infusion, the exposure time on IVIS Spectrum Imaging System is usually set as 1 min. In some cases, the exposure time needs to be reduced to 15 s to avoid saturation of the signal.
8. As a comparison, VR1255 plasmid DNA is also delivered through hydrodynamic injection to the rat liver. After the rat is anesthetized with 2.5–3.5% isoflurane, the hair on left hinder leg is removed with electric clippers. The femoral vein underlying the skin should be visible after the hair is removed. The exposed skin is then disinfected with 70% ethanol and 10% povidone-iodine consecutively. A small midline skin incision (~ 2 cm in length) is made in parallel to the femoral vein. A 27-Gauge infusion set is then connected to a 30-mL plastic syringe preloaded with 20 mL of plasmid DNA solution (20 μg DNA in 20 mL of 5% glucose). The needle is carefully inserted into the femoral vein and the plasmid DNA solution is rapidly injected into femoral vein in about 10 s. Right after injection, the needle is removed from femoral vein and a cotton gauze

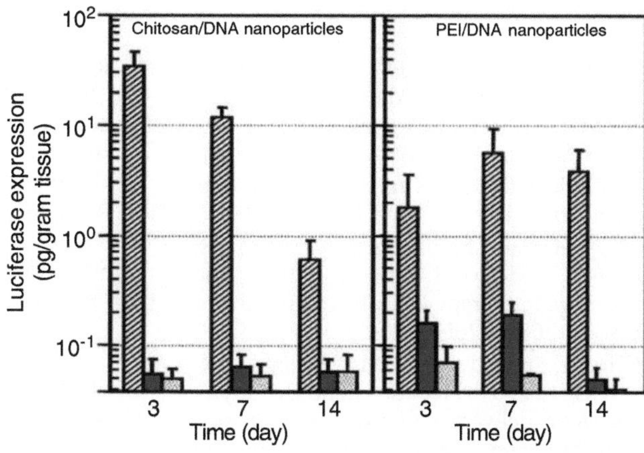

Fig. 4. Luciferase expression in rat liver following RII, intraportal infusion, and tail vein injection of chitosan (390 kDa)–DNA nanoparticles, branched PEI (25 kDa)-DNA nanoparticles, and VR1255 plasmid DNA. Each bar represents average ± standard deviation ($n = 5$). Chitosan–DNA and PEI–DNA nanoparticles were prepared at N/P ratios of 3 and 10, respectively. Nanoparticles and naked DNA were infused to rats at a dose equivalent to 200 μg of plasmid per rat (~0.8 mg/kg of body weight) in 4 mL of medium into the common bile duct (0.1 mL/min), portal vein (1 mL/min), or tail vein (1 mL/min). On days 3, 7, and 14, five rats from each group were sacrificed, and livers were harvested and homogenized in lysis buffer. The luciferase activity was determined as described in Subheading 3.4.1 with a luminometer. Rats receiving 4 mL of PBS infusions were included as the background control that defines the detection limit for this assay at 0.04 pg luciferase/g tissue. The figure was reprinted from International Journal of Nanomedicine, Volume 1, Dai H., Jiang X., Tan G.C., Chen Y., Torbenson M., Leong K.W., Mao H.Q.: Chitosan-DNA nanoparticles delivered by intrabiliary infusion enhance liver-targeted gene delivery., Pages No. 507–522, Copyright (2006), with permission from Dove Medical Press Ltd. (23).

pad is pressed against the injection site for about 30 s to stop bleeding. The skin of the rat is then closed by wound clips. At 4 h after injection, the luciferase expression in the liver is characterized by bioluminescence imaging (BLI) as described in Subheading 3.4.2. In addition to the delivery of plasmid DNA, RII can be used as an effective route for the delivery of polymer/DNA nanoparticles. For liver-targeted gene delivery, RII is a much more efficient route for polymer/DNA nanoparticles than intraportal injection (23). As shown in Fig. 4, the levels of transgene expression in rat liver mediated by chitosan and PEI/DNA nanoparticles after RII were 600- and 11-fold higher than those achieved through intraportal injection, respectively.

References

1. Nguyen TH, Ferry N (2004) Liver gene therapy: advances and hurdles. Gene Ther 11:S76–S84
2. Kren BT, Chowdhury NR, Chowdhury JR, Steer CJ (2002) Gene therapy as an alternative to liver transplantation. Liver Transpl 8:1089–1108
3. Prieto J, Herraiz M, Sangro B, Qian C, Mazzolini G, Melero I, Ruiz J (2003) The promise of gene therapy in gastrointestinal and liver diseases. Gut 52:ii49–ii54
4. Cristiano RJ, Smith LC, Kay MA, Brinkley BR, Woo SLC (1993) Hepatic gene-therapy - efficient gene delivery and expression in primary hepatocytes utilizing a conjugated adenovirus-DNA complex. Proc Natl Acad Sci USA 90:11548–11552
5. Nathwani AC, Davidoff AM, Hanawa H, Hu YN, Hoffer FA, Nikanorov A, Slaughter C, Ng CYC, Zhou JF, Lozier JN, Mandrell TD, Vanin EF, Nienhuis AW (2002) Sustained high-level expression of human factor IX (hFIX) after liver-targeted delivery of recombinant adeno-associated virus encoding the hFIX gene in rhesus macaques. Blood 100:1662–1669
6. Cheung ST, Tsui TY, Wang WL, Yang ZF, Wong SY, Ip YC, Luk J, Fan ST (2002) Liver as an ideal target for gene therapy: expression of CTLA4Ig by retroviral gene transfer. J Gastroenterol Hepatol 17:1008–1014
7. Jooss K, Chirmule N (2003) Immunity to adenovirus and adeno-associated viral vectors: implications for gene therapy. Gene Ther 10:955–963
8. Cichon G, Schmidt HH, Benhidjeb T, Loser P, Ziemer S, Haas R, Grewe N, Schnieders F, Heeren J, Manns MP, Schlag PM, Strauss M (1999) Intravenous administration of recombinant adenoviruses causes thrombocytopenia, anemia and erythroblastosis in rabbits. J Gene Med 1:360–371
9. Treco DA, Selden RF (1995) Nonviral genetherapy. Mol Med Today 1:314–321
10. Herweijer H, Wolff JA (2003) Progress and prospects: naked DNA gene transfer and therapy. Gene Ther 10:453–458
11. Niidome T, Huang L (2002) Gene therapy progress and prospects: nonviral vectors. Gene Ther 9:1647–1652
12. Kawakami S, Yamashita F, Nishida K, Nakamura J, Hashida M (2002) Glycosylated cationic Liposomes for cell-selective gene delivery. Crit Rev Ther Drug Carrier Syst 19:171–190
13. Rogers JV, Bigley NJ, Chiou HC, Hull BE (2000) Targeted delivery of DNA encoding herpes simplex virus type-1 glycoprotein D enhances the cellular response to primary viral challenge. Arch Dermatol Res 292:542–549
14. Reichen J (1999) The role of the sinusoidal endothelium in liver function. News Physiol Sci 14:117–121
15. Otsuka M, Baru M, Delriviere L, Talpe S, Nur I, Gianello P (2000) In vivo liver-directed gene transfer in rats and pigs with large anionic multilamellar liposomes: routes of administration and effects of surgical manipulations on transfection efficiency. J Drug Target 8:267–279
16. Zakim D, Boyer TD (2003) Hepatology: a textbook of liver disease. Saunders, Philadelphia
17. Ludwig J, Ritman EL, LaRusso NF, Sheedy PF, Zumpe G (1998) Anatomy of the human biliary system studied by quantitative computer-aided three-dimensional imaging techniques. Hepatology 27:893–899
18. Bircher J (1999) Oxford textbook of clinical hepatology. Oxford University Press, Oxford
19. Zhang XH, Collins L, Sawyer GJ, Dong XB, Qiu Y, Fabre JW (2001) In vivo gene delivery via portal vein and bile duct to individual lobes of the rat liver using a polylysine-based nonviral DNA vector in combination with chloroquine. Hum Gene Ther 12:2179–2190
20. Xie X, Forsmark CE, Lau JYN (1997) Feasibility of adenoviral-mediated gene delivery through ERCP - the effect or bile and pancreatic juice. Hepatology 26:276–276
21. Xie XM, Forsmark CE, Lau JYN (2000) Effect of bile and pancreatic juice on adenoviral-mediated gene delivery - implications on the feasibility of gene delivery through ERCP. Dig Dis Sci 45:230–236
22. Tominaga K, Kuriyama S, Yoshiji H, Deguchi A, Kita Y, Funakoshi F, Masaki T, Kurokohchi K, Uchida N, Tsujimoto T, Fukui H (2004) Repeated adenoviral administration into the biliary tract can induce repeated expression of the original gene construct in rat livers without immunosuppressive strategies. Gut 53:1167–1173
23. Dai H, Jiang X, Tan GC, Chen Y, Torbenson M, Leong KW, Mao HQ (2006) Chitosan-DNA nanoparticles delivered by intrabiliary infusion enhance liver-targeted gene delivery. Int J Nanomedicine 1:507–522

INDEX

Manfred Ogris and David Oupicky (eds.), *Nanotechnology for Nucleic Acid Delivery: Methods and Protocols*, Methods in Molecular Biology, vol. 948, DOI 10.1007/978-1-62703-140-0, © Springer Science+Business Media, LLC 2013

MIX
Papier aus verantwortungsvollen Quellen
Paper from responsible sources
FSC® C105338

If you have any concerns about our products,
you can contact us on
ProductSafety@springernature.com

In case Publisher is established outside the EU,
the EU authorized representative is:
Springer Nature Customer Service Center GmbH
Europaplatz 3, 69115 Heidelberg, Germany

Printed by Libri Plureos GmbH
in Hamburg, Germany